Lecture Notes in Artificial Intelligence   6703
Edited by R. Goebel, J. Siekmann, and W. Wahlster

Subseries of Lecture Notes in Computer Science

Kishan G. Mehrotra   Chilukuri K. Mohan
Jae C. Oh   Pramod K. Varshney
Moonis Ali (Eds.)

Modern Approaches in Applied Intelligence

24th International Conference
on Industrial Engineering and Other Applications
of Applied Intelligent Systems, IEA/AIE 2011
Syracuse, NY, USA, June 28 – July 1, 2011
Proceedings, Part I

 Springer

Series Editors

Randy Goebel, University of Alberta, Edmonton, Canada
Jörg Siekmann, University of Saarland, Saarbrücken, Germany
Wolfgang Wahlster, DFKI and University of Saarland, Saarbrücken, Germany

Volume Editors

Kishan G. Mehrotra
Chilukuri K. Mohan
Jae C. Oh
Pramod K. Varshney
Syracuse University, Department of Electrical Engineering and Computer Science
Syracuse, NY 13244-4100, USA
E-mail: {mehrotra, mohan, jcoh, varshney}@syr.edu

Moonis Ali
Texas State University San Marcos, Department of Computer Science
601 University Drive, San Marcos, TX 78666-4616, USA
E-mail: ma04@txstate.edu

ISSN 0302-9743 e-ISSN 1611-3349
ISBN 978-3-642-21821-7 e-ISBN 978-3-642-21822-4
DOI 10.1007/978-3-642-21822-4
Springer Heidelberg Dordrecht London New York

Library of Congress Control Number: 2011929232

CR Subject Classification (1998): I.2, H.3-4, F.1-2, C.2, I.4-5, H.2.8

LNCS Sublibrary: SL 7 – Artificial Intelligence

© Springer-Verlag Berlin Heidelberg 2011
This work is subject to copyright. All rights are reserved, whether the whole or part of the material is concerned, specifically the rights of translation, reprinting, re-use of illustrations, recitation, broadcasting, reproduction on microfilms or in any other way, and storage in data banks. Duplication of this publication or parts thereof is permitted only under the provisions of the German Copyright Law of September 9, 1965, in its current version, and permission for use must always be obtained from Springer. Violations are liable to prosecution under the German Copyright Law.
The use of general descriptive names, registered names, trademarks, etc. in this publication does not imply, even in the absence of a specific statement, that such names are exempt from the relevant protective laws and regulations and therefore free for general use.

Typesetting: Camera-ready by author, data conversion by Scientific Publishing Services, Chennai, India

Printed on acid-free paper

Springer is part of Springer Science+Business Media (www.springer.com)

Preface

There has been a steady increase in demand for efficient and intelligent techniques for solving complex real-world problems. The fields of artificial intelligence and applied intelligence cover computational approaches and their applications that are often inspired by biological systems. Applied intelligence technologies are used to build machines that can solve real-world problems of significant complexity. Technologies used in applied intelligence are thus applicable to many areas including data mining, adaptive control, intelligent manufacturing, autonomous agents, bio-informatics, reasoning, computer vision, decision support systems, fuzzy logic, robotics, intelligent interfaces, Internet technology, machine learning, neural networks, evolutionary algorithms, heuristic search, intelligent design, planning, and scheduling.

The International Society of Applied Intelligence (ISAI), through its annual IEA/AIE conferences, provides a forum for international scientific and industrial communities to interact with each other to develop and advance intelligent systems that address such concerns.

The 24th International Conference on Industrial, Engineering and Other Applications of Applied Intelligence Systems (IEA/AIE-2011), held in Syracuse, NY (USA), followed the IEA/AIE tradition of providing an international scientific forum for researchers in the diverse field of applied intelligence. Invited speakers and authors addressed problems we face and presented their solutions by applying a broad spectrum of intelligent methodologies. Papers presented at IEA/AIE-2011 covered theoretical approaches as well as applications of intelligent systems in solving complex real-life problems.

We received 206 papers and selected the 92 best papers for inclusion in these proceedings. Each paper was reviewed by at least three members of the Program Committee. The papers in the proceedings cover a wide number of topics including feature extraction, discretization, clustering, classification, diagnosis, data refinement, neural networks, genetic algorithms, learning classifier systems, *Bayesian* and probabilistic methods, image processing, robotics, navigation, optimization, scheduling, routing, game theory and agents, cognition, emotion, and beliefs.

Special sessions included topics in the areas of incremental clustering and novelty detection techniques and their applications to intelligent analysis of time varying information, intelligent techniques for document processing, modeling and support of cognitive and affective human processes, cognitive computing facets in intelligent interaction, applications of intelligent systems, nature-inspired optimization – foundations and application, chemoinformatic and bioinformatic methods, algorithms and applications.

These proceedings, consisting of 92 chapters authored by participants of IEA/AIE-2011, cover both the theory and applications of applied intelligent systems. Together, these papers highlight new trends and frontiers of applied intelligence and show how new research could lead to innovative applications of considerable practical significance. We expect that these proceedings will provide useful reference for future research.

The conference also invited three outstanding scholars to give plenary keynote speeches. They were Ajay K. Royyuru from IBM Thomas J. Watson Research Center, Henry Kauts from the University of Rochester, and Linderman from Air Force Research Laboratory.

We would like to thank Springer for their help in publishing the proceedings. We would also like to thank the Program Committee and other reviewers for their hard work in assuring the high quality of the proceedings. We would like to thank organizers of special sessions for their efforts to make this conference successful. We especially thank Syracuse University for their generous support of the conference.

We thank our main sponsor, ISAI, as well as our cooperating organizations: Association for the Advancement of Artificial Intelligence (AAAI), Association for Computing Machinery (ACM/SIGART, SIGKDD), Austrian Association for Artificial Intelligence (OeGAI), British Computer Society Specialist Group on Artificial Intelligence (BCS_SGAI), European Neural Network Society (ENNS), International Neural Network Society (INNS), Japanese Society for Artificial Intelligence (JSAI), Slovenian Artificial Intelligence Society (SLAIS), Spanish Society for Artificial Intelligence (AEPIA), Swiss Group for Artificial Intelligence and Cognitive Science (SGAICO), Taiwanese Association for Artificial Intelligence (TAAI), Taiwanese Association for Consumer Electronics (TACE), Texas State University-San Marcos.

Finally, we cordially thank the organizers, invited speakers, and authors, whose efforts were critical for the success of the conference and the publication of these proceedings. Thanks are also due to many professionals who contributed to making the conference successful.

April 2011

Kishan G. Mehrotra
Chilukuri Mohan
Jae C. Oh
Pramod K. Varshney
Moonis Ali

Organization

Program Committee

General Chair Moonis Ali, USA

Program Chairs Kishan G. Mehrotra, USA
Mohan Chilukuri, USA
Jae C. Oh, USA
Pramod K. Varshney, USA

Invited Session Chair Sanjay Ranka, USA

Local Arrangements Chair Thumrongsak Kosiyatrakul, USA

Program Committee

Adam Jatowt, Japan
Ah-Hwee Tan, Singapore
Amruth Kumar, USA
Andres Bustillo, Spain
Anna Fensel, Austria
Antonio Bahamonde, Spain
Azizi Ab Aziz, The Netherlands
Bärbel Mertsching, Germany
Bin-Yih Liao, Taiwan
Bipin Indurkhya, India
Bohdan Macukow, Poland
Bora Kumova, Turkey
C.W. Chan, Hong Kong
Catholijn Jonker, The Netherlands
Cecile Bothorel, France
César García-Osorio, Spain
Changshui Zhang, Canada
Chien-Chung Chan, USA
Chih-Cheng Hung, USA
Chilukuri K. Mohan, USA
Chiung-Yao Fang, Taiwan
Chunsheng Yang, Canada
Chun-Yen Chang, Taiwan
Colin Fyfe, UK
Coral Del Val-Muñoz, Spain

Dan Halperin, Israel
Dan Tamir, USA
Daniela Godoy, Argentina
Dariusz Krol, Poland
David Aha, USA
Djamel Sadok, Brazil
Domingo Ortiz-Boyer, Spain
Don Potter, USA
Don-Lin Yang, Taiwan
Duco Ferro, The Netherlands
Emilia Barakova, The Netherlands
Enrique Frias-Martinez, Spain
Enrique Herrera-Viedma, Spain
Erik Blasch, USA
Fevzi Belli, Germany
Floriana Esposito, Italy
Fran Campa Gómez, Spain
Francois Jacquenet, France
Fred Freitas, Brazil
Gerard Dreyfus, France
Geun-Sik Jo, South Korea
Gonzalo Aranda-Corral, Spain
Gonzalo Cerruela-García, Spain
Greg Lee, Taiwan
Gregorio Sainz-Palmero, Spain

Guillen Quintana, Spain
Guna Seetharaman, USA
Gwo-Jen Hwang, Taiwan
Hamido Fujita, USA
Hans-Werner Guesgen, New Zealand
Hasan Selim, Turkey
Henri Prade, France
Hiroshi Okuno, Japan
Hisao Ishibuchi, Japan
Huey-Ming Lee, Taiwan
Humberto Bustince, Spain
Iris Van De Kieft, The Netherlands
Ishfaq Ahmad, USA
Istenes Zoltán, Hungary
Jae Oh, USA
Jamal Bentahar, Canada
Jan Treur, The Netherlands
Janusz Kacprzyk, Poland
Jason J. Jung, South Korea
Jean-Charles Lamirel, France
Jeffrey Saltz, USA
Jeng-Shyang Pan, Taiwan
Jennifer Golbeck, USA
Jesús Aguilar, Spain
Jesús Maudes Raedo, Spain
Jing Peng, USA
John Dolan, USA
Jorge Romeu, USA
José Francisco Diez-Pastor, Spain
Juan José Rodríguez-Díez, Spain
Judy Qiu, USA
Jun Hakura, Japan
Jyh-Horng Chou, Taiwan
Kaikhah Khosrow, USA
Kaoru Hirota, Japan
Katarzyna Musial, UK
Kazuhiko Suzuki, Japan
Kishan Mehrotra, USA
Krzysztof Juszczyszyn, Poland
Kurosh Madani, France
Kush Varshney, USA
Laszlo Monostori, Hungary
Lav Varshney, USA
Leszek Borzemski, Poland
Ling-Jyh Chen, Taiwan

Lin-Yu Tseng, Taiwan
Lipo Wang, Singapore
Longbing Cao, Australia
Maciej Grzenda, Poland
Man-Kwan Shan, Taiwan
Manton Matthews, USA
Marco Valtorta, USA
Mario Köppen, Japan
Maritza Correa, Spain
Mark Hoogendoorn, The Netherlands
Masaki Kuremtsu, Japan
Matthijs Pontier, The Netherlands
Michal Lower, Poland
Michele Folgheraiter, Germany
Miquel Sánchez-Marré, Spain
Monika Lanzenberger, Austria
Nancy McCraken, USA
Natalie Van Der Wal, The Netherlands
Ngoc-Thanh Nguyen, Germany
Nicolás García-Pedrajas, Spain
Niek Wijngaards, The Netherlands
Nikolay Mirenkov, Japan
Oshadi Alahakoon, Australia
Pascal Wiggers, The Netherlands
Patrick Brézillon, France
Paul Chung, UK
Philipp Baer, Germany
Prabhat Mahanti, Canada
Pramod Varshney, USA
RadosLaw Katarzyniak, Poland
Raja Velu, USA
Rajmohan M., India
Rianne Van Lambalgen,
   The Netherlands
Riichiro Mizoguchi, Japan
Robbert-Jan Beun, The Netherlands
Rocio Romero, Spain
Rodolfo Haber, Spain
Rodrigo Ventura, Portugal
Rung-Ching Chen, Taiwan
Ruppa Thulasiram, Canada
Shaheen Fatima, UK
Shie-Jue Lee, Taiwan
Shogo Okada, Japan
Shusaku Tsumoto, Japan

Shyi-Ming Chen, Taiwan
Simone Mainai, Italy
Srini Ramaswamy, USA
Stefano Ferilli, Italy
Sung-Bae Cho, South Korea
Takayuki Ito, Japan
Tetsuo Kinoshita, Japan
Tibor Bosse, The Netherlands
Tim Hendtlass, Australia
Tim Verwaart, The Netherlands
Thumrongsak Kosiyatrakul, Thailand
Tiranee Achalakul, Thailand
Valery Tereshko, UK
Victor Rayward-Smith, UK
Victor Shen, Taiwan

Viktória Zsók, Hungary
Vincent S. Tseng, Taiwan
Vincenzo Loia, Italy
Walter Potter, USA
Wei-Shinn Ku, USA
Wen-Juan Hou, Taiwan
Wilco Verbeeten, Spain
Yasser Mohammad, Egypt
Ying Han, Spain
Yo-Ping Huang, Taiwan
Youngchul Bae, South Korea
Yu-Bin Yang, Canada
Yukio Ohsawa, Japan
Zia Ul-Qayyum, Pakistan
Zsolt-Janos Viharos, Hungary

Additional Reviewers

Chein-I Chang, USA
Chun-Nan Hsu, Taiwan
John Henry, USA
Jozsef Vancza, Hungary
Michelle Hienkelwinder, USA

Table of Contents – Part I

Section 1: Incremental Clustering and Novelty Detection Techniques and Their Application to Intelligent Analysis of Time Varying Information

Classification Model for Data Streams Based on Similarity 1
  *Dayrelis Mena Torres, Jesús Aguilar Ruiz, and
  Yanet Rodríguez Sarabia*

Comparison of Artificial Neural Networks and Dynamic Principal
Component Analysis for Fault Diagnosis 10
  *Juan Carlos Tudón-Martínez, Ruben Morales-Menendez,
  Luis Garza-Castañón, and Ricardo Ramirez-Mendoza*

Comparative Behaviour of Recent Incremental and Non-incremental
Clustering Methods on Text: An Extended Study 19
  Jean-Charles Lamirel, Raghvendra Mall, and Mumtaz Ahmad

Section 2: Bayesian and Probabilistic Networks

Fault Diagnosis in Power Networks with Hybrid Bayesian Networks
and Wavelets 29
  *Luis Eduarda Garza Castañón, Deneb Robles Guillén, and
  Ruben Morales-Menendez*

Learning Temporal Bayesian Networks for Power Plant Diagnosis 39
  *Pablo Hernandez-Leal, L. Enrique Sucar, Jesus A. Gonzalez,
  Eduardo F. Morales, and Pablo H. Ibarguengoytia*

On the Fusion of Probabilistic Networks 49
  Salem Benferhat and Faiza Titouna

Section 3: Methodologies

Basic Object Oriented Genetic Programming 59
  Tony White, Jinfei Fan, and Franz Oppacher

Inferring Border Crossing Intentions with Hidden Markov Models 69
  *Gurmeet Singh, Kishan.G. Mehrotra, Chilukuri K. Mohan, and
  Thyagaraju Damarla*

A Framework for Autonomous Search in the Ecl^{i}ps^e Solver............ 79
    Broderick Crawford, Ricardo Soto, Mauricio Montecinos,
    Carlos Castro, and Eric Monfroy

Multimodal Representations, Indexing, Unexpectedness and Proteins... 85
    Eric Paquet and Herna Lydia Viktor

A Generic Approach for Mining Indirect Association Rules in Data
Streams 95
    Wen-Yang Lin, You-En Wei, and Chun-Hao Chen

Status Quo Bias in Configuration Systems 105
    Monika Mandl, Alexander Felfernig, Juha Tiihonen, and Klaus Isak

Improvement and Estimation of Prediction Accuracy of Soft Sensor
Models Based on Time Difference 115
    Hiromasa Kaneko and Kimito Funatsu

Network Defense Strategies for Maximization of Network
Survivability 125
    Frank Yeong-Sung Lin, Hong-Hsu Yen, Pei-Yu Chen, and
    Ya-Fang Wen

PryGuard: A Secure Distributed Authentication Protocol for Pervasive
Computing Environment 135
    Chowdhury Hasan, Mohammad Adibuzzaman, Ferdaus Kawsar,
    Munirul Haque, and Sheikh Iqbal Ahamed

Section 4: Feature Extraction, Discretization, Clustering, Quantization, and Data Refinement

A Global Unsupervised Data Discretization Algorithm Based on
Collective Correlation Coefficient.................................... 146
    An Zeng, Qi-Gang Gao, and Dan Pan

A Heuristic Data-Sanitization Approach Based on TF-IDF 156
    Tzung-Pei Hong, Chun-Wei Lin, Kuo-Tung Yang, and
    Shyue-Liang Wang

Discovering Patterns for Prognostics: A Case Study in Prognostics of
Train Wheels 165
    Chunsheng Yang and Sylvain Létourneau

Section 5: Applications of Artificial Intelligence

Automating the Selection of Stories for *AI in the News* 176
    Liang Dong, Reid G. Smith, and Bruce G. Buchanan

Diagnosability Study of Technological Systems 186
    Michel Batteux, Philippe Dague, Nicolas Rapin, and Philippe Fiani

Using Ensembles of Regression Trees to Monitor Lubricating Oil
Quality 199
    Andres Bustillo, Alberto Villar, Eneko Gorritxategi,
    Susana Ferreiro, and Juan J. Rodríguez

Section 6: Image Processing and Other Applications

Image Region Segmentation Based on Color Coherence Quantization ... 207
    Guang-Nan He, Yu-Bin Yang, Yao Zhang, Yang Gao, and Lin Shang

Image Retrieval Algorithm Based on Enhanced Relational Graph 220
    Guang-Nan He, Yu-Bin Yang, Ning Li, and Yao Zhang

Prediction-Oriented Dimensionality Reduction of Industrial Data
Sets 232
    Maciej Grzenda

Informative Sentence Retrieval for Domain Specific Terminologies 242
    Jia-Ling Koh and Chin-Wei Cho

Section 7: Intelligent Techniques for Document Processing

Factoring Web Tables 253
    David W. Embley, Mukkai Krishnamoorthy, George Nagy, and
    Sharad Seth

Document Analysis Research in the Year 2021 264
    Daniel Lopresti and Bart Lamiroy

Markov Logic Networks for Document Layout Correction 275
    Stefano Ferilli, Teresa M.A. Basile, and Nicola Di Mauro

Extracting General Lists from Web Documents: A Hybrid Approach 285
    Fabio Fumarola, Tim Weninger, Rick Barber, Donato Malerba, and
    Jiawei Han

Section 8: Modeling and Support of Cognitive and Affective Human Processes

Towards a Computational Model of the Self-attribution of Agency 295
    Koen Hindriks, Pascal Wiggers, Catholijn Jonker, and
    Willem Haselager

An Agent Model for Computational Analysis of Mirroring
Dysfunctioning in Autism Spectrum Disorders 306
   Yara van der Laan and Jan Treur

Multi-modal Biometric Emotion Recognition Using Classifier
Ensembles 317
   *Ludmila I. Kuncheva, Thomas Christy, Iestyn Pierce, and
Sa'ad P. Mansoor*

Towards a Fully Computational Model of Web-Navigation.............. 327
   Saraschandra Karanam, Herre van Oostendorp, and Bipin Indurkhya

Section 9: Robotics and Navigation

Stairway Detection Based on Single Camera by Motion Stereo 338
   Danilo Cáceres Hernández, Taeho Kim, and Kang-Hyun Jo

Robot with Two Ears Listens to More than Two Simultaneous
Utterances by Exploiting Harmonic Structures....................... 348
   *Yasuharu Hirasawa, Toru Takahashi, Tetsuya Ogata, and
Hiroshi G. Okuno*

Author Index 359

Table of Contents – Part II

Section 1: Cognitive Computing Facets in Intelligent Interaction

Environmental Sound Recognition for Robot Audition Using Matching-Pursuit 1
   Nobuhide Yamakawa, Toru Takahashi, Tetsuro Kitahara, Tetsuya Ogata, and Hiroshi G. Okuno

Cognitive Aspects of Programming in Pictures 11
   Yutaka Watanobe, Rentaro Yoshioka, and Nikolay Mirenkov

An Approach for Smoothly Recalling the Interrupted Tasks by Memorizing User Tasks 21
   Kohei Sugawara and Hamido Fujita

Implementing an Efficient Causal Learning Mechanism in a Cognitive Tutoring Agent 27
   Usef Faghihi, Philippe Fournier-Viger, and Roger Nkambou

Model Checking Commitment Protocols 37
   Mohamed El-Menshawy, Jamal Bentahar, and Rachida Dssouli

Mobile Architecture for Communication and Development of Applications Based on Context 48
   Luis M. Soria-Morillo, Juan A. Ortega-Ramírez, Luis González-Abril, and Juan A. Álvarez-García

A Simplified Human Cognitive Approach for Supporting Crowd Modeling in Tunnel Fires Emergency Simulation 58
   Enrico Briano, Roberto Mosca, Roberto Revetria, and Alessandro Testa

Model Checking Epistemic and Probabilistic Properties of Multi-agent Systems 68
   Wei Wan, Jamal Bentahar, and Abdessamad Ben Hamza

Modeling Users of Crisis Training Environments by Integrating Psychological and Physiological Data 79
   Gabriella Cortellessa, Rita D'Amico, Marco Pagani, Lorenza Tiberio, Riccardo De Benedictis, Giulio Bernardi, and Amedeo Cesta

Personality Estimation Based on Weblog Text Classification 89
   Atsunori Minamikawa and Hiroyuki Yokoyama

Design of an Optimal Automation System: Finding a Balance between
a Human's Task Engagement and Exhaustion 98
    Michel Klein and Rianne van Lambalgen

A Cognitive Agent Model Using Inverse Mirroring for False Attribution
of Own Actions to Other Agents 109
    Jan Treur and Muhammad Umair

Explaining Negotiation: Obtaining a Shared Mental Model of
Preferences.. 120
    Iris van de Kieft, Catholijn M. Jonker, and M. Birna van Riemsdijk

A Computational Model of Habit Learning to Enable Ambient Support
for Lifestyle Change 130
    Michel C.A. Klein, Nataliya Mogles, Jan Treur, and
    Arlette van Wissen

Section 2: Applications of Intelligent Systems

An Intelligent Method to Extract Characters in Color Document with
Highlight Regions 143
    Chun-Ming Tsai

Development of Technological System Structure for Threaded
Connections Assembly under Conditions of Uncertainty 153
    Roman Chumakov

Automatic Vehicle Identification by Plate Recognition for Intelligent
Transportation System Applications 163
    Kaushik Deb, My Ha Le, Byung-Seok Woo, and Kang-Hyun Jo

Intelligent Page Recommender Agents: Real-Time Content Delivery for
Articles and Pages Related to Similar Topics 173
    Robin M.E. Swezey, Shun Shiramatsu, Tadachika Ozono, and
    Toramatsu Shintani

Meta-learning Based Optimization of Metabolic Pathway Data-Mining
Inference System.. 183
    Tomás V. Arredondo, Wladimir O. Ormazábal,
    Diego C. Candel, and Werner Creixell

Section 3: Optimization, Scheduling, and Routing

Multiple Pickup and Delivery TSP with LIFO and Distance Constraints:
A VNS Approach 193
    Xiang Gao, Andrew Lim, Hu Qin, and Wenbin Zhu

Distributed Learning with Biogeography-Based Optimization 203
   Carre Scheidegger, Arpit Shah, and Dan Simon

Scheduling a Single Robot in a Job-Shop Environment through
Precedence Constraint Posting.................................... 216
   *Daniel Díaz, M. Dolres R-Moreno, Amendo Cesta, Angelo Oddi,
and Riccardo Rasconi*

An Intelligent Framework to Online Bin Packing in a Just-In-Time
Environment 226
   Sergey Polyakovskiy and Rym M'Hallah

A Greedy Heuristic for Airline Crew Rostering: Unique Challenges in a
Large Airline in China.. 237
   Qiao Chen, Andrew Lim, and Wenbin Zhu

Optimal Algorithms for Two-Dimensional Box Placement Problems 246
   Wenbin Zhu, Wee-Chong Oon, Yujian Weng, and Andrew Lim

An Algorithm for the Freight Allocation Problem with All-Units
Quantity-Based Discount 256
   Xiang Gao, Andrew Lim, Wee-Chong Oon, and Hu Qin

A Distributed, Heterogeneous, Target-Optimized Operating System for
a Multi-robot Search and Rescue Application 266
   Karl Muecke and Brian Powell

A Heuristic for the Multiple Container Loading Cost Minimization
Problem 276
   Chan Hou Che, Weili Huang, Andrew Lim, and Wenbin Zhu

A Skyline-Based Heuristic for the 2D Rectangular Strip Packing
Problem 286
   Lijun Wei, Andrew Lim, and Wenbin Zhu

Real-Time Resource Allocation Co-processor 296
   Stuart W. Card

A Hybrid Search Strategy to Enhance Multiple Objective
Optimization 302
   Li Ma and Babak Forouraghi

Section 4: Nature Inspired Optimization – Foundations and Applications

Forest Planning Using Particle Swarm Optimization with a Priority
Representation 312
   Philip W. Brooks and Walter D. Potter

Fuzzy Robot Controller Tuning with Biogeography-Based
Optimization 319
  George Thomas, Paul Lozovyy, and Dan Simon

Development of a Genetic Fuzzy Controller for an Unmanned Aerial
Vehicle 328
  Yan Qu, Swetha Pandhiti, Kalesha S. Bullard, Walter D. Potter,
  and Karl F. Fezer

Toward Evolving Self-organizing Software Systems: A Complex System
Point of View 336
  Liguo Yu, David Threm, and Srini Ramaswamy

Evolving Efficient Sensor Arrangement and Obstacle Avoidance Control
Logic for a Miniature Robot 347
  Muthukumaran Chandrasekaran, Karthik Nadig, and Khaled Rasheed

Section 5: Chemoinformatic and Bioinformatic Methods, Algorithms, and Applications

Feature Selection for Translation Initiation Site Recognition 357
  Aida de Haro-García, Javier Pérez-Rodríguez, and
  Nicolás García-Pedrajas

DTP: Decision Tree-Based Predictor of Protein Contact Map 367
  Cosme Ernesto Santiesteban-Toca and Jesus Salvardor Aguilar-Ruiz

Translation Initiation Site Recognition by Means of Evolutionary
Response Surfaces 376
  Rafael del Castillo-Gomariz and Nicolás García-Pedrajas

An Evolutionary Algorithm for Gene Structure Prediction 386
  Javier Pérez-Rodríguez and Nicolás García-Pedrajas

Prediction of Drug Activity Using Molecular Fragments-Based
Representation and RFE Support Vector Machine Algorithm 396
  Gonzalo Cerruela García, Irene Luque Ruiz, and
  Miguel Ángel Gómez-Nieto

Section 6: Neural Network, Classification, and Diagnosis

A Hybrid Video Recommendation System Using a Graph-Based
Algorithm 406
  Gizem Öztürk and Nihan Kesim Cicekli

A Diagnostic Reasoning Approach to Defect Prediction 416
  Rui Abreu, Alberto Gonzalez-Sanchez, and Arjan J.C. van Gemund

Multiple Source Phoneme Recognition Aided by Articulatory
Features 426
  Mark Kane and Julie Carson-Berndsen

Plan Recommendation for Well Engineering 436
  Richard Thomson, Stewart Massie, Susan Craw, Hatem Ahriz, and
  Ian Mills

Lung Cancer Detection Using Labeled Sputum Sample: Multi Spectrum
Approach 446
  Kesav Kancherla and Srinivas Mukkamala

Section 7: Neural Network and Control

Improvement of Building Automation System....................... 459
  Mark Sh. Levin, Aliaksei Andrushevich, and Alexander Klapproth

Efficient Load Balancing Using the Bees Algorithm................. 469
  Anabela Moreira Bernardino, Eugénia Moreira Bernardino,
  Juan Manuel Sánchez-Pérez, Juan Antonio Gómez-Pulido, and
  Miguel Angel Vega-Rodríguez

Predicting the Distribution of Thermal Comfort Votes 480
  Anika Schumann and Nic Wilson

Section 8: Agents, Game Theory, and Bidding

Strategic Bidding Methodology for Electricity Markets Using Adaptive
Learning 490
  Tiago Pinto, Zita Vale, Fátima Rodrigues, Hugo Morais, and
  Isabel Praça

Compromising Strategy Based on Estimated Maximum Utility for
Automated Negotiation Agents Competition (ANAC-10) 501
  Shogo Kawaguchi, Katsuhide Fujita, and Takayuki Ito

Negotiating Privacy Preferences in Video Surveillance Systems 511
  Mukhtaj Singh Barhm, Nidal Qwasmi, Faisal Z. Qureshi, and
  Khalil el-Khatib

The Bayesian Pursuit Algorithm: A New Family of Estimator Learning
Automata... 522
  Xuan Zhang, Ole-Christoffer Granmo, and B. John Oommen

A Two-Armed Bandit Based Scheme for Accelerated Decentralized
Learning 532
  Ole-Christoffer Granmo and Sondre Glimsdal

Specification of Interlevel Relations for Agent Models in Multiple
Abstraction Dimensions..................................... 542
  Jan Treur

Section 9: Cognition, Emotion, Psychology, and Beliefs

An Argumentation Framework for Deriving Qualitative Risk Sensitive
Preferences.. 556
  Wietske Visser, Koen V. Hindriks, and Catholijn M. Jonker

Agent-Based Analysis of Patterns in Crowd Behaviour Involving
Contagion of Mental States.................................. 566
  *Tibor Bosse, Mark Hoogendoorn, Michel C.A. Klein,
  Jan Treur, and C. Natalie van der Wal*

Author Index... 579

Classification Model for Data Streams Based on Similarity

Dayrelis Mena Torres[1], Jesús Aguilar Ruiz[2], and Yanet Rodríguez Sarabia[3]

[1] University of Pinar del Río "Hermanos Saíz Montes de Oca", Cuba
dayro@info.upr.edu.cu.cu
[2] University "Pablo de Olavide", Spain
jsagurui@upo.es
[3] Central University of Las Villas "Marta Abreu", Cuba
yrsarabia@uclv.edu.cu

Abstract. Mining data streams is a field of study that poses new challenges. This research delves into the study of applying different techniques of classification of data streams, and carries out a comparative analysis with a proposal based on similarity; introducing a new form of management of representative data models and policies of insertion and removal, advancing also in the design of appropriate estimators to improve classification performance and updating of the model.

Keywords: classification, data streams, similarity.

1 Introduction

Mining data streams has attracted the attention of the scientific community in recent years with the development of new algorithms for data processing and classification. In this area it is necessary to extract the knowledge and the structure gathered in data, without stopping the flow of information. Examples include sensor networks, the measure of electricity consumption, the call log in mobile networks, the organization and rankings of e-mails, and records in telecommunications and financial or commercial transactions.

Incremental learning techniques have been used extensively in these issues. A learning task is incremental if the training set is not available completely at the beginning, but generated over time, the examples are being processed sequentially forcing the system to learn in successive episodes.

Some of the main techniques used in incremental learning algorithms for solving problems of classification, are listed below:

- Artificial Neural Networks. (Fuzzy-UCF [1] LEARN [2])
- Symbolic Learning (Rules). (Facil [3], FLORA [4], STAGGER [5], LAIR [6])
- Decision trees. (ITI [7], VFDT [8], NIP [9], IADEM [10])
- Instance-based learning. (IBL-DS [11], TWF [12], LWF [12], SlidingWindows [13], DROP3 [12], ICF [14])

Instance-based learning (IBL) works primarily through the maintenance of the representative examples of each class and has three main features [15]: the similarity function, the selection of the prototype instances and the classification function. The main difficulty of this technique is to combine optimally these three functions.

This paper introduces a new algorithm for classification in data streams based on similarity, called Dynamic Model Partial Memory-based (DMPM), where a representative set of current information are maintained.

The paper is structured as follows: section 2 presents a description of the proposal and its formalization; Section 4 presents a comparative analysis, introducing other relevant techniques, then the parameters defined for experiments and finally the results; Section 5 presents conclusions and some future work.

2 Description of the Proposal

To develop the proposed classifier implies the design of three methods:

Build Classifier: structuring the model with the first 100 generated instances, which are stored as a group for each class.

Then, depending on the type of experiment which is performed (cross validation or hold out), the algorithm classifies and updates the model, or only classifies.

Classify Instance: Returns the class that has been assigned.

Basically this classification functions as a standard kNN (with k = 1, this parameter can also be defined by the user) using the heterogeneous distance measure and updateable HVDM [16].

Update Classifier: Update the model with a new example. It works as follows:

When a new instance arrives, the distances to the centers of the groups are calculated, and if it is closer to another class, a new group is created, otherwise it is included in the closest group with same class.

As our approach uses a buffer of 400 instances (defined by the user), when a new instance must be included another instance must also be deleted. When a new group is created, the farthest neighbor to its center is removed, the entropy is updated, and the buffer restructured, removing the oldest group if necessary.

3 Formalization of the Proposal

To construct the classifier, the instance base (B) is empty, and the model is initially built with the first 100 arriving instances, in the procedure *buildClassifier*. To create the instance base, a TreeMap structure is used for each class, composed by groups of instances (g), a mean instance (m) and the age of the group.

Once the model is built, new instances are classified, taken into account the type of experiment being performed.

The objective of the procedure of updating the model (*UpdateClassifier*) is the proper selection of each instance (e_i) according to the proposed criteria. The procedure works as follows:

Algorithm 1. UpdateClassifier

    input : $e_i = (x_i, c)$: instance; d: distanceFunction; n: max number of instance in g
    output: B: base of instance

Update $d(e_i)$
if $c \notin B$ **then**
    CreateNewClassGroup(c)
    Update B
    UpdateEntropy
else
    $g(m, c_i) \longleftarrow \min d(e_i, c_i)$
    if $c_i \neq c$ **then**
        CreateNewInstanceGroup(e_i)
        Update B
        UpdateEntropy
        EditGroup
    else
        if $instanceCount(g) \leq n$ **then**
            Add(e_i, g)
        else
            EditGroup

When an instance (e_i) is coming, whose class (c) is not in the set, a new group for this class is created, and it is inserted in B in an organized manner.

In case it already exists, the proximity of the new instance to groups of other class is checked. If it is closer to another class, another group is created, containing the instance, the mean instance (firstly, the only instance), and the age of the group (initially 1). Otherwise, if the new instance is closer to groups of its class, it is added to the group in question.

It may happen that the class in question has a maximum permitted number of instances (n), because the whole set has a limited number of examples, in this case the procedure *EditGroup* is called, which is responsible for adding the new instance and defining which of the remaining will be removed, following a policy of inclusion - deletion that is aided of different estimators. The process of editing the groups works as illustrate the algorithm 2.

The strategy of insertion and deletion of instances in a group (*EditGroup*) is carried out mainly using two estimators: the mean instance or center of the group (m) and a parameter defined as a threshold of gradual change (b), which is updated with each insertion and removal of the buffer (instance base). This threshold is defined for each group (g) stored and obtained by calculating the distance from the center of the group to its farthest neighbor.

The procedure for *UpdateEntropy* is called every time there is a substantial change in the instance base. The actions defined as substantial changes, are the creating of new groups, the removing groups and the elimination of instances in

Algorithm 2. EditGroup

input : $e_i = (x_i, c)$: instance; d: distanceFunction; $g(m, c_i)$: instance group
output: B: base of instance

$nearestNeighbor \longleftarrow NearestNeighbor(g, m)$
$farthestNeighbor \longleftarrow FarthestNeighbor(g, m)$
$b \longleftarrow d(m, farthestNeighbor)$
if $d(m, x_i) < b$ **then**
  Delete $nearestNeighbor$
  Add e_i Update B
else
  Delete $farthestNeighbor$
  Add e_i Update B
  UpdateEntropy

a group when the instance removed is the farthest neighbor of the center of the group. In these cases, entropy is calculated for all classes, taking into account the number of instances of each class, and adjusts the number of examples for each class, ensuring that there is a proper balance with respect to stored examples of each class.

This adjustment might lead to increasing the number of instances to be stored, or to decreasing it by eliminating groups whose contribution to classification of new instances has been poor.

4 Experimentation

In this section we analyze the algorithms that have been selected for the experimentation.

IBL-DS [11]: This algorithm is able to adapt to concept changes, and also shows high accuracy for data streams that do not have this feature. However, these two situations depend on the size of the case base; if the change of concept is stable, the classification accuracy increases with the size of the case base. On the other hand, a large database of cases turns out to be unfavorable in cases where occur concept changes. It establishes a replacement policy based on:

Temporal Relevance: Recent observations are considered more useful than other, less recent ones.
Space Relevance: A new example in a region of space occupied by other examples of instances is less relevant than an example of a sparsely occupied region.
Consistency: An example should be removed if it is inconsistent with the current concept.

As a function of distance SVDM is used which is a simplified version of the VDM distance measure, which implements two strategies that are used in combination to successfully face the gradual or abrupt concept changes. IBL-DS is relatively robust and produces good results when uses the default parameters.

LWF (Locally Weighted Forgetting) [12]: One of the best adaptive learning algorithms. It is a technique that reduces the weight of the k nearest neighbors (in increasing order of distance) of a new instance. An example is completely eliminated if its weight is below a threshold. To keep the size of the case base, the parameter k is adapted, taking into account the size of the case base. As an obvious alternative to LWF, it was considered the TWF (Time Weighted Forgetting) algorithm that determines the weight of the cases according to their age.

In previous approaches, strategies for adapting the size window, if any, are mostly heuristic in nature. In the algorithm Sliding Windows [13] the authors propose to adapt the window size in such a way as to minimize the estimated generalization error of the learner trained on that window. To this end, they divide the data into batches and successively (that is, for k = 1; 2 ...) test windows consisting of batches t..k; t..k+1 ... t. On each of these windows, they train a classifier (in this case a support vector machine) and estimate its predictive accuracy (by approximating the leave-one-out error). The window/model combination that yields the highest accuracy is finally selected. In [17], this approach is further generalized by allowing the selection of arbitrary subsets of batches instead of only uninterrupted sequences. Despite the appealing idea of this approach to window (training set) adjustment, the successive testing of different window lengths is computationally expensive and therefore not immediately applicable in a data stream scenario with tight time constraints.

4.1 Parameters

Experiments were performed with the following setting parameter:
IBL-DS, $k = 1$.
Single sliding windows of size 400.
LWF, $\beta = 0.8$.
TWF, $w = 0.996$
DMPM, k = 1, change in entropy = 0.98.
To simulate data streams generators, were used Agrawal, Led24, RDG1 and RandomRBF included in Weka and also were incorporated a total of 8 different data streams found in the literature that have been used in multiple experiments [18] [19] [11]. They are: Gauss, Sine, Distrib, Random, Stagger, Mixed, Hyperplane and Means. Each set with a maximum of 20000 instances, adding 5% of noise, generating an initial training set of 100 examples and each algorithm with a window of 400 instances. Table 1 shows the attribute information, classes and types of concept (presence of drift) for each of the test data.

4.2 Results

For the evaluation of the proposed model, 4 algorithms were used: IBL-DS [11], TWF [12], LWF [12], and Win400 (variant of Sliding Windows with size 400) [13].

Table 1. Data Streams features

Data Streams	No. Att	Type	Class	Drift
Agrawal	9	Mix	2	-
LED24	24	Nom	10	-
Gauss	2	Num	2	X
Sine	2	Num	2	X
Distrib	2	Num	2	X
Random	2	Num	2	X
Stagger	3	Nom	2	X
Mixed	4	Mix	2	X
Hyperplane	2-5	Num	5	X
Means	2-5-15	Num	5	X
RDG1	10	Nom	2	-
RandomRBF	10	Num	2	-

Two performance measures were taking into account in this comparatives analysis:
- Absolute Accuracy: The ratio between the number of examples classified correctly and the total number of examples up to time t.
- Accuracy of Streaming: It is the proportion of examples correctly classified within the last 100 instances.

The results obtained for absolute accuracy, are shown in Table 2.

Table 2. Absolute Precision

Data Streams	DMPM	IBL-DS	LWF	TWF	Win400
Agrawal	.80	.62	.85	.79	.55
LED24	.65	.39	.61	.54	.36
Gauss	.62	.80	.55	.54	.54
Sine	.61	.83	.67	.67	.67
Distrib	.82	.91	.88	.86	.86
Random	.50	.64	.68	.63	.62
Stagger	.67	.85	.72	.72	.69
Mixed	.59	.83	.76	.75	.75
Hyperplane	.69	.84	.89	.78	.79
Means	.23	.38	.31	.23	.22
RDG1	.86	.88	.81	.91	.89
RandomRBF	.83	.80	.86	.91	.91

The results show that the algorithm DMPM shows better results with the data stream Led24, of nominal attributes and higher dimensionality.

Table 3 shows the values of standard deviation (σ) and mean ($\bar{x}$) values obtained with the measure of absolute accuracy.

The experiment shows that the average performance of the proposed model is good. Showing an average of 65% accuracy and a low standard deviation, giving the model a desired behavior.

Table 3. Standard Deviation and Means of the Absolute Precision

	DMPM	IBL-DS	LWF	TWF	Win400
σ	.17	.18	.16	.19	.21
$\bar{x}$	.65	.73	.71	.69	.65

The results obtained with the precision of the streaming, are shown in Table 4.

Table 4. Precision of the streaming

Data Streams	DMPM	IBL-DS	LWF	TWF	Win400
Agrawal	.80	.64	.84	.81	.80
LED24	.72	.39	.62	.64	.72
Gauss	.78	.66	.43	.55	.78
Sine	.79	.85	.59	.80	.79
Distrib	.89	.96	.79	.96	.89
Random	.52	.51	.56	.52	.52
Stagger	.76	.80	.70	.87	.76
Mixed	.76	.77	.69	.81	.76
Hyperplane	.66	.80	.91	.87	.66
Means	.21	.26	.34	.17	.21
RDG1	.90	.84	.80	.91	.90
RandomRBF	.84	.80	.77	.97	.84

In this performance measure DMPM algorithm shows the best result for the data stream Led24 and Gauss.

Table 5 shows the values of standard deviation (σ) and mean ($\bar{x}$) values obtained with the measurement accuracy of the streaming assessment.

With this measure, the proposed model achieved superior results, showing a mean accuracy of 71%, and a very low standard deviation, which shows the accuracy of the algorithm in the classification of the 100 instances last seen at any time of learning.

Fig 1. contains the results of all performance measures and all the algorithms for data stream LED24, the highest dimensionality of nominal attributes, without concept change. The algorithm proposed in this paper shows the best results with this data set, because the model presented does not have a policy to handle concept drift.

Table 5. Standard Deviation and Means of the Precision of the streaming

	DMPM	IBL-DS	LWF	TWF	Win400
σ	.19	.20	.16	.23	.23
$\bar{x}$	.71	.69	.67	.74	.66

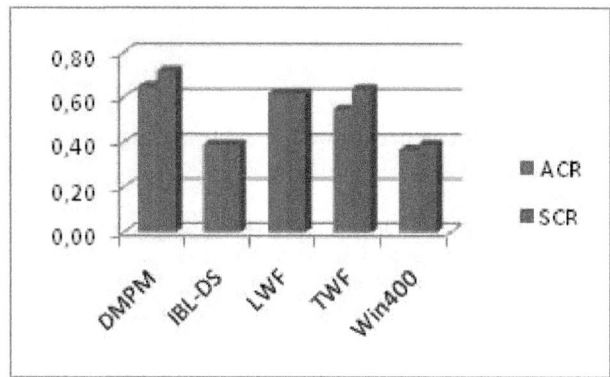

Fig. 1. Results with the data streams LED24

It should be noted that the proposed algorithm does not treat abrupt concept change, and it is being compared to others that are prepared for this situation. Taking into account these criteria, the results obtained for the measures of efficiency are comparable to those obtained with other algorithms.

5 Conclusions

This paper presents a Dynamic Model Partial Memory based (DMPM) for classification data streams, based on similarity, introducing new proposals for managing data models and policies of insertion and removal.

The proposed model shows better results in data streams with high dimensionality and no concept changes.

Experiments were conducted considering two performance measures, noting that the model proposed gives results comparable to other classifiers from literature, taking opportunities to improve both response time and precision.

References

1. Orriols-puig, A., Casillas, J., Bernado, E.: Fuzzy-UCS: A Michigan-style Learning Fuzzy-Classifier System for Supervised Learning. Transactions on Evolutionary Computation, 1–23 (2008)

2. Polikar, R., Udpa, L., Udpa, S.S., Honavar, V.: LEARN ++: an Incremental Learning Algorithm For Multilayer Perceptron Networks. IEEE Transactions on System, Man and Cybernetics (C), Special Issue on Knowledge Management, 3414–3417 (2000)
3. Ferrer, F.J., Aguilar, J.S., Riquelme, J.C.: Incremental Rule Learning and Border Examples Selection from Numerical Data Streams. Journal of Universal Computer Science, 1426–1439 (2005)
4. Widmer, G.: Combining Robustness and Flexibility in Learning Drifting Concepts. Machine Learning, 1–11 (1994)
5. Schlimmer, J.C., Granger, R.H.: Incremental learning from noisy data. Machine Learning 1(3), 317–354 (1986)
6. Watanabe, L., Elio, R.: Guiding Constructive Induction for Incremental Learning from Examples. Knowledge Acquisition, 293–296 (1987)
7. Kolter, J.Z., Maloof, M.A.: Dynamic Weighted Majority: A New Ensemble Method for Tracking Concept Drift. In: Proceedings of the Third International IEEE Conference on Data Mining, pp. 123–130 (2003)
8. Domingos, P., Hulten, G.: Mining High-Speed Data Streams. In: Proceedings of the Sixth ACM SIGKDD International Conference on Knowledge Discovery and Data Mining, pp. 71–80 (2000)
9. Jin, R., Agrawal, G.: Efficient Decision Tree Construction on Streaming Data. In: Proceedings of the Ninth ACM SIGKDD International Conference on Knowledge Discovery and Data Mining, pp. 571 – 576(2003)
10. Ramos-jim, G., Jos, R.M.-b., Avila, C.: IADEM-0: Un Nuevo Algoritmo Incremental, pp. 91–98 (2004)
11. Beringer, J., Hullermeier, E.: Efficient Instance-Based Learning on Data Streams. Intelligent Data Analysis, 1–43 (2007)
12. Salganicoff, M.: Tolerating Concept and Sampling Shift in Lazy Learning Using Prediction Error Context Switching. Artificial Intelligence Review, 133–155 (1997)
13. Klinkenberg, R., Joachims, T.: Detecting Concept Drift with Support Vector Machines. In: Proceedings of the Seventeenth International Conference on Machine Learning (ICML), pp. 487–494 (2000)
14. Mukherjee, K.: Application of the Gabriel Graph to Instance Based Learning Algorithms. PhD thesis, Simon Fraser University (2004)
15. Aha, D.W., Kibler, D., Albert, M.K.: Instance-Based Learning Algorithms. Machine Learning 66, 37–66 (1991)
16. Randall Wilson, D., Martinez, T.R.: Improved Heterogeneous Distance Functions. Artificial Intelligence 6, 1–34 (1997)
17. Stanfill, C., Waltz, D.: Toward memory-based reasoning. Communications of the ACM 29(12), 1213–1228 (1986)
18. Gama, J., Medas, P., Rocha, R.: Forest Trees for On-line Data. In: Proceedings of the 2004 ACM Symposium on Applied Computing, pp. 632–636 (2004)
19. Gama, J., Rocha, R., Medas, P.: Accurate Decision Trees for Mining High-speed Data Streams. In: Proc. SIGKDD, pp. 523–528 (2003)

Comparison of Artificial Neural Networks and Dynamic Principal Component Analysis for Fault Diagnosis

Juan C. Tudón-Martínez, Ruben Morales-Menendez, Luis Garza-Castañón, and Ricardo Ramirez-Mendoza

Tecnológico de Monterrey, Av. E. Garza Sada 2501,
64849, Monterrey N.L., México
{A00287756,rmm,legarza,ricardo.ramirez}@itesm.mx

Abstract. Dynamic Principal Component Analysis ($DPCA$) and Artificial Neural Networks (ANN) are compared in the fault diagnosis task. Both approaches are process history based methods, which do not assume any form of model structure, and rely only on process historical data. Faults in sensors and actuators are implemented to compare the online performance of both approaches in terms of quick detection, isolability capacity and multiple faults identifiability. An industrial heat exchanger was the experimental test-bed system. Multiple faults in sensors can be isolated using an individual control chart generated by the principal components; the error of classification was 15.28% while ANN presented 4.34%. For faults in actuators, ANN showed instantaneous detection and 14.7% lower error classification. However, $DPCA$ required a minor computational effort in the training step.

Keywords: $DPCA$, Artificial Neural Network, Fault Detection, Fault Diagnosis.

1 Introduction

Any abnormal event in an industrial process can represent financial deficit for the industry in terms of loss of productivity, environmental or health damage, etc. Fault Detection and Diagnosis (FDD) task is more complicated when there are many sensors or actuators in the process such as chemical processes. Advanced methods of FDD can be classified into two major groups [1]: process history-based methods and model-based methods.

Most of the existing Fault Detection and Isolation (FDI) approaches tested on Heat Exchangers (HE) are based on model-based methods. For instance, fuzzy models based on clustering techniques are used to detect leaks in a HE [2]. In [3], an adaptive observer is used to estimate model parameters, which are used to detect faults in a HE. In [4], particle filters are proposed to predict the fault probability distribution in a HE. These approaches require a process model with high accuracy; however, due to permanent increase in size and complexity of chemical processes, the modeling and analysis task for their monitoring and control have become exceptionally difficult. For this reason, process history based-methods are gaining on importance.

In [5] an Artificial Neural Network (*ANN*) is proposed to model the performance of a *HE* including fault detection and classification. In [6] a fast training *ANN* based on a Bayes classifier is proposed, the *ANN* can be online retrained to detect and isolate incipient faults. On the other hand, in [7] and [8], different adaptive approaches based on Dynamic Principal Component Analysis (*DPCA*) have been proposed to detect faults and avoid normal variations. A comparative analysis between *DPCA* and Correspondence Analysis (*CA*) is presented in [9], *CA* showed best performance but needs greater computational effort. Another comparison between two *FDI* systems is shown in [10], *DPCA* could not isolate sequential failures using the multivariate T^2 statistic while the *ANN* did it.

This paper presents an extended and improved version of [10] in terms of the quantitative comparison between *DPCA* and *ANN*. Moreover, this work proposes the use of an individual control chart with higher sensitivity to multiple abnormal deviations, instead of using the multivariate T^2 statistic used in [10]. The comparative analysis, based on the same experimental data provided from an industrial *HE*, is divided into two major groups: (1) detection stage in terms of detection time and probabilities of detection and false alarms and (2) diagnosis stage based on the percentage of classification error.

The outline of this paper is as follows: in the next section, *DPCA* formulation is presented. Section 3 describes the *ANN* design. Section 4 shows the experimentation. Section 5 and 6 present the results and discussion respectively. Conclusions are presented in section 7.

2 *DPCA* Formulation

Process data in the normal operating point must be acquired and standardized. The variability of the process variables and their measurement scales are factors that influence in the *PCA* performance [11]. In chemical processes, serial and cross-correlations among the variables are very common. To increase the decomposition in the correlated data, the column space of the data matrix X must be augmented with past observations for generating a static context of dynamic relationships:

$$X_D(t) = [X_1(t), X_1(t-1), \ldots, X_1(t-w), \ldots, X_n(t), X_n(t-1), \ldots, X_n(t-w)] \ . \quad (1)$$

see variable definition in Table 1. The main objective of *DPCA* is to get a set of a smaller number ($r < n$) of variables; r must preserve most of the information given by the correlated data. *DPCA* formulation can be reviewed in detail in [12]. Once the scaled data matrix $\bar{X}$ is projected by a set of orthogonal vectors, called loading vectors (P), a new and smaller data matrix T is obtained. Matrix T can be back-transformed into the original data coordination system as, $X^* = TP'$.

Normal operating conditions can be characterized by T^2 statistic [13] based on the first r Principal Components (*PC*). Using the multivariate Hotelling statistic, it is possible to detect multiple faults; however, the fault isolation is not achieved [10]. Therefore, the individual control chart of the T^2 statistic is proposed in this work for diagnosing multiple faults. In this case, the variability of all measurements is amplified by the *PC*, i.e. the statistic is more sensitive to any abnormal deviation once the measurement vector is filtered by *P*. Using the X^* data matrix, the Mahalanobis distance between a variable and its mean is:

$$T_i^2(t) = (x_i^*(t) - \mu_i) * s_i^{-1} * (x_i^*(t) - \mu_i) \ . \tag{2}$$

Since each T^2 statistic follows a *t*-distribution, the threshold is defined by:

$$T^2(\alpha) = t_{m-1}\left(\frac{\alpha}{2n}\right) \ . \tag{3}$$

A sensor or actuator fault is correctly detected and isolated when its T^2 statistic overshoots the control limit. All variables are defined in Table 1.

Table 1. Definition of variables

Variable	Description
X	Data matrix
X_D	Augmented data matrix
$\bar{X}$	Scaled data matrix
X^*	Back-transformed data matrix
w	Number of time delays to include in n process variables.
n	Number of process variables
m	Number of observations
r	Number of variables associated to the *PC*
T^2	Statistic of Hotelling (multivariate or univariate)
T	Scores data matrix
P	Loading vectors (matrix of eigenvectors)
μ_i	Mean of the variable i used in the *DPCA* training.
s_i	Standard deviation of the variable i used in the in the *DPCA* training.
t_{m-1}	*t*-distribution with $m-1$ degrees of freedom.
α	Significance value.

3 *ANN* Formulation

An *ANN* is a computational model capable to learn behavior patterns of a process, it can be used to model nonlinear, complex and unknown dynamic systems, [15]. A Multilayer Perceptron (*MLP*) network, which corresponds to a *feedforward* system, has been proposed for sensor and actuator fault diagnosis. For this research, *ANN* inputs correspond directly to the process measurements, and *ANN* outputs generate a fault signature which must be codified into pre-defined operating states. The *ANN* training algorithm was *backpropagation*; [14].

The codifier of the output indicates a fault occurrence; when the *ANN* output is zero, the sensor or actuator is free of faults; otherwise, it is faulty. A fault signature is used for identification of individual or multiple faults. The trained network can be subsequently validated by using unseen process data (*fresh data*). Crossed validation was used to validate the results.

4 Experimental System

Heat Exchangers (*HE*) are widely used in industry for both cooling and heating large scale industrial processes. An industrial shell-tube *HE* was the test bed, for this research. The experimental set up has all desirable characteristics for testing a fault detection scheme due to its industrial components. Fig. 1 shows a photo of the system; while right picture displays a conceptual diagram of the main instrumentation: 2 temperature sensors (TT_1, TT_2), 2 flow sensors (FT_1, FT_2) and their control valves (FV_1, FV_2). A data acquisition system (NI USB-6215) communicates the process with a computer.

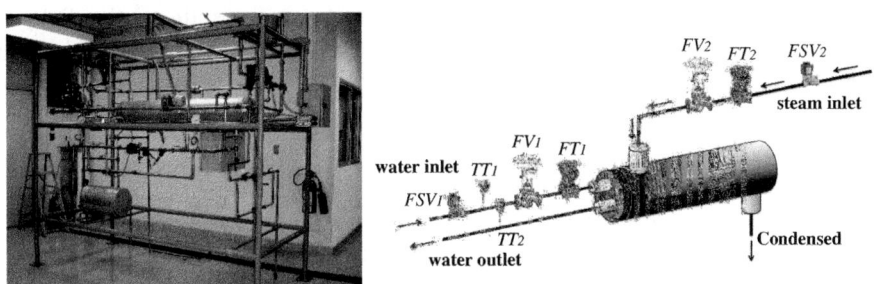

Fig. 1. Experimental System. An industrial *HE* is used as test bed, water runs through the tubes whereas low pressure steam flows inside the shell.

Design of Experiments. Faults in sensors and actuators, called *soft faults*, have been implemented in additive form. Sensor faults, which simulate transmitter biases, have been implemented in sequential form. The fault magnitude for flow sensors were 5σ and for temperature sensors were 8σ. For actuator faults, 4 fault cases have been implemented in the water (cases 1 and 2) and steam (cases 3 and 4) control valves. An actuator fault is considered as a change in the pressure of the pneumatic valve, this is equivalent to a change in the opening percentage of the valve ($\pm 10\%$). The normal operating point (case 0) was: 70% steam valve and 38% water valve.

Implementations. In the training stage of *DPCA*, 1,900 experimental data of each sensor have been used. The measurement vector is $x(t) = [FT_2\ FT_1\ TT_1\ TT_2]$. All data vectors are considered in the normal operating point. The *ANN* design is based on the minimal dimensions criterion [14], which selects the possible lowest number of hidden layers with the possible lowest number of neurons. A *MLP* was trained with 3,000 data vectors for diagnosing sensor faults, for diagnosing actuator faults 1,000 data vectors. Normal and faulty operating conditions were learned.

5 Results

Both approaches have been compared under same experimental conditions.

DPCA approach. Taking 1 as delay ($w = 1$), it is possible to explain a high quantity of variance including the possible auto and cross correlations. 5 PC are enough for explaining the 99.97% of the data variance. Left plot in Fig. 2 clearly shows that multiple failures in sensors can not be isolated ($t > 1,800$ sec.) using the multivariate T^2 statistic; while, the individual control charts allow the fault location (right plots in Fig. 2). The fault in the steam flow sensor (FT_2 signal) is the most sensitive for being detected. Similarly results were obtained for faults in actuators. In this case, the univariate statistic allows the FDD in parallel form, while the multivariate statistic requires another tool for achieving the fault isolation, for example: contribution plots [10].

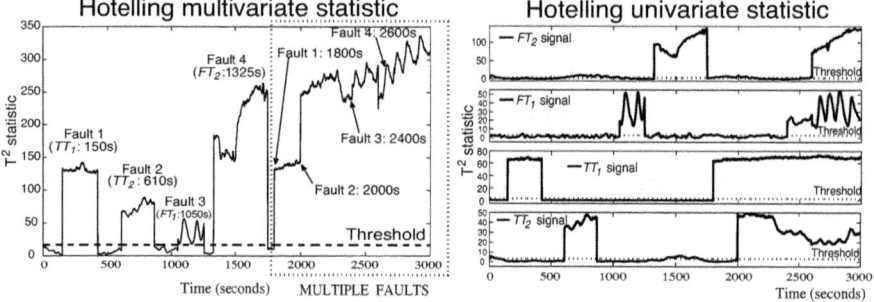

Fig. 2. FDD analysis for sensor faults. Multiple faults can not be isolated using the multivariate T^2 statistic (*left*); while, the univariate statistic can do it (*right*). The univariate statistics are obtained from decomposing the multivariate statistic, this allows to detect and isolate the fault sequence simultaneously.

ANN approach. Fig. 3 shows the real faulty condition and ANN estimation for faults in actuators. Since 5 different operating points were designed for actuator faults, a codifier is used to translate a set of binary fault signature to specific operating case. The FDD analysis for sensor faults using ANN can be reviewed in detail in [10]; the ANN estimation is binary (0 is considered free of faults and 1 as faulty condition), the inlet temperature sensor (TT_1) showed the greatest false alarms rate.

6 Discussion

The comparative analysis is quantified in terms of detection time, probability of false alarms and percentage of classification error.

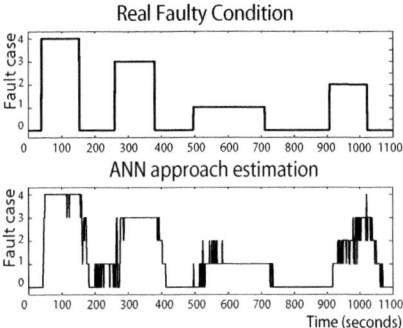

Fig. 3. *FDD* analysis for actuator faults using *ANN*. Several times the fault case 1 and 3 are erroneously detected instead the real faulty case.

A Receiver Operating Characteristic (*ROC*) curve is used to present the probability of right detection (p_d) and its respective false alarm probability (p_{fa}), [16]. The detection probability considers only the *FDI* property of indicating an abnormal event without considering the fault isolation. For sensor and actuator faults, both approaches showed similar probabilities of detection and false alarms, practically the same performance; however, the probability p_{fa} is greater for faults in actuators, Fig. 4. Therefore, faults in actuators are more complicated to detect because two or more process measurements can be affected in different magnitude or the fault incidence is not instantaneous.

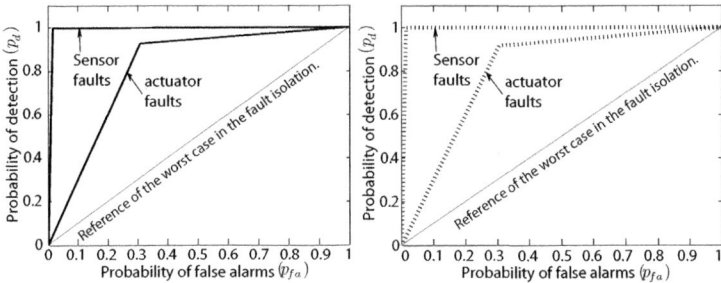

Fig. 4. *ROC* curve. The relation between the right fault detection and false alarms is presented by the *ROC* curve. Both applications show similar probabilities for detecting sensor and actuator faults; *DPCA* (right plot) and *ANN* (left plot). However, the detection of faults in actuators is worse than the fault detection in sensors.

For testing the isolation property of both approaches, a confusion matrix is used to quantify the relative and total error of fault classification. For constructing the confusion matrix, it is necessary to accommodate all classified data into the cells of the matrix. The column sum represents the real cases of a process condition; while, the row sum is the number of cases in which a process condition was classified. The elements of the principal diagonal indicates the cases which have been classified correctly, while the elements out of the diagonal are misclassified events. The last column contains 2 numbers: the right classification and relative false alarm based on the estimated number of faults in each faulty condition; whereas, the last row represents the percentage of right classification in terms of the real number of faults in each faulty condition and relative error respectively. Finally, the last cell indicates the global performance of classification: total right classification and total error in the isolation step.

Fig. 5 shows the isolation results for faults in actuators using $DPCA$ (left) and ANN (right), both confusion matrices have the same number of samples. A fault in the valve FV_1 has the greatest error (40.8%) of isolation using $DPCA$, while the case free of faults is erroneously classified several times (78+135) instead a real faulty condition (false alarm rate of 37.7%). In the ANN approach, the normal operating condition has the lowest percentage of right classification (68.6%). In general, ANN shows a lower total error of classification (23.32%) than $DPCA$ (27.17%). Similarly, for sensor faults, the ANN has the lowest total error of fault classification than $DPCA$: 4.34% versus 15.28%.

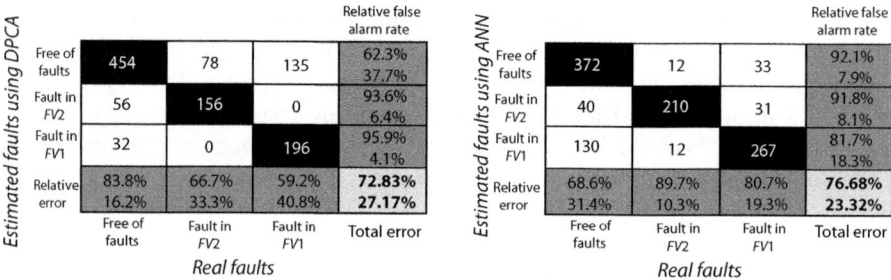

Fig. 5. Classification results. $DPCA$ approach shows a greater value in the total error of classification than ANN approach for faults in actuators.

Considering the real time necessary for fault detection, the detection time is greater in $DPCA$ approach even when these faults were abrupt; whereas, ANN shows an instantaneous detection. Table 2 shows a summary of the FDD performance in both approaches.

Table 2. Comparison of $DPCA$ and ANN approaches. ANN shows a better performance in detection and isolation of faults.

Approach	Location	Detection Time (s)	Isolability	p_d	p_{fa}	Error of Classification
DPCA	Actuators	9-18	✓	0.9168	0.3044	27.17%
	Multiple Sensors	Instantaneous	✓	0.9883	0.0142	15.28%
ANN	Actuators	Instantaneous	✓	0.9256	0.3081	23.32%
	Multiple Sensors	Instantaneous	✓	0.9966	0.0157	4.34%

According to computational requirements, the *ANN* training needs greater computational resources since historical data of the normal and faulty conditions must be known. The latest condition is an important drawback. While, the *DPCA* training is quickly executed; it only demands historical data of the normal operating point. Although both approaches are based on historical data, they have different frameworks. *DPCA* is a multivariate statistical method; whereas, *ANN* is based on pattern recognition.

When unknown soft faults are presented, the performance of *ANN* can be deteriorated; whereas, *DPCA* does not suffer this limitation. Although, the *ANN* has the extrapolation property in nonlinear systems as well as the capacity to interpolate unknown results, a new *ANN* training can be necessary under these new malfunctions. However, for online applications, a fast training algorithm must be taking account [6]. *DPCA* does not require more training effort because only deviations from the nominal point are considered. Furthermore, in the presence of a new source of variance caused by a new fault, it is possible to use a recursive *DPCA* algorithm that models the relationships between the measurements and their past history with the minimum computational effort [7].

Both methods are deteriorated when the process is time-variant; however, the training stage of both methods (retraining) can be easily applied every specific time window. If an adaptive training algorithm is not used above these methods, a change in the normal operating point can be interpreted as a fault.

The *MLP* used in this work requires and external classifier; however, the false alarm rate is similar to the obtained in [5], who proposes a self-organizing-map network that operates like a fault classifier by itself.

Both methods can detect and isolate multiple faults; while in [9] and [10], a fault is related to 2 or more variables using the multivariate T^2 statistic. Moreover, in [9] and [10], another tool (contribution plots) is used to isolate faults, while the univariate statistic allows to detect and isolate faults simultaneously.

7 Conclusions

A comparison between Dynamic Principal Component Analysis (*DPCA*) and Artificial Neural Networks (*ANN*) under the same experimental data generated by an industrial heat exchanger is presented. Both approaches can be used to detect and isolate abnormal conditions in nonlinear processes without considering a reliable model structure of the process, moreover the *FDD* task can be achieved when the industrial process has many sensors/actuators. The *ANN* showed the best performance in the *FDD* task with greater detection probability (1%), lower error of fault isolation (14.7% for actuator faults and 71.6% for sensor faults) and instantaneous detection.

The individual control chart generated by the principal components is proposed for fault diagnosis in *DPCA*, the method can isolate multiple faults due to its greater sensitivity to the data variability instead of using the simultaneous T^2 statistic. For *ANN*, there is not problem in the fault isolation task if all faulty conditions are known. New and unknown faults can be diagnosed by *DPCA*; while, *ANN* must be trained again including all faulty behaviors.

References

1. Venkatasubramanian, V., Rengaswamy, R., Kavuri, S., Yin, K.: A Review of Process Fault Detection and Diagnosis Part I Quantitative Model-Based Methods. Computers and Chemical Eng. 27, 293–311 (2003)
2. Habbi, H., Kinnaert, M., Zelmat, M.: A Complete Procedure for Leak Detection and Diagnosis in a Complex Heat Exchanger using Data-Driven Fuzzy Models. ISA Trans. 48, 354–361 (2008)
3. Astorga-Zaragoza, C.M., Alvarado-Martínez, V.M., Zavala-Río, A., Méndez-Ocaña, R., Guerrero-Ramírez, G.V.: Observer-based Monitoring of Heat Exchangers. ISA Trans. 47, 15–24 (2008)
4. Morales-Menendez, R., Freitas, N.D., Poole, D.: State Estimation and Control of Industrial Processes using Particles Filters. In: IFAC-ACC 2003, Denver Colorado U.S.A, pp. 579–584 (2003)
5. Tan, C.K., Ward, J., Wilcox, S.J., Payne, R.: Artificial Neural Network Modelling of the Thermal Performance of a Compact Heat Exchanger. Applied Thermal Eng. 29, 3609–3617 (2009)
6. Rangaswamy, R., Venkatasubramanian, V.: A Fast Training Neural Network and its Updation for Incipient Fault Detection and Diagnosis. Computers and Chemical Eng. 24, 431–437 (2000)
7. Perera, A., Papamichail, N., Bârsan, N., Weimar, U., Marco, S.: On-line Novelty Detection by Recursive Dynamic Principal Component Analysis and Gas Sensor Arrays under Drift Conditions. IEEE Sensors J. 6(3), 770–783 (2006)
8. Mina, J., Verde, C.: Fault Detection for MIMO Systems Integrating Multivariate Statistical Analysis and Identification Methods. In: IFAC-ACC 2007, New York U.S.A, pp. 3234–3239 (2007)
9. Detroja, K., Gudi, R., Patwardhan, S.: Plant Wide Detection and Diagnosis using Correspondance Analysis. Control Eng. Practice 15(12), 1468–1483 (2007)
10. Tudón-Martínez, J.C., Morales-Menendez, R., Garza-Castañón, L.: Fault Diagnosis in a Heat Exchanger using Process History based-Methods. In: ESCAPE 2010, Italy, pp. 169–174 (2010)
11. Peña, D.: Análisis de Datos Multivariantes. McGrawHill, España (2002)
12. Tudón-Martínez, J.C., Morales-Menendez, R., Garza-Castañón, L.: Fault Detection and Diagnosis in a Heat Exchanger. In: 6th ICINCO 2009, Milan Italy, pp. 265–270 (2009)
13. Hotelling, H.: Analysis of a Complex of Statistical Variables into Principal Components. J. Educ. Psychol. 24 (1993)
14. Freeman, J.A., Skapura, D.M.: Neural Networks: Algorithms, Applications and Programming Techniques. Adisson-Wesley, Reading (1991)
15. Korbicz, J., Koscielny, J.M., Kowalczuk, Z., Cholewa, W.: Fault Diagnosis Models, Artificial Intelligence, Applications. Springer, Heidelberg (2004)
16. Woods, K., Bowyer, K.W.: Generating ROC Curves for Artificial Neural Networks. IEEE Trans. on Medical Imaging 16(3), 329–337 (1997)

Comparative Behaviour of Recent Incremental and Non-incremental Clustering Methods on Text: An Extended Study

Jean-Charles Lamirel[1], Raghvendra Mall[2], and Mumtaz Ahmad[1]

[1] LORIA, Campus Scientifique,
BP 239, Vandœuvre-lès-Nancy, France
{jean-charles.lamirel,mumtaz.ahmad}@inria.fr
http://www.loria.fr,
[2] Center of Data Engineering, IIIT Hyderabad,
NBH-61, Hyderabad, Andhra Pradesh, India
raghvendra.mall@research.iiit.ac.in
http://www.iiit.ac.in

Abstract. This paper represents an attempt to throw some light on the quality and on the defects of some recent clustering methods, either they are incremental or not, on "real world data". An extended evaluation of the methods is achieved through the use of textual datasets of increasing complexity. The third test dataset is a highly polythematic dataset that figures out a static simulation of evolving data. It thus represents an interesting benchmark for comparing the behaviour of incremental and non incremental methods. The focus is put on neural clustering methods but the standard K-means method is included as reference in the comparison. Generic quality measures are used for quality evaluation.

1 Introduction

Most of the clustering methods show reasonable performance on homogeneous textual dataset. However, the highest performance on such datasets are generally obtained by neural clustering methods [6]. The neural clustering methods are based on the principles of neighbourhood relation between clusters, either they are preset (fixed topology), like the "Self-Organizing Maps" also named SOM [4], or dynamic (free topology), like static "Neural Gas" (NG) [10] or "Growing Neural Gas" (GNG) [3]. As compared to usual clustering method, like K-means [9], this strategy makes them less sensitive to the initial conditions, which represents an important asset within the framework of data analysis of highly multidimensional and sparse data, like textual data.

The most known neural clustering method is the SOM method which is based on a mapping of the original data space onto a two dimensional grid of neurons. The SOM algorithm is presented in details in [4]. It consists of two basic procedures: (1) selecting a winning neuron on the grid and (2) updating weights of the winning neuron and of its neighbouring neurons.

In the NG algorithm [10], the weights of the neurons are adapted without any fixed topological arrangement within the neural network. Instead, this algorithm utilizes a neighbourhood ranking process of the neuron weights for a given input data. The weight changes are not determined by the relative distances between the neuron within a topologically pre-structured grid, but by the relative distance between the neurons within the input space, hence the name "neural gas" network.

The GNG algorithm [3] solves the static character of the NG algorithm bringing out the concept of evolving network. Hence, in this approach the number of neuron is adapted during the learning phase according to the characteristics of the data distribution. The creation of the neurons is made only periodically (each T iteration or time period) between the two neighbour neurons that accumulated the most important error for representing the data.

Recently, an incremental growing neural gas algorithm or (IGNG) [12] has been proposed by Prudent and Ennaji for general clustering tasks. It represents an adaptation of the GNG algorithm that relaxes the constraint of periodical evolution of the network. Hence, in this algorithm a new neuron is created each time the distance of the current input data to the existing neuron is greater than a prefixed threshold σ. The σ value is a global parameter that corresponds to the average distance of the data to the center of the dataset. Prudent and Ennaji have proved that their method produces better results than the existing neural and the non-neural methods on standard test distributions.

More recently, the results of the IGNG algorithm have been evaluated on heterogeneous datasets by Lamirel & al. [6] using generic quality measures and cluster labeling techniques. As the results have been proved to be insufficient for such data, a new incremental growing neural gas algorithm using the cluster label maximization (IGNG-F) has been proposed by the said authors. In this strategy the use of a standard distance measure for determining a winner is completely suppressed by considering the label maximization approach as the main winner selection process. Label maximization approach is sketched at section 2 and it is more precisely detailed in [6]. One if its important advantage is that it provides the IGNG method with an efficient incremental character as it becomes independent of parameters.

In order to throw some light on the quality and on the defects of above mentioned clustering methods on "real world data", we propose hereafter a extended evaluation of those methods through the use of 3 textual datasets of increasing complexity. The third dataset which we exploit is a highly polythematic dataset that figures out a static simulation of evolving data. It thus represents an interesting benchmark for comparing the behaviour of incremental and non incremental methods. Generic quality measures like *Micro-Precision, Micro-Recall, Cumulative Micro-Precision* and cluster labeling techniques are exploited for comparison. In the next section, we give some details on these clustering quality measures. The third section provides a detailed analysis of the datasets and on their complexity. Following which we present the results of our experiments on our 3 different test datasets.

2 Clustering Quality Evaluation

An inherent problem of cluster quality evaluation persists when we try to compare various clustering algorithms. It has been shown in [5] that the inertia measures, or their adaptations [1], which are based on cluster profiles are often strongly biased and highly dependent on the clustering method. A need thus arised for such quality metrics which validate the intrinsic properties of the numerical clusters. We have thus proposed in [5] unsupervised variations of the recall and precision measures which have been extensively used in IR systems for evaluating the clusters.

For such purpose, we transform the recall and precision metrics to appropriate definitions for the clustering of a set of documents with a list of labels, or properties. The set of labels S_c that can be attributed to a cluster c are those which have maximum value for that cluster, considering an unsupervised $Recall - Precision$ metric [7] [1]. The greater the unsupervised precision, the nearer the intentions of the data belonging to the same cluster will be with respect to each other. In a complementary way, the unsupervised recall criterion measures the exhaustiveness of the contents of the obtained clusters, evaluating to what extent single properties are associated with single clusters.

Global measures of Precision and Recall are obtained in two-ways. *Macro-Precision* and *Macro-Recall* measures are generated by averaging Recall and Precision of the cluster labels at the cluster level, in a first step, and by averaging the obtained values between clusters, in a second step. *Micro-Precision* and *Micro-Recall* measures are generated by averaging directly Recall and Precision of the cluster labels at a global level. Comparison of *Macro* measures and *Micro* measures makes it possible to identify heterogeneous results of clustering [8].

It is possible to refer not only to the information provided by the indices *Micro-Precision* and *Micro-Recall*, but to the calculation of the *Micro-Precision* operated cumulatively. In the latter case, the idea is to give a major influence to large clusters which are most likely to repatriate the heterogeneous information, and therefore, by themselves, lowering the quality of the resulting clustering. This calculation can be made as follows:

$$CMP = \frac{\sum_{i=|c_{inf}|,|c_{sup}|} \frac{1}{|C_{i+}|^2} \sum_{c \in C_{i+}, p \in S_c} \frac{|c_p|}{|c|}}{\sum_{i=|c_{inf}|,|c_{sup}|} \frac{1}{C_{i+}}} \quad (1)$$

where C_{i+} represents the subset of clusters of C for which the number of associated data is greater than i, and:

$$inf = argmin_{c_i \in C}|c_i|, sup = argmax_{c_i \in C}|c_i| \quad (2)$$

---

[1] The IGNG-F algorithm uses this strategy as a substitute for the classical distance based measure which provides best results for homogeneous datasets [6].

3 Dataset Analysis

For the experimental purpose we use 3 different datasets namely the Total-Use, PROMTECH and Lorraine dataset. The clustering complexity of each dataset is more than the previous one. We provide detailed analysis of the datasets and try to estimate their complexity and heterogeneity.

The **Total-Use** dataset consisted of 220 patent records related to oil engineering technology recorded during the year 1999. This dataset contains information such as the relationship between the patentees, the advantages of different type of oils, what are the technologies used by patentees and the context of exploitation of their final products. Based on the information present in the dataset, the labels belonging to the datasets have been categorized by the domain experts into four viewpoints or four categories namely Patentees, Title, Use and Advantages. The task of extracting the labels from the dataset is divided into two elementary steps. At the step 1, the rough index set of each specific text sub field is constructed by the use of a basic computer-based indexing tool. At the step 2, the normalization of the rough index set associated to each viewpoint is performed by the domain expert in order to obtain the final index sets. In our experiment, we solely focus on the labels belonging to the Use viewpoint. Thus, the resulting corpus can be regarded as a homogeneous dataset as soon as it covers an elementary description field of the patents with a limited and contextual standardized vocabulary of 234 keywords or labels spanning over 220 documents.

The **PROMTECH** dataset is extracted from the original PROMTECH dataset of the PROMTECH project which has been initially constituted by the use of INIST PASCAL database and relying on its classification plan with the overall goal of analysing the dynamics of the various identified topics. For building up this dataset, a simple search strategy, consisting in the selection of the bibliographic records having at the same time a code in Physics and a code corresponding to a technological scope of application has been firstly employed. The selected applicative field was the Engineering. By successive selections, combining statistical techniques and expert approaches, 5 promising sets of themes have been released [11]. The final choice was to use the set of themes of the optoelectronic devices here because this field is one of the most promising of last decade. 3890 records related to these topics were thus selected in the PASCAL database. The corpus has then been cut off in two periods that are (1996-1999: period 1) and (2000-2003: period 2), to carry out for each one an automatic classification by using the description of the content the bibliographic records provided by the indexing keywords. Only data associated to first period has been used in our own experiment. In such a way, our second experimental dataset finally consisted of 1794 records indexed by 903 labels.

The **Lorraine** dataset is also build up from a set of bibliographic records resulting from the INIST PASCAL database and covering all the research activity performed in the Lorraine region during the year 2005. The structure of the records makes it possible to distinguish the titles, the summaries, the indexing labels and the authors as representatives of the contents of the information published in the corresponding article. In our experiment, the research topics associated with the labels field are solely considered. As soon as these labels cover themselves a large set of different topics (as far one to another as medicine from structural physics or forest cultivation, etc.), the resulting experimental dataset can be considered as a highly heterogeneous dataset. The number of records is 1920. A frequency threshold of 2 is applied on the initial label set resulting in a description space constituted of 3556 labels. Although the keyword data do not come directly from full text, we noted that the distribution of the terms took the form of a Zipf law (linear relation between the log of the frequencies of terms and the log of their row) characteristic of the full text data. The final keyword or label set also contains a high ratio of polysemic terms, like age, system, structure, etc.

3.1 Estimation of Complexity and Heterogeneity of Datasets

The clustering complexity of a dataset is intrinsically inter-related with its different kinds of heterogeneity. We follow the principle that a dataset heterogeneity is mainly related to the diversity of the topics represented in its data and to the ratio and the diversity of the information shared by the said data. The latter kind of heterogeneity that we have called "overlapping heterogeneity" is particularly critical for clustering as soon as it will lead the clustering methods to mix information belonging to different clusters together.

At a first glance, we can observe that for Total-Use dataset all the labels are related to one viewpoint, which is a first indication of its topic homogeneity. Conversely, for PROMTECH and Lorraine datasets, the labels cover various research themes over a period of time, providing heterogeneity of topics. Table 1 presents basic statistics on the datasets. It highlights that the average labels per document for the Total-Use dataset is as low as 3.268 and out of the total 234 labels only 62 are overlapping, with quite low average overlapping labels per document, suggesting that this dataset is relatively easy to cluster. Table 1 also highlights that PROMTECH and Lorraine datasets have higher but nearly similar values for average number of labels per document. But the average number of overlapping labels per document is high (0.503) for PROMTECH dataset indicating its overlapping heterogeneity and is extremely high (1.852) for Lorraine dataset, which makes this latter an even more complex dataset for the task of clustering.

As soon as our approach seemed not sufficient to clearly highlight the clustering complexity of the PROMTECH and the Lorraine datasets, we thus apply a second methodology. We construct a label×document matrix in order to properly estimate the diversity of the labels co-occurring in the documents. This matrix provides a list of all the labels which are occurring in the document couples. We then determine for each label the maximum sized list in which it has occurred. This enables us to assess the maximum length of the common label patterns for which this label was contributing. So, it becomes easier to estimate the number of distinct labels that are participating for common label patterns of different sizes. We normalize the results using the total number of documents

Table 1. Summary of datasets Information

Dataset	Keywords	Documents	Frequency Threshold	Average Label nb. per Document	Total Overlapping Labels	Average Label nb. per Document
Total	234	220	2	3.268	62	0.28
PROMTECH	903	1794	3	8.12	903	0.503
Lorraine	3556	1920	2	8.99	3556	1.852

in the dataset to have average fraction of distinct terms occurring in common label patterns of various lengths. Figure 1 represents the trends observed for the 3 datasets. The Total-Use dataset has only 62 distinct overlapping label for common label patterns of length 1. This helps to confirm that the Total-Use dataset is relatively easier to cluster as it has less diverse overlapping terms. For the PROMTECH dataset the fraction of distinct terms is greater than for Total-Use dataset and follows a relatively smooth decreasing curve (0.503 for common label patterns of length 1, 0.501 for common label patterns of size 2, etc.). As the length of the common label patterns increases, the number of distinct participating labels decreases gradually which suggests that the overlapping labels are nearly similar but present in more quantity and more extent than the Total-Use dataset. However, the Lorraine dataset has extremely high average values with all the 3556 distinct labels participating for common label patterns of length 1. It also has very high value (i.e. 1.36) for common label patterns of size 2, etc. This indicates that the number of distinct bridging terms is very high even for large sized common label patterns, leading to the highest complexity of clustering, as it will be confirmed by our experiments.

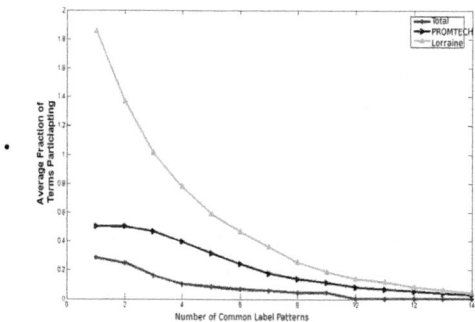

Fig. 1. Average fraction of distinct labels occurring in common label patterns

4 Results

For each method, we do many different experiments letting varying the number of clusters in the case of the static methods and the neighbourhood parameters in the case the incremental ones (see below). We have finally kept the best clustering results for each method regarding to the value of *Recall-Precision F-measure* and the *Cumulative Micro-Precision*.

We first conducted our experiments on the Total-Use dataset which is homogeneous by nature. Figure 2 shows the *Macro-F-Measure, Micro F-Measure* and the *Cumulative Micro-Precision(CMP)* for the dataset. We see that the *Macro F-Measure* and *Micro F-Measure* values are nearly similar for the different clustering approaches, although in the case of the SOM approach more than half of the clusters are empty. However, the *CMP* value shows some difference [2]. The NG and GNG algorithms have good *Macro* and *Micro F-Measure* but lower *CMP* than the IGNG approaches. We can thus conclude that the best results are obtained for IGNG and IGNG-F methods. The dataset is homogeneous by nature, so it is possible to reach such high precision values. Thus small, distinct and non-overlapping clusters are formed with the best methods. Its lowest *CMP* values also highlights that standard K-Means approach produces the worst results on the first dataset.

Clustering Method	No of clusters	F-Measure Macro	F-Measure Micro	Cumulative Micro-Precision	MSE
SOM	90 (196)	0.766719	0.692499	0.530187	0.346252
K-Means	53 (55)	0.758615	0.594195	0.291856	0.139917
NG	82(96)	0.815711	0.714835	0.6	0.225764
GNG	86 (93)	0.784056	0.755068	0.670653	0.244798
IGNG	86 (87)	0.837917	0.796471	0.729932	0.259827
IGNG-F	68 (81)	0.83054	0.745341	0.690571	0.270658

Fig. 2. Clustering quality results of various algorithms on Total-Use homogeneous dataset

We next conduct our experiments on the PROMTECH dataset. The dataset is heterogeneous by nature and is marginally complex dataset. Figure 3 presents the results for the said dataset. The SOM approach fails to identify the optimal model and has low *Micro F-Measure* values. This can be attributed to its fixed topology principle. On their own side, the free topology based approaches got high but nearly similar *Micro* and *Macro F-Measure* values, so we cannot distinguish the approaches based on these measures. It is the *CMP* measure which helps to determine the best results. We see that *CMP* values are high for SOM, K-Means and GNG but the maximum value is attained by the GNG algorithm (0.488) which is far better than the other approaches. One of the main reasons is that GNG applies an exhaustive error accumulation procedure and so

---

[2] In the figures, the main number of clusters represents the actual number of non-empty clusters among the total number of clusters used for clustering the dataset, which is mentioned into ().

it traverses the entire dataset. The K-Means method obtained the second best *CMP* value (0.35). However, its *Micro-Precision* value is less illustrating lower global average performance than all the other methods. IGNG-F has moderate *CMP* value (0.25) but has high *Micro-Precision*, so the average clustering quality is quite good even though there might be some cluster for this method which is still agglomerating the labels. The neural NG method has been left out from this experiment and from the next one because of its too high computation time.

Clustering Method	No of clusters	F-Measure Macro	F-Measure Micro	Cumulative Micro-Precision	MSE
SOM	311 (361)	0.368056	0.353343	0.309484	0.80597
K-Means	186 (185)	0.41429	0.375741	0.356678	0.4697
GNG	215 (215)	0.440858	0.407637	0.48383	0.62427
IGNG	296 (300)	0.465731	0.416523	0.242226	0.932746
IGNG-F	300 (300)	0.490176	0.430829	0.249885	0.922738

Fig. 3. Clustering quality results of various algorithms on PROMTECH dataset

Even if it embeds stable topics, the Lorraine dataset is a very complex heterogeneous dataset as we have illustrated earlier. In a first step we restricted our experiment to 198 clusters as beyond this number, the GNG approach went to an infinite loop (see below). A first analysis of the results on this dataset within this limit shows that most of the clustering methods have huge difficulties to deal with it producing consequently very bad quality results, even with such high expected number of clusters, as it is illustrated at Figure 4 by the very low CMP values. It indicates the presence of degenerated results including few garbage clusters attracting most of the data in parallel with many chunks clusters representing either marginal groups or unformed clusters. This is the case for K-Means, IGNG, IGNG-F methods and at a lesser extent for GNG method.

This experiment also highlights the irrelevance of Mean Square Error (MSE) (or distance-based) quality indexes for estimating the clustering quality in complex cases. Hence, the K-Means methods that got the lowest MSE practically produces the worth results. This behaviour can be confirmed when one looks more precisely to the cluster content for the said method, using the methodology that is described in section 2. It can be thus highlighted that K-means method mainly produced a garbage cluster with very big size (1722 data or more than 80% of the dataset) attracting (i.e. maximising) many kinds of different labels (3234 labels among 3556), figuring out a degenerated clustering result. Conversely, despite of its highest MSE, the correct results of the SOM method can also be confirmed in the same way. Hence, cluster labels extraction clearly highlights that this latter method produces different clusters of similar size attracting semantically homogeneous labels groups which figure out the main research topics covered by the analysed dataset. The grid constrained learning of the SOM method seems to be a good strategy for preventing to produce too bad results in such a critical context. Hence, it enforces the homogeneity of the results by splitting both data and noise on the grid.

Clustering Method	No of clusters	F-Measure Macro	F-Measure Micro	Cumulative Micro-Precision	MSE
SOM	216 (225)	0.35142	0.295278	0.136492	0.913648
K-Means	198 (400)	0.392339	0.288297	0.027891	0.582161
GNG	136 (136)	0.327476	0.256538	0.07864	0.73263
IGNG	198 (198)	0.339541	0.279246	0.028195	1.351955
IGNG-F	198 (198)	0.351322	0.280931	0.030203	1.334549

Fig. 4. Clustering quality results of various algorithms on Lorraine heterogeneous dataset (within 198 clusters limit)

As mentioned earlier, for higher number of clusters than 198 the GNG method does not provide any results on this dataset because of its incapacity to escape from an infinite cycle of creation-destruction of neurons (i.e. clusters). Moreover, the CMP value for GNG approach was surprisingly greater for 136 clusters than for 198 clusters (see figure 4). Thus, increasing the expected number of clusters is not helpful to the method to discriminate between potential data groups in the Lorraine dataset context. At the contrary, it even leads the method to increase its garbage agglomeration effect. However, we found that as we increase the number of clusters beyond 198 for the other methods the actual peaks value for the SOM, IGNG and IGNG-F methods are reached, and the static SOM method and the incremental IGNG-F method reach equivalently the best clustering quality. Thus, only these two methods are really able to appropriately cluster this highly complex dataset. Optimal quality results are reported at figure 5.

Clustering Method	No of clusters	F-Measure Macro	F-Measure Micro	Cumulative Micro-Precision	MSE
SOM	283 (441)	0.419857	0.335212	0.185579	0.814
K-Means	198 (400)	0.392339	0.288297	0.027891	0.582161
IGNG	298 (298)	0.426154	0.365496	0.118324	1.1865
IGNG-F	270 (270)	0.413174	0.336916	0.17449	1.209

Fig. 5. Optimal clustering quality results on Lorraine heterogeneous dataset

5 Conclusion

Clustering algorithms show reasonable performance in the usual context of the analysis of homogeneous textual dataset. This is especially true for the recent adaptive versions of neural clustering algorithms, like the incremental neural gas algorithm (IGNG) or the incremental neural gas algorithm based on label maximization (IGNG-F). Nevertheless, using a stable evaluation methodology and dataset of increasing complexity, this paper highlights clearly the drastic decrease of performance of most of these algorithms, as well as the one of more classical non neural algorithms, when a very complex heterogeneous textual dataset, figuring out a static simulation of evolving data, is considered as an input. If incrementality is considered as a main constraint, our experiment also showed that only non standard distance based methods, like the IGNG-F method, can produce reliable results in such case. Such a method also presents the advantage to produce the most stable and reliable results in the different experimental contexts, unlike the other neural and non-neural approaches which have highly varying on the datasets. However, our experiments also highlighted that one of the problems IGNG-F faces in some case is that it can associate a data with labels completely different from the ones existing in the prototypes. This leads to significant decrease in performance in such case as labels belonging to different clusters are clubbed together. We are thus investigating to use a distance based criteria to limit the number of prototypes which are considered for a new upcoming data point. This will allows to set a neighbourhood threshold and focus for each new data point, which is lacking in the IGNG-F approach.

References

1. Davies, D., Bouldin, W.: A cluster separation measure. IEEE Transaction on Pattern Analysis and Machine Intelligence 1, 224–227 (1979)
2. Dempster, A., Laird, N., Rubin, D.: Maximum likelihood for incomplete data via the em algorithm. ournal of the Royal Statistical Society, B 39, 1–38 (1977)
3. Frizke, B.: A growing neural gas network learns topologies. Advances in neural Information processing Systems 7, 625–632 (1995)
4. Kohonen, T.: Self-organized formation of topologically correct feature maps. Biological Cybernetics 43, 56–59 (1982)
5. Lamirel, J.-C., Al-Shehabi, S., Francois, C., Hofmann, M.: New classification quality estimators for analysis of documentary information: application to patent analysis and web mapping. Scientometrics 60 (2004)
6. Lamirel, J.-C., Boulila, Z., Ghribi, M., Cuxac, P.: A new incremental growing neural gas algorithm based on clusters labeling maximization: application to clustering of heterogeneous textual data. In: The 22th Int. Conference on Industrial, Engi- neering and Other Applications of Applied Intelligent Systems (IEA-AIE), Cordoba, Spain (2010)
7. Lamirel, J.-C., Phuong, T.A., Attik, M.: Novel labeling strategies for hierarchical representation of multidimensional data analysis results. In: IASTED International Conference on Artificial Intelligence and Applications (AIA), Innsbruck, Austria (February 2008)
8. Lamirel, J.-C., Ghribi, M., Cuxac, P.: Unsupervised recall and precision measures: a step towards new efficient clustering quality indexes. In: Proceedings of the 19th Int. Conference on Computational Statistics (COMPSTAT 2010), Paris, France (August 2010)
9. MacQueen, J.: Some methods of classifcation and analysis of multivariate observations. In: Proc. 5th Berkeley Symposium in Mathematics, Statistics and Probability, vol. 1, pp. 281–297. Univ. of California, Berkeley (1967)
10. Martinetz, T., Schulten, K.: A neural gas network learns topologies. Articial Neural Networks, 397–402 (1991)
11. Oertzen, J.V.: Results of evaluation and screening of 40 technologies. Deliverable 04 for Project PROMTECH, 32 pages + appendix (2007)
12. Prudent, Y., Ennaji, A.: An incremental growing neural gas learns topology. In: 13th European Symposium on Artificial Neural Networks, Bruges, Belgium (April 2005)

Fault Diagnosis in Power Networks with Hybrid Bayesian Networks and Wavelets

Luis E. Garza Castañón, Deneb Robles Guillén,
and Ruben Morales-Menendez

Tecnológico de Monterrey, Campus Monterrey
Department of Mechatronics and Automation
Av. Eugenio Garza Sada Sur No. 2501
Monterrey, N.L. 64489 México
{legarza,A00507182,rmm}@itesm.mx

Abstract. A fault diagnosis framework for electrical power transmission networks, which combines Hybrid Bayesian Networks (HBN) and Wavelets is proposed. HBN is a probabilistic graphical model in which discrete and continuous data are analyzed. In this work, power network's protection breakers are modeled as discrete nodes, and information extracted from voltages measured in every electrical network node represent the continuous nodes. Protection breakers are devices with the function to isolate faulty nodes by opening the circuit, and are considered to be working in one of three states: OK, OPEN, and FAIL. On the other hand, node voltages data are processed with wavelets, delivering specific coefficients patterns which are encoded into probability distributions of continuous HBN nodes. Experimental tests show a good performance of the diagnostic system when simultaneous multiple faults are simulated in a 24 nodes electrical network, in comparison with a previous approach in the same domain.

Keywords: Fault Diagnosis, Hybrid Bayesian Networks, Wavelets, Power Networks.

1 Introduction

The monitoring and supervision of power networks plays a very important role in modern societies, because of the close dependency of almost every system to electricity supply. In this domain, failures in one area perturb other neighbor areas, and troubleshooting is very difficult due to excess of information, cascade effects, and noisy data. Early detection of network misbehavior can help to avoid major breakdowns and incidents, such as those reported in the northeast of USA and south of Canada in August 2003, and in 18 states in Brasil and Paraguay in November 2009. These events caused heavy economic loses and millions of affected users. Therefore, an alternative to increase the efficiency of electrical distribution systems, is the use of automated tools, which could help the operator to speed up the process of system diagnosis and recovery.

In order to tackle those problems, fault detection and system diagnosis has been a very active research domain since a few decades ago. Recently, the need to develop more powerful methods has been recognized, and approaches based on Bayesian Networks (BN), able to deal very efficiently with noise and the modeling of uncertainties have been developed. For instance, in [1], a discrete BN with noisy-OR and noisy-AND nodes, and a parameter-learning algorithm similar to the used by artificial neural networks (ANN), is used to estimate the faulty section in a power system. In [2] and [3], discrete BN are used to handle discrete information coming from the protection breakers. Continuous data voltages coming from nodes are handled in [2], with dynamic probabilistic maximum entropy models, whereas in [3] ANN are used. A Bayesian selectivity technique and Discrete Wavelet Transform are used in [4], to identify the faulty feeder in a compensated medium voltage network. Although BN are used in several processes for fault diagnosis and support decision making, none of the surveyed papers deal with multiple simultaneous faults, and none reports the use of HBN to diagnose faults in power systems.

In this paper is proposed a hybrid diagnostic framework, which combines the extraction of relevant features from continuous data with wavelet theory and the modeling of a complete 24-nodes power network with HBN. Wavelets are used to analyze and extract the characteristic features of the voltage data obtained from electrical network nodes measurements, creating specific coefficients fault patterns for each node. These coefficients are encoded as probability distributions in the continuous nodes of HBN model, which in turn help to identify the fault type of the component. On the other hand, the operation of protection breakers is modeled with discrete nodes encoding the probability of the status of breakers. The HBN model of the system states probabilistic relationships between continuous and discretes system's components. Promising results of three different evaluations made from simulations performed with a power network composed by 24 nodes and 67 breakers and a comparison to other approach. The organization of the paper is as follows: Section 2 reviews the fundamentals of the wavelets theory. Section 3 presents the BN framework. Section 4 gives the general framework description. Section 5 shows the case study, and finally, Section 6 concludes the paper.

2 Wavelets

Wavelet Transform (WT) methods have been effectively used for multi-scale representation and analysis of signals. The WT decomposes signal transients into a series of wavelet components, each of which corresponds to a time-domain signal that covers a specific frequency band containing more detailed information. Wavelets localize the information in the time-frequency plane, and they are capable of trading one type of resolution for another, which makes them especially suitable for the analysis of non-stationary signals. Also, wavelet analysis can analyze appropriately rapid changes in transients of signals. The main strength of wavelet analysis is its ability to demonstrate the local feature of a particular area of a large signal.

In a Discrete Wavelet Transform (DWT), a time-scale representation of a digital signal is obtained using digital filtering techniques. In this case, filters of different cut-off frequencies are used to analyze the signal at different scales. The signal (S) is decomposed into two types of components: approximation (A) and detail (D). The approximation is the high scale, low-frequency component of the signal. The detail is the low scale, high-frequency component. These decompositions are achieved through the iterated filter bank technique which is elaborated in the following way: First, the original signal $x[n]$ is passed through a high-pass filter $g[n]$ and a low-pass filter $h[n]$, which divides the frequency bandwidth by the half, reducing the highest frequency. This two steps constitute the first level of decomposition, and after this, the signal can be down-sampled by two [5]. This decomposition decreases the time resolution and double the frequency resolution; the frequency uncertainties diminishes also by the half and it must be said, that this can be realized as many times as the decomposition of the signal is needed, presenting in each decomposition the same described characteristics. The prominent frequencies in the original signal will appear with high amplitudes in the regions that include this particular frequencies in the DWT. The temporal localization will have a resolution that depends on the level in which they appear.

3 Bayesian Networks

Bayesian Networks (BN) are a representation for probability distributions over complex domains. Traditional BN can handle only discrete information, but Hybrid Bayesian Networks (HBN) contain both, discrete data and continuous conditional probability distributions (CPDs) as numerical inputs. The main advantages of these models are that they can infer probabilities based on continuous and discrete information, and also they can deal with a great amount of data, uncertainty and noise.

The Bayesian inference is a kind of statistical inference in which the evidence or observations are employed to update or infer the probability of a hypothesis to be true [6]. The name Bayesian comes from the frequent use of the Bayes' Theorem (Eqn. 1) during the inference process. The basic task of a probabilistic inference system is calculating the subsequent probability distributions for a group of variables, given an observed event.

$$P(b \mid a) = \frac{P(a \mid b)P(b)}{P(a)} \qquad (1)$$

In an HBN the conditional distribution of continuous variables is given by a linear Gaussian model:

$$P(x_i | I = i, Z = z) = N(\alpha(i) + \beta(i) * z, \gamma(i)), \qquad (2)$$

where Z and I are the set of continuous and discrete parents of x_i respectively and $N(\mu, \sigma)$ is a multivariate normal distribution. The network represents a joint distribution over all its variables given by a product of all its CPDs.

4 Framework

A portion of the HBN model developed for the power system is shown in Figure 1. Each node in the power network is represented with a model of this type; it is composed of discrete nodes, representing the status of protection breakers, and continuous nodes representing wavelets features extracted from voltage measurements at power network nodes. The model has three parts:

1. **Upper part:** To manage the discrete data, the breaker status. The main function of this part is to establish the status of the power network node. The status node (faulty or not) is known, but not the type.
2. **Bottom part:** The main function is the analysis of continuous data. In this way, this information can be communicated with the other parts of the network. The inserted data in this section are the wavelets coefficients computed from measurements of node voltages. The main contribution is the confirmation and determination of which possible fault may have the device.
3. **Middle part:** It is probably the most important of the whole structure, because here the weighting probabilities of continuous and discrete data are communicated. Faults can be isolated and known.

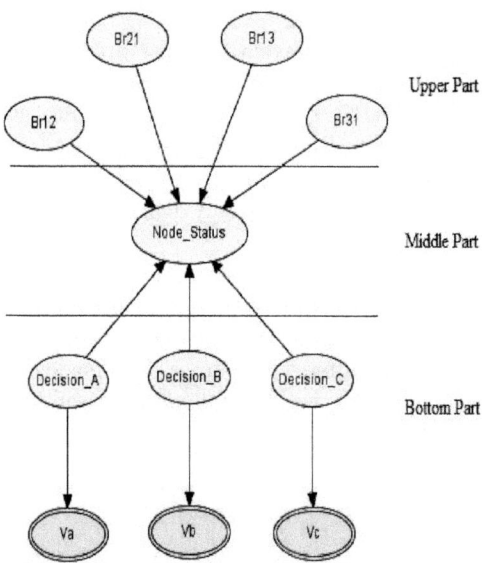

Fig. 1. Representation of one power network node in HBN model

The parameters for probability distributions of continuous nodes of HBN are learned by simulating all possible types of faults in each node and their subsequent treatment with wavelet filtering.

In order to perform the system diagnosis, the first step is to obtain discrete and continuous observations. The discrete observations will reflect in some way the status of system's components (node status). It is assumed that every breaker has a finite set of states indicating normal or faulty operation. Protection breakers help to isolate faulted nodes by opening the circuit, and are considered to be working in one of three states: OK, OPEN, and FAIL. The OK status means the breaker remains closed and has not detected any problem. The OPEN status is related to the opening of breaker when a fault is detected. The FAIL status means the breaker malfunctioned and did not open when a fault event was detected. The inference task over discrete information will be accomplished by applying the HBN upper-part of the model for each node, Figure 1. The output is a possible detection of a faulty or non-faulty node consistent with discrete observations.

In the second step, continuous data coming from all the system components is analyzed. First, features are extracted through a bank of iterated filters according to the wavelet theory, in which data is decomposed in approximated and detailed information, generating specific patterns for each component. A signature of the different possible states or fault types of the system's components is considered. The first decomposition is chosen to be the representative pattern of each signal, because it gives a general view of the analyzed wave and unlike the detailed information it does not present too much sensibility in presence of simultaneous multiple faults. In order to make an adequate diagnosis, the patterns here obtained, are inserted in the HBN bottom-part of the model of every node (continuous part), which with the CPD will elaborate a weighted magnitude. Finally, in the HBN middle-part of the model, with both of the weighted magnitude parts (discrete and continuous), the real power network node status will be calculated.

5 Case Study

The system under testing is a network that represents an electrical power system with dynamic load changes and it was taken from [7], Figure 2. The system has 24 nodes, and every node is protected by several breakers devices, which will be active when a malfunction is detected. For every node on the system, there is a three-phases continuous voltage measurement, and for every breaker there is a discrete signal of the state of the device.

The complete HBN model is shown in Figure 3. Each node of the power network (Figure 2) is modeled individually with the proposed HBN in Figure 1. The 24 nodes of the power network in the complete HBN model are only communicated through the discrete nodes, which represents the breakers between two electrical nodes.

The methodology proposed is applied as follows:

1. Collect discrete observations from protection breakers and continuous data from the devices (nodes' voltages) of the power network.

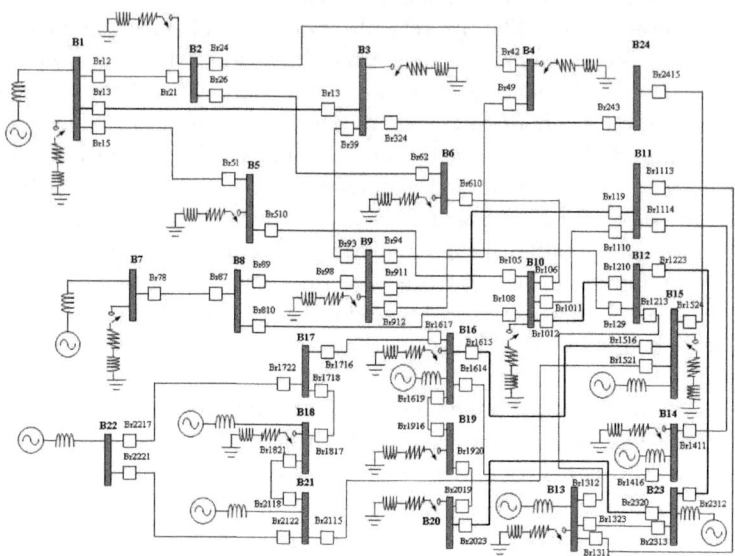

Fig. 2. Power network test system, [7]

2. Continuous data must be first treated, so, this information is analized by wavelets theory through the iterated filter bank technique.
3. Once the signal is decomposed, only the first decomposition is selected.
4. The sum of the values obtained in the first decomposition of the signal is taken as the pattern coefficient of that particular node.
5. The pattern coefficients and discrete data are inserted in the final model.
6. The diagnosis of each node being monitored is calculated. If there is a fault in a specific node, the type and location of it are given, else NO FAULT is printed.

There are 3 different evaluations:

1. First evaluation was made only to test and validate the HBN model in 25 scenarios with faults at randomly selected nodes (five simulations with one fault, five with two simultaneous faults, and so on, till five simultaneous faults).
2. Second and third evaluation were made for comparison purposes. 24 fault simulations for each evaluation were included, with faults at 4 test nodes (3, 9, 10 and 13). The results of the second evalutaion were obtained through the correct diagnose of one test node at a time, while in the third evaluation the results were obtained through the correct diagnose of the 4 test nodes in each scenario. The results were compared with the investigation presented in [3]. In this article, a 2-phase model for fault detection is elaborated; in the first phase, they used a discrete BN for generating a group of suspicious fault nodes, and in the second phase, the eigenvalues of the correlation matrix of

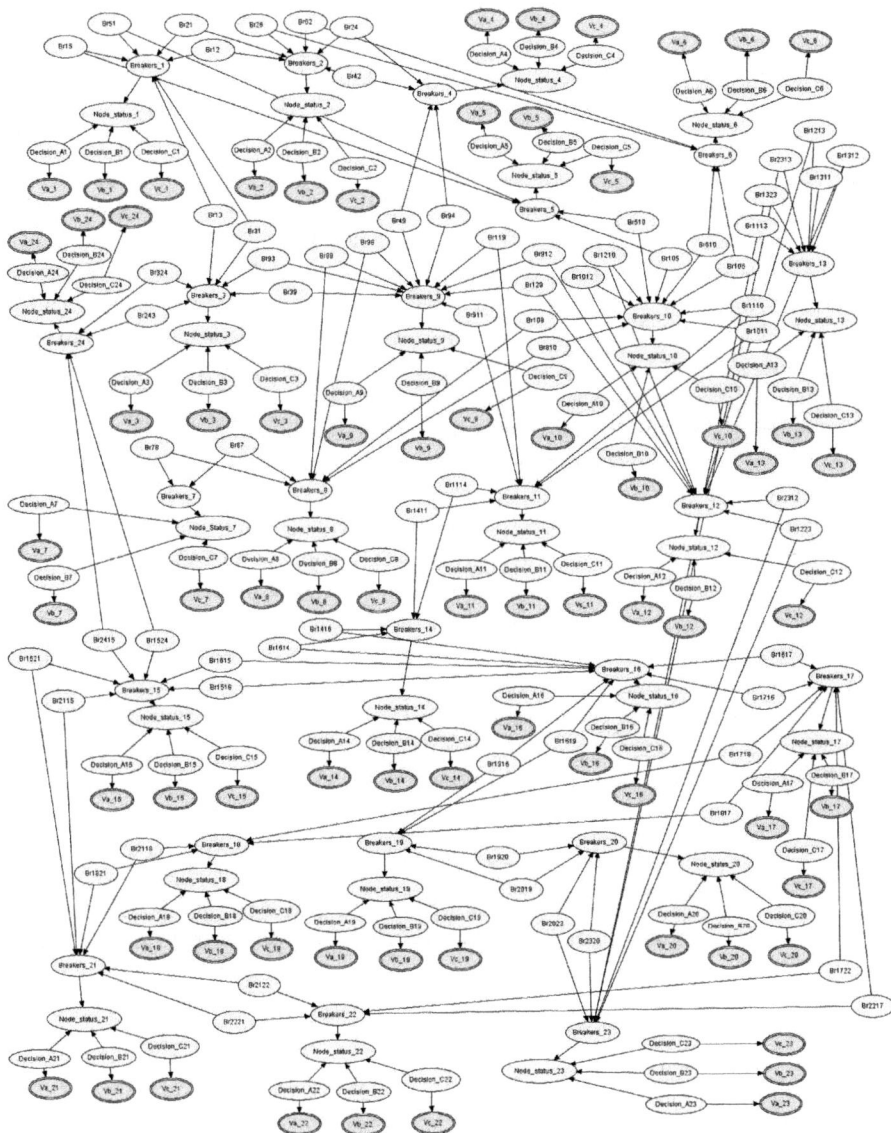

Fig. 3. Complete HBN model of power network

the continuous data from the suspicious nodes are computed, and inserted in a neural network, to confirm which of this suspicious nodes are faulty.

Continuous data voltages were taken from simulations of electrical network in the MicroTran software. This software is a reliable tool for the simulation of electromagnetic transients in power systems [8]. Examples of simulated faults are: one line to ground (A GND), two lines to ground (A-B GND), three lines to ground (A-B-C GND), or faults between two lines (A-B or B-C). Data where none of the nodes of the system were under a faulty situation (NO FAULT) are included. For every fault scenario, the discrete response of breakers associated to faulty nodes were simulated. The response of a breaker is given by one of three states: *OK*, which means not affected by a fault, *OPEN*, which means opening the circuit to isolate a fault, and *FAIL*, which means that the breaker malfunctioned and the breaker did not open when the event was detected.

The three evaluations were tested in our network under ideal conditions (no uncertainty), and the first evaluation also under non-ideal conditions (with uncertainty). In the first case there was no missing information or uncertainty in discrete data, and in the second case we simulated wrong reading on the status of protection breakers that can mislead the diagnosis.

The efficiency of the first evaluation can be seen in Table 1. The term efficiency refers to the percentage of the correct identification of nodes' status in the different scenarios. The results presented by the model with ideal conditions have high values of efficiency, even in scenarios with 5 simultaneous faults. The lowest efficiency is presented in scenarios with 3 and 4 faults. A possible explanation for this behavior is that the randomly faults were presented in nodes that are not directly connected to a generator, such as node 8, 9, 10 (see Figure 2), which may not recover as quickly from a failure. The results obtained with the model in non-ideal conditions have also high values of efficiency, which means that the scenarios were correctly diagnosed, except for the scenarios with 5 simultaneous faults, that presents a degradation level due to the interconnection effects of the faulted nodes in the electrical network.

Table 1. Test and Validation of the model

Number of Faults	efficiency (%) without uncertainty	efficiency (%) with uncertainty
1	100.0	100.0
2	90.0	80.0
3	86.67	80.0
4	85.0	80.0
5	96.0	56.0

In Table 2, the results of the second evaluation are presented. In this evaluation, 1 to 4 faults in random nodes were simulated in every scenario, but the results were only based on the correct determination of the status of 4 test nodes (3,9,10 and 13). The efficiency is almost perfect with ideal conditions, while with non-ideal conditions there is a decrease in the nodes with line-to-line faults and faults in neighbor nodes.

In Tables 3 and 4, a comparison of the results of our work with the research in [3] are shown. In this research, three evaluations with different number of fault samples in the same scenarios were made. In case 1 they use 75 % of faulty

Table 2. Evaluation 2 by type of fault

Fault type	efficiency (%) without uncertainty	efficiency (%) with noise
A-B-C-GND	100.0	100.0
A-B-GND	100.0	100.0
A-GND	80.0	80.0
A-B	100.0	75.0
B-C	100.0	50.0
NO FAULT	100.0	100.0

samples, in case 2 50 %, and in case 3 just 25 % of faulty samples. The scenarios contained from 1 to 4 faults in the specified test nodes (3, 9, 10 and 13).

Table 3. Evaluation by type of fault

Fault type	Case 1	Case 2	Case 3	HBN model without uncertainty
A-B-C-GND	100	100	100	100
A-B-GND	100	100	100	100
A-GND	100	85.7	92.9	100
A-B	100	83.3	50	88.9
B-C	100	68.8	68.8	81.3
NO FAULT	54.2	58.3	79.2	100

Table 4. Evaluation by node

Node number	Case 1	Case 2	Case 3	HBN Model without uncertainty
3	83.3	83.3	83.3	100
9	79.2	75	70.8	91.7
10	91.2	87.5	62.5	87.5
13	100	95.8	100	100

Tables 3 and 4 give a summary of the obtained percentages for each of the three cases considered in [3] and the work here presented, according to the type of fault in table 3, and according to the node in table 4. There exists degradation in the investigation by [3], mainly due to the similarity of some of their processed data, when there are more normal operation data than faulted (Case 2 and 3). According these 2 tables the proposal is competitive with the work done in [3], it even presents a better performance obtaining a higher efficiency of up to 15.65%, but with the advantage that there are no restrictions regarding the number of fault samples analyzed.

6 Conclusions

A fault detection framework for power systems, based on Hybrid Bayesian Networks and wavelets was presented. The wavelet processing analyses the continuous data of the power network, to extract the relevant fault features in pattern coefficients. These patterns, will then be used in the HBN model. This model combines the discrete observations of the system's components and the results of the application of wavelets in continuous data, and determines the status of the node (location), and the type of fault. The experiments have shown that the type of fault in most cases can be identified. The proposal was compared with a previous approach in the same domain [3], and have shown a better performance obtaining a higher efficiency of up to 15.65%.

Future work may include the extended version of the model here presented, to include the degradation of the power network (devices) with the time, to develop a more realistic project through the use of Dynamic HBN models.

References

1. Yongli, Z., Limin, H., Jinling, L.: Bayesian Networks-based Approach for Power Systems Fault Diagnosis. IEEE Trans. on Power Delivery 2005 21(2), 634–639 (2006)
2. Garza Castañón, L., Acevedo, P.S., Cantú, O.F.: Integration of Fault Detection and Diagnosis in a Probabilistic Logic Framework. In: Garijo, F.J., Riquelme, J.-C., Toro, M. (eds.) IBERAMIA 2002. LNCS (LNAI), vol. 2527, pp. 265–274. Springer, Heidelberg (2002)
3. Garza Castañón, L., Nieto, G.J., Garza, C.M., Morales, M.R.: Fault Diagnosis of Industrial Systems with Bayesian Networks and Neural Networks. In: Gelbukh, A., Morales, E.F. (eds.) MICAI 2008. LNCS (LNAI), vol. 5317, pp. 998–1008. Springer, Heidelberg (2008)
4. Elkalashy, N.I., Lehtonen, M., Tarhuni, N.G.: DWT and Bayesian technique for enhancing earth fault protection in MV networks. In: Power Systems Conference and Exposition, PSCE 2009, IEEE/PES, Power Syst.& High Voltage Eng., Helsinki Univ. of Technol., Helsinki, April 23, pp. 89–93 (2009)
5. Valens, C., (Copyright Valens, C., 1999-2004).: A Really Friendly Guide to Wavelets. PolyValens, http://www.polyvalens.com/ (recovered February 1, 2010)
6. Jensen, F.V.: Bayesian Networks and Influence Diagrams. Aalborg University. Department of Mathematics and Computer Science, Denmark
7. Reliability Test System Task Force, Application of Probability Methods Subcomitee. IEEE Reliability Test System. IEEE Transactions on Power Apparatus and Systems 98(6), 2047–2054 (1979)
8. MicroTran official webpage, http://www.microtran.com/

Learning Temporal Bayesian Networks for Power Plant Diagnosis

Pablo Hernandez-Leal[1], L. Enrique Sucar[1], Jesus A. Gonzalez[1], Eduardo F. Morales[1], and Pablo H. Ibarguengoytia[2]

[1] National Institute of Astrophysics, Optics and Electronics
Tonantzintla, Puebla, Mexico
{pablohl,esucar,jagonzalez,emorales}@inaoep.mx
[2] Electrical Research Institute
Cuernavaca, Morelos, Mexico
pibar@iie.org.mx

Abstract. Diagnosis in industrial domains is a complex problem because it includes uncertainty management and temporal reasoning. Dynamic Bayesian Networks (DBN) can deal with this type of problem, however they usually lead to complex models. Temporal Nodes Bayesian Networks (TNBNs) are an alternative to DBNs for temporal reasoning that result in much simpler and efficient models in certain domains. However, methods for learning this type of models from data have not been developed. In this paper we propose a learning algorithm to obtain the structure and temporal intervals for TNBNs from data. The method has three phases: (i) obtain an initial interval approximation, (ii) learn the network structure based on the intervals, and (iii) refine the intervals for each temporal node. The number of possible sets of intervals is obtained for each temporal node based on a clustering algorithm and the set of intervals that maximizes the prediction accuracy is selected. We applied this method to learn a TNBN for diagnosis and prediction in a combined cycle power plant. The proposed algorithm obtains a simple model with high predictive accuracy.

1 Introduction

Power plants and their effective operation are vital to the development of industries, schools, and even for our houses, for this reason they maintain strict regulations and quality standards. However, problems may appear and when these happen, human operators have to take decisions relying mostly on their experience to determine the best recovery action with very limited help from the system. In order to provide useful information to the operator, different models have been developed that can deal with industrial diagnosis. These models must manage uncertainty because real world information is usually imprecise, incomplete, and with errors (noisy). Furthermore, they must manage temporal reasoning, since the timing of occurrence of the events is an important piece of information.

Bayesian Networks [9] are an alternative to deal with uncertainty that has proven to be successful in various domains. Nevertheless, these models cannot deal with temporal information. An extension of BNs, called Dynamic Bayesian Networks (DBNs), can deal with temporal information. DBNs can be seen as multiple slices of a *static* BN over time, with links between adjacent slices. However, these models can become quite complex, in particular, when only a few important events occur over time.

Temporal Nodes Bayesian Networks (TNBNs) [1] are another extension of Bayesian Networks. They belong to a class of temporal models known as *Event Bayesian Networks* [5]. TNBNs were proposed to manage uncertainty and temporal reasoning. In a TNBN, each Temporal Node has intervals associated to it. Each node represents an event or state change of a variable. An arc between two Temporal Nodes corresponds to a causal–temporal relation. One interesting property of this class of models, in contrast to Dynamic Bayesian Networks, is that the temporal intervals can differ in number and size.

TNBNs have been used in diagnosis and prediction of temporal faults in a steam generator of a fossil power plant [1]. However, one problem that appears when using TNBNs is that no learning algorithm exists, so the model has to be obtained from external sources (i.e., a domain expert). This can be a hard and prone to error task. In this paper, we propose a learning algorithm to obtain the structure and the temporal intervals for TNBNs from data, and apply it to the diagnosis of a combined cycle power plant.

The learning algorithm consists of three phases. In the first phase, we obtain an approximation of the intervals. For this, we apply a clustering algorithm. Then we convert these clusters into initial intervals. In the second phase, the BN structure is obtained with a structure learning algorithm [2]. The last step is performed to refine the intervals for each Temporal Node. Our algorithm obtains the number of possible sets of intervals for each configuration of the parents by clustering the data based on a Gaussian mixture model. It then selects the set of intervals that maximizes the prediction accuracy. We applied the proposed method to fault diagnosis in a subsystem of a power plant. The data was obtained from a power plant simulator. The structure and intervals obtained by the proposed algorithm are compared to a uniform discretization and a k-means clustering algorithm; the results show that our approach creates a simpler TNBN with high predictive accuracy.

2 Related Work

Bayesian Networks (BN) have been applied to industrial diagnosis [6]. However, *static* BNs are not suited to deal with temporal information. For this reason Dynamic Bayesian Networks [3] were created. In a DBN, a copy of a base model is created for each time stage. These copies are linked via a transition network which is usually connected through links only allowing connections between consecutive stages (Markov property). The problem is that DBNs can become very complex; and this is unnecessary when dealing with problems for which there are only few changes for each variable in the model. Moreover, DBNs are not capable of managing different levels of time granularity. They have a fixed time interval between stages.

In TNBNs, each variable represents an event or state change. So, only one (or a few) instance(s) of each variable is required, assuming there is one (or a few) change(s) of a variable state in the temporal range of interest. No copies of the model are needed, and no assumption about the Markovian nature of the process is made. TNBNs can deal with multiple granularity because the number and the size of the intervals for each node can be different.

There are several methods to learn BNs from data [8]. Unfortunately, the algorithms used to learn BNs cannot deal with the problem of learning temporal intervals, so these cannot be applied directly to learn TNBNs. To the best of our knowledge, there is only one previous work that attempts to learn a TNBN. Liu et al. [7] proposed a method to build a TNBN from a *temporal probabilistic database*. The method obtains the structure from a set of temporal dependencies in a *probabilistic* temporal relational model (PTRM). In order to build the TNBN, they obtain a variable ordering that maximizes the set of conditional independence relations implied by a dependency graph obtained from the PTRM. Based on this order, a directed acyclic graph corresponding to the implied independence relations is obtained, which represents the structure of the TNBN. The previous work assumes a known probabilistic temporal–relational model from the domain of interest, which is not always available. Building this PTRM can be as difficult as building a TNBN. In contrast, our approach constructs the TNBN directly from data, which in many applications is readily available or can be generated, for instance, using a simulator.

3 Temporal Nodes Bayesian Networks

A Temporal Nodes Bayesian Network (TNBN) [1,5] is composed by a set of Temporal Nodes (TNs). TNs are connected by edges. Each edge represents a causal-temporal relationship between TNs. There is at most one state change for each variable (TN) in the temporal range of interest. The value taken by the variable represents the interval in which the event occurs. Time is discretized in a finite number of intervals, allowing a different number and duration of intervals for each node (multiple granularity). Each interval defined for a child node represents the possible delay between the occurrence of one of its parent events (cause) and the corresponding child event (effect). Some Temporal Nodes do not have temporal intervals. These correspond to Instantaneous Nodes. Formally:

Definition 1. *A TNBN is defined as a pair* $B = (G, \Theta)$. *G is a Directed Acyclic Graph*, $G = (\mathbf{V}, \mathbf{E})$. *G is composed of* $\mathbf{V}$, *a set of Temporal and Instantaneous Nodes;* $\mathbf{E}$ *a set of edges between Nodes. The Θ component corresponds to the set of parameters that quantifies the network. Θ contains the values* $\Theta_{v_i} = P(v_i | Pa(v_i))$ *for each* $v_i \in \mathbf{V}$; *where* $Pa(v_i)$ *represents the set of parents of* v_i *in* G.

Definition 2. *A Temporal Node, v_i, is defined by a set of states $\mathbf{S}$, each state is defined by an ordered pair $S = (\lambda, \tau)$, where λ is the value of a random variable and $\tau = [a, b]$ is the interval associated, with initial value a and final value b, that corresponds to the time interval in which the variable state changes to λ. In addition, each Temporal Node contains an extra default state $s = ($ 'no change', $\emptyset)$, which has no interval associated. If a Node has no intervals defined for all its states then it receives the name of Instantaneous Node.*

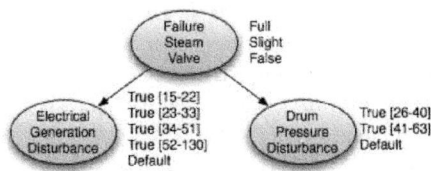

Fig. 1. The TNBN for Example 1. Each oval represents a node. The Failure Steam Valve is an Instantaneous Node, so it does not have temporal intervals. The Electrical Generation Disturbance and Drum Pressure Disturbance are Temporal Nodes. Therefore, they have temporal intervals associated with their values.

Example 1. Assume that at time $t = 0$, a Failure in a Steam Valve occurs. This kind of failure can be classified as *Full, Slight* and *False*. To simplify the model we will consider only two immediate consequences in the plant process, the Disturbance in Electrical Generation (DEG) and the Disturbance in the Drum Pressure (DDP). These events are not immediate, they depend on the severity of the accident, therefore, have temporal intervals associated. For the DEG node four intervals are defined $[15-22], [23-33], [34-51], [52-130]$, for the DDP node two intervals are defined $[26-40], [41-63]$. These intervals represent that the state of the node changed during that period of time. A TNBN for this simple example is shown in Figure 1.

4 Learning Algorithm

First, we present the interval learning algorithm for a TN, assuming that we have a defined structure, and later we present the whole learning algorithm.

4.1 Interval Learning

Initially, we will assume that the events follow a known distribution. With this idea, we can use a clustering algorithm with the temporal data. Each cluster corresponds, in principle, to a temporal interval. The algorithm is presented first by ignoring the values of the parent nodes (first approximation). Later we refine the method by incorporating the parent nodes configurations.

4.2 First Approximation: Independent Variables

Our approach uses a Gaussian mixture model (GMM) to perform an approximation of the data. Therefore, we can use the Expectation-Maximization (EM) algorithm [4]. EM works iteratively using two steps: (i) The E-step tries to *guess* the parameters of the Gaussian distributions, (ii) the M-step updates the parameters of the model based on the previous step. By applying EM, we obtain a number of Gaussians (clusters), specified by their mean and variance. For now, assume that the number of temporal intervals (Gaussians), k, is given. For each TN, we have a dataset of points over time and these are clustered using GMM, to obtain k Gaussian distributions. Based on the parameters of each Gaussian, each temporal interval is initially defined by: $[\mu - \sigma, \mu + \sigma]$.

Now we will deal with the problem of finding the number of intervals. The ideal solution has to fulfill two conditions: (i) The number of intervals must be small, in order to reduce the complexity of the network, and (ii) the intervals should yield good estimations when performing inference over the network. Based on the above, our approach uses the EM algorithm with the parameter for the number of clusters in the range from 1 to ℓ, where ℓ is the highest value (for the experiments in this paper we used $\ell = 3$).

To select the best set of intervals an evaluation is performed over the network, which is an indirect measure of the quality of the intervals. In particular, we used the *Relative Brier Score* to measure the predictive accuracy of the network. The selected set of intervals for each TN are those that maximize the Relative Brier Score. The Brier Score is defined as $BS = \sum_{i=1}^{n}(1-P_i)^2$, where P_i is the marginal posterior probability of the correct value of each node given the evidence. The maximum brier score is $BS_{max} = \sum_n 1^2$. The Relative Brier Score (RBS) is defined as: RBS $(in\ \%) = (1 - \frac{BS}{BS_{max}}) \times 100$. This RBS is used to evaluate the TNBN instantiating a random subset of variables in the model, predicting the unseen variables and obtaining the RBS for these predictions.

4.3 Second Approximation: Considering the Network Topology

Now we will construct a more precise approximation. For this, we use the configurations (combination of the values of the nodes) of the parent nodes. The number of configurations of each node i is $q_i = \prod_{x \in Pa(i)} |s_x|$ (the product of the number of states of the parents nodes).

Formally, we construct partitions of the data (disjoint sets of values), one partition for each configuration. Then we get the combinations taking 2 partitions p_i, p_j from the total. This yields $q(q-1)/2$ different combinations of partitions. For p_i and p_j, we apply the first approximation and obtain ℓ sets of intervals for each partition. Then, we obtain the combination of these sets of intervals, that yield ℓ^2 sets of final intervals for each p_i, p_j. For example, if a node has parents: X (with states a, b) and Y (with states c, d), there are 4 partitions in total. We select two partitions out of those four and apply the first approximation to each of them to obtain different intervals. After this process is applied, we have different sets of intervals, that we need to adjust using Algorithm 1.

Algorithm 1. Algorithm to adjust the intervals
Require: Array of intervals sets
Ensure: Array of intervals adjusted
 1: **for** each set of intervals s **do**
 2:    sortIntervalsByStart(s)
 3:    **while** Interval i is contained in Interval j **do**
 4:       tmp=AverageInterval(i,j)
 5:       s.replaceInterval(i,j,tmp)
 6:    **end while**
 7:    **for** $k = 0$ to number of intervals in set s-1 **do**
 8:       Interval[k].end=(Interval[k].end + Interval[k+1].start)/2
 9:    **end for**
10: **end for**

Algorithm 1 performs two nested loops, for each set of intervals, we sort the intervals by their starting point. Then checks if there is an interval contained in another interval. While this is true, the algorithm obtains an average interval, taking the average of the start and end points of the intervals and replacing these two intervals with the new one. Next, it refines the intervals to be continuous by taking the mean of two adjacent values.

As in the first approximation, the best set of intervals for each TN is selected based on the predictive accuracy in terms of their RBS. However, when a TN has as parents other Temporal Nodes (an example of this situation is illustrated in Figure 3), the state of the parent nodes is not initially known. So, we cannot directly apply the second approximation. In order to solve this problem, the intervals are selected sequentially in a top–down fashion according to the TNBN structure. That is, we first select the intervals for the nodes in the second level of the network (the root nodes are instantaneous by definition in a TNBN [1]). Once these are defined, we know the values of the parents of the nodes in the 3rd level, so we can find their intervals; and so on, until the leaf nodes are reached.

4.4 Pruning

Taking the combinations and joining the intervals can become computationally too expensive, the number of sets of intervals for node is in $O(q^2 \ell^2)$ where q is the number of configurations and ℓ is the maximum number of clusters for the GMM. For this reason we used two pruning techniques for each TN to reduce the computation time.

The first pruning technique discriminates the partitions that contain few instances. For this, we count the number of instances in each partition, and if it is greater than a value $\beta = \frac{\text{Number of instances}}{\text{Number of partitions} \times 2}$ the configuration is used, otherwise it is discarded. A second technique is applied when the intervals for each combination are being obtained. If the final set of intervals contains only one interval (no temporal information) or more than α (producing a complex network), the set of intervals is discarded. For our experiments we used $\alpha = 4$.

4.5 Structural Learning

Now we present the complete algorithm that learns the structure and the intervals of the TNBN. First we perform an initial discretization of the temporal variables based on a clustering algorithm (k-means), the obtained clusters are converted into intervals according to the process shown in Algorithm 2. With this process, we obtain an initial approximation of the intervals for all the Temporal Nodes and we can perform a standard BN structural learning. We used the K2 algorithm [2]. This algorithm has as a parameter an ordering of the nodes. For learning TNBN, we can exploit this parameter and define an order based on the temporal domain information.

When a structure is obtained, we can apply the interval learning algorithm described in Section 4.1. Moreover, this process of alternating interval learning and structure learning may be iterated until convergence.

Algorithm 2. Algorithm to obtain the initial intervals
Require: Sorted Points *point* obtained by k-means algorithm, An array of continuous values *data* from Node n, *min* minimum value of *data*, *max* maximum value of *data*.
Ensure: Array of intervals for a Node n.
1: Interval[0].start=min,Interval[0].end=average(point[0],point[1])
2: **for** i=0 to size(*point*)-1 **do**
3:     Interval[i+1].start=average(point[i],point[i+1]) ,
4:     Interval[i+1].end=average(point[i+1],point[i+2])
5: **end for**
6: Interval[i].start=average(point[i],point[i+1]),Interval[i].end=max

5 Application to Power Plant Diagnosis

The proposed algorithm was tested on a subsystem of a combined cycle power plant. A simplified diagram is shown in Figure 2. In the process, a signal exceeding its specified limit of normal functioning is called an event.

5.1 Application Domain

A power plant mainly consists of three equipments: the steam generator (HRSG), the steam turbine and the electric generator. The steam generator, with the operation of burners, produces steam from the feed-water system. After the steam is superheated, it is introduced to the steam turbine to convert the energy carried out by the steam in work and finally in electricity through the corresponding steam generator.

The HRSG consists of a huge boiler with an array of tubes, the drum and the circulation pump. The feed-water flows through the tubes receiving the heat provided by the gases from the gas turbine and the burners. Part of the water

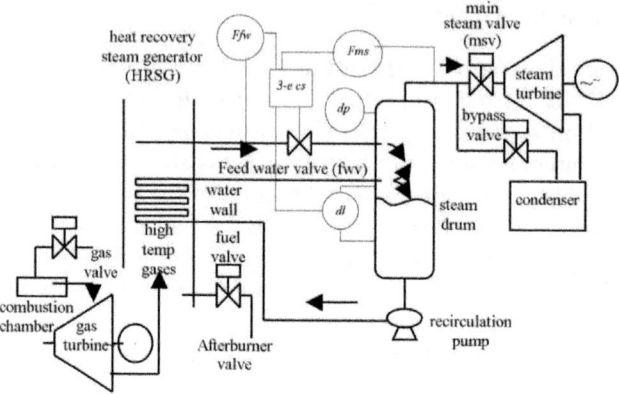

Fig. 2. Schematic description of a power plant showing the feedwater and main steam subsystems. Ffw refers to feedwater flow, Fms refers to main stream flow, dp refers to drum pressure, dl refers to drum level.

mass becomes steam and is separated from the water in a special tank called the drum. Here, water is stored to provide the boiler tubes with the appropriate volume of liquid that will be evaporated and steam is stored to be sent to the steam turbine.

From the drum, water is supplied to the rising water tubes called water walls by means of the water recirculation pump, where it will be evaporated, and water-steam mixture reaches the drum. From here, steam is supplied to the steam turbine. The conversion of liquid water to steam is carried out at a specific saturation condition of pressure and temperature. In this condition, water and saturated steam are at the same temperature. This must be the stable condition where the volume of water supply is commanded by the feed-water control system. Furthermore, the valves that allow the steam supply to the turbine are controlled in order to manipulate the values of pressure in the drum. The level of the drum is one of the most important variables in the generation process. A decrease of the level may cause that not enough water is supplied to the rising tubes and the excess of heat and lack of cooling water may destroy the tubes. On the contrary, an excess of level in the drum may drag water as humidity in the steam provided to the turbine and cause a severe damage in the blades. In both cases, a degradation of the performance of the generation cycle is observed.

Even with a very well calibrated instrument, controlling the level of the drum is one of the most complicated and uncertain processes of the whole generation system. This is because the mixture of steam and water makes very difficult the reading of the exact level of mass.

5.2 Experiments and Results

For obtaining the data used in the experiments, we used a full scale simulator of the plant, then we simulate two failures randomly: failure in the Water Valve and failure in the Steam Valve. These types of failures are important because, they may cause disturbances in the generation capacity and the drum.

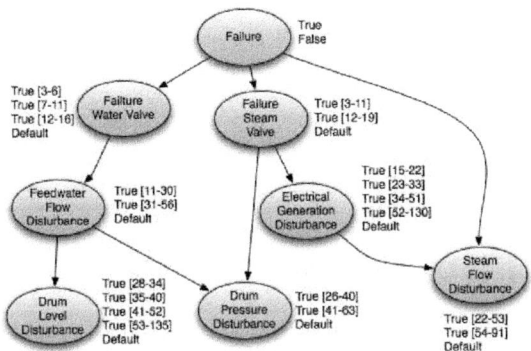

Fig. 3. The learned TNBN for a subsystem of a combined cycle power plant. For each node the obtained temporal intervals are shown. The TNBN presents the possible effects of the failure of two valves over different important components.

In order to evaluate our algorithm, we obtained the structure and the intervals for each Temporal Node with the proposed algorithm. In this case, we do not have a reference network, so to compare our method, we used as baselines an equal-width discretization (EWD) and a K-means clustering algorithm to obtain the intervals for each TN. We evaluated the model using three measures: (i) the predictive accuracy using RBS, (ii) the error in time defined as the difference between the real event and the expected mean of the interval, and (iii) the number of intervals in the network. The best network should have high predictive RBS, low error in time and low complexity (reduced number of intervals).

We performed three experiments varying the number of cases. First, we generate the data with the simulator, then we learned the structure and the intervals. Finally, we used the learned network to compare the results with the original data. The results are presented in Table 1. The network obtained with the proposed algorithm with higher accuracy is presented in Figure 3.

The following observations can be obtained from these results. In all the experiments, our algorithm obtained the best RBS score and the lowest number of intervals. The K-means and EW discretization obtained the best score in time error. However, this happens because they produced a high number of intervals of smaller size, which decreases the difference between the mean of an interval and the real event. Even though our algorithm does not obtain the best time error, it is not far from the other algorithms. It is important to note that our algorithm obtains the highest accuracy with a simpler model.

Table 1. Evaluation on the power plant domain. We compare the proposed algorithm (Prop), K-means clustering and equal-width discretization (EWD) in terms of predictive accuracy (RBS), time error and number of intervals generated.

Num. of Cases	Algorithm	RBS (Max 100)	Time Error	Average num. intervals
50	Prop.	**93.26**	18.02	**16.25**
50	K-means	83.57	**15.6**	24.5
50	EWD	85.3	16.5	24.5
75	Prop.	**93.7**	17.8	**16**
75	K-means	85.7	**16.3**	24.5
75	EWD	86.9	17.2	24.5
100	Prop.	**93.37**	17.7	**17**
100	K-Means	90.4	17.1	24.5
100	EW D	91.9	**15.29**	24.5

6 Conclusions and Future Research

We have developed a method for learning both the structure and the temporal intervals for a TNBN from data. The method generates initially a set of candidate intervals for each Temporal Node based on a Gaussian clustering algorithm, and then the best intervals are selected based on predictive accuracy. We evaluated our method with data generated by a power plant simulator. The proposed method produces a simpler (low number of intervals) and better (high predictive accuracy) model than EWD and K-means clustering. As future work we propose to evaluate our model with a larger industrial case and apply the algorithm in other domains such as a medical case.

References

1. Arroyo-Figueroa, G., Sucar, L.E.: A temporal Bayesian network for diagnosis and prediction. In: Proceedings of the 15th UAI Conference, pp. 13–22 (1999)
2. Cooper, G.F., Herskovits, E.: A bayesian method for the induction of probabilistic networks from data. Machine learning 9(4), 309–347 (1992)
3. Dagum, P., Galper, A., Horvitz, E.: Dynamic network models for forecasting. In: Proc. of the 8th Workshop UAI, pp. 41–48 (1992)
4. Dempster, A.P., Laird, N.M., Rubin, D.B.: Maximum likelihood from incomplete data via the EM algorithm. Journal of the Royal Statistical Society 39(1), 1–38 (1977)
5. Galán, S.F., Arroyo-Figueroa, G., Díez, F.J., Sucar, L.E.: Comparison of two types of Event Bayesian Networks: A case study. Applied Artif. Intel. 21(3), 185 (2007)
6. Knox, W.B., Mengshoel, O.: Diagnosis and Reconfiguration using Bayesian Networks: An Electrical Power System Case Study. In: SAS 2009, p. 67 (2009)
7. Liu, W., Song, N., Yao, H.: Temporal Functional Dependencies and Temporal Nodes Bayesian Networks. The Computer Journal 48(1), 30–41 (2005)
8. Neapolitan, R.E.: Learning Bayesian Networks. Pearson Prentice Hall, London (2004)
9. Pearl, J.: Probabilistic reasoning in intelligent systems: networks of plausible inference. Morgan Kaufmann, San Francisco (1988)

On the Fusion of Probabilistic Networks

Salem Benferhat[1] and Faiza Titouna[2]

[1] Université Lille-Nord de France Artois, F-62307 Lens, CRIL, F-62307 Lens
CNRS UMR 8188, F-62307 Lens
benferhat@cril.fr
[2] Université Badji Mokhtar, Annaba, Algeria

Abstract. This paper deals with the problem of merging multiple-source uncertain information in the framework of probability theory. Pieces of information are represented by probabilistic (or bayesian) networks, which are efficient tools for reasoning under uncertainty. We first show that the merging of probabilistic networks having the same graphical (DAG) structure can be easily achieved in polynomial time. We then propose solutions to merge probabilistic networks having different structures. Lastly, we show how to deal with the sub-normalization problem which reflects the presence of conflicts between different sources.

1 Introduction

This paper addresses the problem of fusion of information multi-source represented in the framework of probability theory. Uncertain pieces of information are assumed to be represented by probabilistic networks. Probabilistic networks [3] are important tools proposed for an efficient representation and analysis of uncertain information. They are directed acyclic graphs (DAG), where each node encodes a variable and every edge represents a causal or influence relationship between two variables. Uncertainties are expressed by means of conditional probability distributions for each node in the context of its parents.

Several works have been proposed to fuse propositional or weighted logical knowledge bases issued from different sources (e.g., [1], [6]). However there are few works on the fusion of belief networks in agreement with the combination laws of probability distributions, see [7], [8], [4] and [5]. In these existing works fusion is based on exploiting intersections or unions of independence relations induced by individual graphs. In this paper, we are more interested in computing the counterparts of fusing probability distributions, associated with bayesian networks. Results of this paper can be viewed as a natural counterpart of the ones recently developed in a possibility theory for merging possibilistic networks [2]. In fact, all main steps proposed in [2] have natural counterparts. More precisely, we propose an approach to merge n probabilistic networks. The obtained bayesian network is such that its associated probability distribution, is a function of probability distributions associated with initial bayesian networks to merge. The merging operator considered in this paper is the product operator. We study, in particular, the problem of sub-normalization that concerns the conflict between information sources.

After giving a brief background on probabilistic networks, we present the fusion of probabilistic networks having same graphical structures. Then we discuss the problem of fusing probabilistic networks having different structures. Next we handle the subnormalization problem which reflects the presence of conflicts between sources. Last section concludes the paper.

2 Probabilistic Networks

Let $V = \{A_1, A_2, ..., A_N\}$ be a set of variables. We denote by $D_A = \{a_1, ..., a_n\}$ the domain associated with the variable A. By a we denote any instance of A. $\Omega = \times_{A_i \in V} D_{A_i}$ denotes the universe of discourse, which is the cartesian product of all variable domains in V. Each element $\omega \in \Omega$ is called a state or an interpretation or a solution. Subsets of Ω are simply called events. This section defines quantitative probabilistic graphs. A *probabilistic network* on a set of variables V, denoted by $\mathbb{B} = (p_\mathbb{B}, G_\mathbb{B})$, consists of:

- a *graphical component*, denoted by $G_\mathbb{B}$, which is a DAG (Directed Acyclic Graph). Nodes represent variables and edges encode the influence between variables. The set of parent of a node A is denoted by U_A and μ_A denotes an instance of parents of A. We say that C is a child of A if there exists an edge from A to C. We define descendants of a variable A as the set of nodes obtained by applying the transitivity closure of relation child.
- a *numerical component*, denoted by $p_\mathbb{B}$, which quantifies different links of the network. For every root node A uncertainty is represented by a conditional probability degree $p_\mathbb{B}(a \mid u_A)$ of each instances $a \in D_A$ and $u_A \in D_{U_A}$.

In the following, probability distributions $p_\mathbb{B}$, defined on nodes level, are called local probability distributions. From the set of local conditional probability distributions, one can define a unique global joint probability distribution.

Definition 1. *Let $\mathbb{B} = (p_\mathbb{B}, G_\mathbb{B})$ be a quantitative probabilistic network. The joint or global probability distribution associated with $\mathbb{B}$ and denoted by $p_\mathbb{B}^J$, is expressed by the following quantitative chain rule:*

$$p_\mathbb{B}^J(A_1, .., A_N) = \prod_{i=1..N} p(A_i \mid U_{A_i}). \tag{1}$$

2.1 Probabilistic Merging

One of the important aim of merging uncertain information is to exploit complementarities between the sources in order to get a complete, precise and global point of view on a given problem. Given a set of probability distributions $p_i's$, the combination mode considered here is the product of probability distributions, namely

$$\forall \omega, p_{\oplus(\omega)} = \prod_{i=1,n} p_i(\omega). \tag{2}$$

Note that $p_{\oplus(\omega)}$ is generally subnormalized. Therefore, the fused probability distribution should be normalized.

2.2 Aims of the Work

For a simplicity and clarity sakes, we restrict ourselves to the case of the fusion of two probabilistic networks. Since the product operator is symmetric and associative, fusion methods presented in this paper can be easily extended to merging n probabilistic networks. Let $\mathbb{B}1$ and $\mathbb{B}2$ be two probabilistic networks. Our aim consists of computing directly from $\mathbb{B}1$ and $\mathbb{B}2$ a new probabilistic network, denoted by $\mathbb{B}\oplus$. The new probabilistic network represents the result of fusion of the probabilistic networks $\mathbb{B}1$ and $\mathbb{B}2$, using the product operator.

3 Fusion of Same-Structure Networks

This section presents a procedure for merging probabilistic networks having a same DAG structure. Namely, the probabilistic networks to merge, denoted by $\mathbb{B}1$ and $\mathbb{B}2$, only differ on conditional probability distributions assigned to variables. The following definition and proposition show that merging networks having a same structure is immediate.

Definition 2. *Let* $\mathbb{B}1 = (p_{\mathbb{B}1}, G_{\mathbb{B}1})$ *and* $\mathbb{B}2 = (p_{\mathbb{B}2}, G_{\mathbb{B}2})$ *be two probabilistic networks such that* $G_{\mathbb{B}1} = G_{\mathbb{B}2}$. *The result of merging* $\mathbb{B}1$ *and* $\mathbb{B}2$ *is a probabilistic network denoted by* $\mathbb{B}\oplus = (p_{\mathbb{B}\oplus}, G_{\mathbb{B}\oplus})$, *where :*

- $G_{\mathbb{B}\oplus} = G_{\mathbb{B}1} = G_{\mathbb{B}2}$ *and*
- *The local conditional probability distributions* $p_{\mathbb{B}\oplus}$'s *are defined by:*
  $\forall A, p_{\mathbb{B}\oplus}(A|U_A) = p_{\mathbb{B}1}(A|U_A) * p_{\mathbb{B}2}(A|U_A)$, *where A is a variable and* U_A *is the set of parents of A.*

Namely, the result of merging $\mathbb{B}1$ and $\mathbb{B}2$ is a probabilistic network such that its DAG is the one of $\mathbb{B}1$ and $\mathbb{B}2$, and its local probability distributions are the product of local probability distributions of $\mathbb{B}1$ and $\mathbb{B}2$. The following proposition shows that $\mathbb{B}\oplus$ is exactly the result of merging $\mathbb{B}1$ and $\mathbb{B}2$. In other words, the joint probability distributions associated with $\mathbb{B}\oplus$ is equal to product of joint probability distributions associated with $\mathbb{B}1$ and $\mathbb{B}2$.

Proposition 1. *Let* $\mathbb{B}1 = (p_{\mathbb{B}1}, G_{\mathbb{B}1})$ *and* $\mathbb{B}2 = (p_{\mathbb{B}2}, G_{\mathbb{B}2})$ *be two probabilistic networks having exactly the same associated DAG. Let* $\mathbb{B}\oplus = (p_{\mathbb{B}\oplus}, G_{\mathbb{B}\oplus})$ *be the result of merging* $\mathbb{B}1$ *and* $\mathbb{B}2$ *using the above definition. Then, we have :*

$$\forall \omega \in \Omega, p_{\mathbb{B}\oplus}^J(\omega) = p_{\mathbb{B}1}^J(\omega) * p_{\mathbb{B}2}^J(\omega),$$

where $p_{\mathbb{B}\oplus}^J, p_{\mathbb{B}1}^J, p_{\mathbb{B}2}^J$ *are respectively the probability distributions associated with* $\mathbb{B}\oplus, \mathbb{B}1, \mathbb{B}2$ *using Definition 1.*

4 Fusion of U-Acyclic Networks

The above section has shown that the fusion of probabilistic networks can be easily achieved if they share the same DAG structure.

This section considers the case when the networks to merge have not the same structure. However we assume that their union does not contain a cycle.

A union of two DAGs (G_1, G_2) is a graph where:

- The set of its variables is the union of the sets of variables in G_1 and in G_2.
- For each variable A, its parents are those in G_1 and G_2.

If the union of G_1 and G_2 does not contain cycles, we say that G_1 and G_2 are U-acyclic networks. In this case the fusion can be easily obtained. The fusion of networks which is not U-acyclic is left for a future work. But we first need to introduce some intermediaries results concerning the addition of variables and arcs to a given probabilistic network.

4.1 Adding Variables and Arcs

The following proposition shows how to add new variables to a probabilistic network without changing its joint probability distribution:

Proposition 2. *Let* $\mathbb{B} = (p_\mathbb{B}, G_\mathbb{B})$ *be a probabilistic network defined on a set of variables V. Let A be a new variable which does not belong to V. Let* $\mathbb{B}1 = (p_{\mathbb{B}1}, G_{\mathbb{B}1})$ *be a new probabilistic network such that:*

- $G_{\mathbb{B}1}$ *is equal to* $G_\mathbb{B}$ *plus the node A,*
- $p_{\mathbb{B}1}$ *is identical to* $p_\mathbb{B}$ *for all the variables in V, and equal to a uniform probability distribution on the node A (namely,* $\forall a \in D_A$, $p_{\mathbb{B}1}(a) = 1 < \div \mid D_A \mid$, *where* $\mid D_A \mid$ *represents the number of elements in* D_A.).

Then, we have:

$$\forall \omega \in \times_{A_i \in V} D_{A_i}, p_\mathbb{B}^J(\omega) = \sum_{a \in D_A} p_{\mathbb{B}1}^J(a\omega). \tag{3}$$

Where $p_\mathbb{B}^J$ *and* $p_{\mathbb{B}1}^J$ *are respectively the probability distributions associated with* $\mathbb{B}$ *and* $\mathbb{B}1$ *using (1).*

The above proposition is essential for a fusion process since it allows, if necessary, to equivalently increase the size of all the probabilistic networks to merge, such that they will use the same set of variables.

The following proposition shows how to add links to a probabilistic network without modifying its probability distributions.

Proposition 3. *Let* $\mathbb{B} = (p_\mathbb{B}, G_\mathbb{B})$ *be a probabilistic network. Let A be a variable, and let* U_A *be the set of parents of A in* $G_\mathbb{B}$. *Let* $B \notin U_A$ *and B is not a descendant of A. Let* $\mathbb{B}1 = (p_{\mathbb{B}1}, G_{\mathbb{B}1})$ *be a new probabilistic network obtained from* $\mathbb{B}$ *by adding a link from B to A. The new conditional probability distributions associated with* $\mathbb{B}1$ *are:*

- at the level of the node A:
 $\forall a \in D_A, \forall b \in D_B, \forall u_a \in D_{U_A}, p_{\mathbb{B}1}(a \mid u_a b) = p_\mathbb{B}(a \mid u_a)$.
- for the other nodes:
 $\forall C, C \neq A, \forall c \in D_C, \forall u_c \in D_{U_C}, p_{\mathbb{B}1}(c \mid u_c) = p_\mathbb{B}(c \mid u_c)$.

Then: $\forall \omega, p_\mathbb{B}^J(\omega) = p_{\mathbb{B}1}^J(\omega)$.

4.2 Computing the Fused Network

Using results of the previous section and Proposition 1, the fusion of two U-acyclic networks, is immediate. Let $G_{\mathbb{B}\oplus}$ be the union of $G_{\mathbb{B}1}$ and $G_{\mathbb{B}2}$. Then the fusion of $\mathbb{B}1$ and $\mathbb{B}2$ can be obtained using the following two steps:

Step 1. Using Proposition 2 and 3, expand $\mathbb{B}1$ and $\mathbb{B}2$ such that $G_{\mathbb{B}1} = G_{\mathbb{B}2} = G_{\mathbb{B}\oplus}$. Extension consists of adding variables and arcs.

Step 2. Apply Proposition 1 on the probabilistic networks obtained from Step 1 (since the two networks have now the same structure).

Example 1. Let us consider two probabilistic networks $\mathbb{B}1$ and $\mathbb{B}2$, where their DAG are given by Figure 1. These two DAG's have a different structure. The conditional probability distributions associated with $\mathbb{B}1$ and $\mathbb{B}2$ are given below. The DAG $G_{\mathbb{B}\oplus}$ is given by Figure 2 which is simply the union of the two graphs of Figure 1.

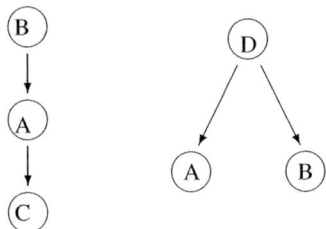

Fig. 1. G_1 and G_2 : Example of U-acyclic networks

Initial conditional probability distributions $p_{\mathbb{B}1}$ and $p_{\mathbb{B}2}$

B	$p_{\mathbb{B}1}(B)$	A B	$p_{\mathbb{B}1}(A\mid B)$	C A	$p_{\mathbb{B}1}(C\mid A)$	D	$p_{\mathbb{B}2}(D)$	A D	$p_{\mathbb{B}2}(A\mid D)$	B D	$p_{\mathbb{B}2}(B\mid D)$
b_1	.8	$a_1\ b_1$	.3	$c_1\ a_1$	.5	d_1	.9	$a_1\ d_1$	.6	$b_1\ d_1$	.3
b_2	.2	$a_1\ b_2$	1	$c_1\ a_2$	0	d_2	.1	$a_1\ d_2$	1	$b_1\ d_2$	.8
		$a_2\ b_1$	.7	$c_2\ a_1$	.5			$a_2\ d_1$	.4	$b_2\ d_1$	.7
		$a_2\ b_2$	0	$c_2\ a_2$	1			$a_2\ d_2$	0	$b_2\ d_2$	.2

Now we transform both of $G_{\mathbb{B}1}$ and $G_{\mathbb{B}2}$ to the common graph $G_{\mathbb{B}\oplus}$ by adding the required variables and links for each graph. In our example we apply the following steps:

- For $G_{\mathbb{B}1}$ we add:
  - a new variable D with a uniform conditional probability distribution
  - a link from the variable D to the variable B in the graph. The new conditional probability distribution on the node B becomes:
    $\forall d \in D_D, \forall b \in D_B, p_{\mathbb{B}1}(b\mid d) = p_{\mathbb{B}1}(b)$.
  - a link from the variable D to the variable A in the graph. The new conditional probability distribution on the node A becomes:
    $\forall d \in D_D, \forall b \in D_B, \forall a \in D_A, p_{\mathbb{B}1}(a\mid b, d) = p_{\mathbb{B}1}(a\mid b)$.
- For $G_{\mathbb{B}2}$ we proceed similarly, namely we add:

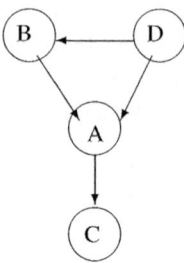

Fig. 2. DAG $G_{\mathbb{B}\oplus}$

- a new variable C, and a link from the variable A to the variable C, with a uniform conditional probability distribution, namely:
  $\forall c \in D_C,, \forall a \in D_A, p_{\mathbb{B}2}(c \mid a) = .5$.
- a link from the variable B to the variable A, and the new conditional probability distribution on the node A becomes:
  $\forall d \in D_D, \forall b \in D_B, \forall a \in D_A, p_{\mathbb{B}2}(a|b,d) = p_{\mathbb{B}2}(a|d)$.

The conditional probability distributions associated with the DAG representing the result of fusion are represented by the following Table.

Conditional probability distributions associated with $p_{\mathbb{B}\oplus}$

D	$p_{\mathbb{B}\oplus}(D)$	C	A	$p_{\mathbb{B}\oplus}(C \mid A)$	B	D	$p_{\mathbb{B}\oplus}(B \mid D)$	A	B	D	$p_{\mathbb{B}\oplus}(A \mid B \wedge D)$
d_1	.9	c_1	a_1	.25	b_1	d_1	.24	a_1	b_1	d_1	.18
d_2	.1	c_1	a_2	0	b_1	d_2	.64	a_1	b_1	d_2	.3
		c_2	a_1	.25	b_2	d_1	.14	a_1	b_2	d_1	.6
		c_2	a_2	.5	b_2	d_2	.4	a_1	b_2	d_2	1
								a_2	b_1	d_1	.42
								a_2	b_1	d_2	.28
								a_2	b_2	d_1	0
								a_2	b_2	d_2	0

From these different tables of conditional probability distributions, we can easily show that the joint probability distribution $p^J_{\mathbb{B}\oplus}$ computed by using chain rule (Definition 1), is equal to the product of $p^J_{\mathbb{B}1}$ and $p^J_{\mathbb{B}2}$. For instance, let $\omega = a_1 b_1 c_1 d_1$ be a possible situation. Using the probabilistic chain rule we have:
$p^J_{\mathbb{B}1}(a_1 b_1 c_1 d_1) = .3 * .5 * .8 = .12$.
$p^J_{\mathbb{B}2}(a_1 b_1 c_1 d_1) = .9 * .6 * .3 = .162$.
$p^J_{\mathbb{B}\oplus}(a_1 b_1 c_1 d_1) = p_{\mathbb{B}\oplus}(c_1 \mid a_1) * p_{\mathbb{B}\oplus}(a_1 \mid b_1 d_1) * p_{\mathbb{B}\oplus}(b_1 \mid d_1) * p_{\mathbb{B}\oplus}(d_1) = 0.5 * .18 * .24 * .9 = .01944$

It is very important to note that local probability distributions are subnormalized. Hence the joint probability distribution is also subnormalized.

5 Handling Sub-normalized Probabilistic Networks

One of important problem when fusing multiple sources information is how to deal conflits. When uncertain information are represented by probabilistic networks, the presence of conflits is materialized by a subnormalized joint probability distribution. Namely, even if initial conditional and joint probability distributions are normalized, it may happen that conditional and joint distributions associated with the result (of fusion) are subnormalized.

In probabilistic networks when local conditional probability distributions are normalized then the joint global distribution is also normalized. However if one of the local probability distribution is subnormalized then the global probability distribution is also subnormalized.

The aim of this section is to propose an approach for normalizing subnormalized probability distributions associated with a fused network. More precisely, let $\mathbb{B}$ be a probabilistic network where local probability distributions are subnormalized. Our goal consists in computing a probabilistic network denoted by $\mathbb{B}1$ such that:

$$\forall \omega, p_{\mathbb{B}1}^J(\omega) = \frac{p_{\mathbb{B}}^J(\omega)}{h(p_{\mathbb{B}}^J)},$$

where, $h(p_{\mathbb{B}}^J) = \sum_{\omega \in \Omega} p_{\mathbb{B}1}^J(\omega)$. Namely, $p_{\mathbb{B}1}^J$ is the result of normalizing $p_{\mathbb{B}}^J$.

The idea of the normalization procedure for $\mathbb{B}$, is to compute a network $\mathbb{B}1$ such that all its local probability distributions are normalized. $\mathbb{B}1$ is obtained by progressively normalizing local distributions for each variable. We first study the case of a probabilistic network where the probability distributions at a root variable is subnormalized.

Proposition 4. *Let $\mathbb{B}$ be a probabilistic network. Let A be a root variable where:* $\sum_{a \in D_A}(p(a)) = \alpha < 1$. *Let us define $\mathbb{B}1$ such that:*
- $G_{\mathbb{B}1} = G_{\mathbb{B}}$,
- $\forall X, X \neq A, p_{\mathbb{B}1}(X \mid U_X) = p_{\mathbb{B}}(X \mid U_X)$,
- $\forall a \in D_A, p_{\mathbb{B}1}(a) = p_{\mathbb{B}}(a)/\alpha$.

Therefore: $\forall \omega, p_{\mathbb{B}1}^J(\omega) = p_{\mathbb{B}}^J(\omega)/\alpha$.

This proposition allows to normalize root variables.

A corollary of the above proposition is the situation where only the probability distributions of a root variable is subnormalized (namely all other variables are normalized), then the probabilistic network $\mathbb{B}1$ is in fact the result of normalizing $\mathbb{B}$. Namely, $p_{\mathbb{B}1} = \frac{p_{\mathbb{B}}}{h(p_{\mathbb{B}})}$.

The next important issue is how to deal with a variable A which is not root. In this situation, we solve the problem by modifying local probability distributions associated with the parents of the variable A. This modification does not change joint global probability distributions. However, the result may produces, as a bord effect, subnormalized probability distribution on parents of A. Hence, the normalization process should be repeated from leaves to the root variables. When we reach roots, it is enough to apply Proposition 6 to get a probabilistic network with normalized local probability distributions. Let us first consider, the case of a variable A, where its local probability distribution is subnormalized and it only admits one parent. The normalization of such variable is given by next proposition.

Proposition 5. *Let $\mathbb{B}$ be a probabilistic network. Let A be a variable which has only one parent B. Assume that there is an instance 'b' of B such that: $\sum_{a \in D_A} p_\mathbb{B}(a \mid b) = \alpha \neq 1$. Let us define $\mathbb{B}1$ where its graph is the same as the one of $\mathbb{B}$ and its local probability distributions associated with $\mathbb{B}1$ are defined as:*

1. $\forall C \neq A, \forall C \neq B, p_{\mathbb{B}1}(C \mid \mu_C) = p_\mathbb{B}(C \mid \mu_C)$.
2.
$$\forall a_j, \forall b_i, \quad p_{\mathbb{B}1}(a_j \mid b_i) = \begin{cases} \frac{p_\mathbb{B}(a_j \mid b_i)}{\alpha} & if\ b_i = b \\ p_\mathbb{B}(a_j \mid b_i) & otherwise \end{cases}$$

3.
$$\forall b_i, \forall \mu_{b_i}, \quad p_{\mathbb{B}1}(b_i \mid \mu_{b_i}) = \begin{cases} p_\mathbb{B}(b_i \mid \mu_{b_i}) * \alpha & if\ b_i = b \\ p_\mathbb{B}(b_i \mid \mu_{b_i}) & otherwise \end{cases}$$

Then

$$\forall \omega,\ p_{\mathbb{B}1}^J(\omega) = p_\mathbb{B}^J(\omega).$$

Let us explain the different conditions enumerated in the above proposition. These conditions allow to normalize local probability distributions at node A level which has one parent B. The first condition says that probability distributions associated with variables, different from A and B, remain unchanged. The second condition specifies that normalization only affects the variable A (variable concerned by the normalization problem). Lastly, the third condition applies on the variable B (parent of A) the reverse operation of normalization, which ensures the equivalence between joint distributions.

Example 2. Let us consider again the previous example. The conditional probability distribution on the node C are sub-normalized. C has only one parent. Hence, we can apply the above proposition. The result is given in Table 4, where the distribution on the node C is now normalized. We also give the distribution on the node A which has been changed.

Conditional distributions after normalizing the node C

C	A	$p_{\mathbb{B}1}(C \mid A)$	A	B	D	$p_{\mathbb{B}1}(A \mid B \wedge D)$
c_1	a_1	.5	a_1	b_1	d_1	.09
c_1	a_2	0	a_1	b_1	d_2	.15
c_2	a_1	.5	a_1	b_2	d_1	.3
c_2	a_2	1	a_1	b_2	d_2	.5
			a_2	b_1	d_1	.21
			a_2	b_1	d_2	.14
			a_2	b_2	d_1	0
			a_2	b_2	d_2	0

The above proposition allows to normalize local distributions of variables having one parent, without modifying the joint global distribution. Let us now generalize the above proposition.

In the rest of paper, we define an ordering relation between variables, $A \leq B$ if B is not a descendant of A. Since DAGs are the graphs associated with probabilistic networks, then if A and B are two variables, we always have either $A \leq B$ or $B \leq A$.

Definition 3. *Let* $\mathbb{B} = (p_\mathbb{B}, G_\mathbb{B})$ *be a probabilistic network. Let A be a variable where its local probability distribution is subnormalized. Namely, there exists μ_a an instance of U_A such that:* $\sum_{a \in D_A}(p(a \mid \mu_a)) = \alpha < 1$. *Let* $B \in U_A$ *be such that* $\forall C \neq B \in U_A, B \leq C$. *The probabilistic network* $\mathbb{B}1 = (p_{\mathbb{B}1}, G_{\mathbb{B}1})$ *representing the result of the normalization of* $\mathbb{B}$, *is such that:*

1. $G_{\mathbb{B}1} = G_\mathbb{B} \cup \{C \to B : C \in U_A, C \neq B\}$.
2. *Local probability distributions are defined by:*
   (a) $\forall X \neq A, \forall X \neq B, p_{\mathbb{B}1}(X \mid U_X) = p_\mathbb{B}(X \mid U_X)$.
   (b) $\forall a \in D_A, \forall \mu_A \in D_{U_A}$,

$$p_{\mathbb{B}1}(a \mid \mu_a) = \begin{cases} \frac{p_\mathbb{B}(a|\mu_A)}{\alpha} & if \ (\sum_a(p_\mathbb{B}(a \mid \mu_A)) = \alpha) \\ p_\mathbb{B}(a \mid \mu_A) \ otherwise \end{cases}$$

(c) $\forall b \in D_B, \forall \mu'_{B}$,

$$p_{\mathbb{B}1}(b \mid \mu'_B) = \begin{cases} p_\mathbb{B}(b \mid \mu_B) * \alpha, & if \ \mu'_B = \mu_B \times (\mu_a - \{b\}) \\ & and \ (\sum_a(p_\mathbb{B}(a \mid \mu_A)) = \alpha) \\ p_\mathbb{B}(b \mid \mu_B) & otherwise. \end{cases}$$

We first recall that this definition concerns the normalization of local probability distribution associated with a variable which has at least two parents. Note that in Definition 3 the variable B always exists (otherwise the graph contains a cycle). The first condition says that $G_\mathbb{B}$ is a subgraph of $G_{\mathbb{B}1}$. Namely, $G_{\mathbb{B}1}$ is obtained from $G_\mathbb{B}$ by adding new links from each variable C (which is different form B and parent of A) to the variable B. We now explain local probability distributions associated with the probabilistic network $\mathbb{B}1$.

Condition (a) says probability distributions, associated with variables which are different from A and B, are exactly the same as ones of the initial network. Condition (b) normalizes the probability distribution associated with the variable A.

The condition (c) applies a normalization inverse operation on the variable B in order to preserve the joint distribution. μ'_B denotes an instance of parents of B in the new network $\mathbb{B}1$, and where μ_B (resp. μ_A) denotes an instance of parents of B(resp. of parents of A) in the initial network $\mathbb{B}$.

Proposition 6. *Let* $\mathbb{B} = (p_\mathbb{B}, G_\mathbb{B})$ *be a probabilistic network, such that there exists a variable having at least two parents and where its local probability distribution is subnormalized. Let* $\mathbb{B}1 = (p_{\mathbb{B}1}, G_{\mathbb{B}1})$ *be the probabilistic network obtained using Definition 3. Then:*

$$\forall \omega, p^J_{\mathbb{B}1}(\omega) = p^J_\mathbb{B}(\omega).$$

Based on the previous propositions, the normalization procedure may be defined as follows: Order the set of variables $V = \{A_1, \ldots, A_n\}$, such that parents of A_i (may be empty) are included in $V' = \{A_{i+1}, \ldots, A_n\}$. The we succesively normalize distributions of nodes from leafs to roots. If A_i is a root then apply Proposition 4. If A_i has one parent then apply Proposition 5, otherwise apply Proposition 6.

Example 3. Let us continue above example. After normalizing C, we need to normalize A, by applying proposition 8. Then the node B is normalized using Proposition 5. Lastly, the node D (a root node) is normalized using Proposition 4. The obtained result is a normalized probabilistic network. In Table 5, we do not simplify computations, in order to make explicit the propagation of normalization degrees from leaves to roots.

Conditional probability distributions after normalizing all nodes

D	$p_{\mathbb{B}\oplus}(D)$	B	D	$p_{\mathbb{B}\oplus}(B \mid D)$	A	B	D	$p_{\mathbb{B}\oplus}(A \mid B \wedge D)$
					a_1	b_1	d_1	.09 /.3
d_1	.9 *0114/0.14116	b_1	d_1	.24 *.3/.0114	a_1	b_1	d_2	.15 /.29
d_2	.1 *0.3856/0.14116	b_1	d_2	.64 *.29/0.3856	a_1	b_2	d_1	.3 /.3
		b_2	d_1	.14 *.3 /.0114	a_1	b_2	d_2	.5 /.5
		b_2	d_2	.4 *.5 /0.3856	a_2	b_1	d_1	.21 /.3
					a_2	b_1	d_2	.14 /.29
					a_2	b_2	d_1	0
					a_2	b_2	d_2	0

6 Conclusions

This paper has proposed the fusion of probabilistic networks. We first showed that when the probabilistic networks have the same structure then the fusion can be realized efficiently. We then studied the fusion of networks having different structure. Lastly, we proposed an approach to handle the sub-normalization problem induced by the fusion of probabilistic networks.

References

1. Baral, C., Kraus, S., Minker, J., Subrahmanian, V.S.: Combining knowledge bases consisting in first order theories. Computational Intelligence 8(1), 45–71 (1992)
2. Benferhat, S., Titouna, F.: Fusion and normalization of quantitative possibilistic networks. Appl. Intell. 31(2), 135–160 (2009)
3. Darwiche, A.: Modeling and Reasoning with Bayesian Networks. Cambridge University Press, Cambridge (2009)
4. de Oude, P., Ottens, B., Pavlin, G.: Information fusion with distributed probabilistic networks. Artificial Intelligence and Applications, 195–201 (2005)
5. JoseDel, S., Serafin, M.: Qualitative combination of bayesian networks. International Journal of Intelligent Systems 18(2), 237–249 (2003)
6. Konieczny, S., Perez, R.: On the logic of merging. In: Proceedings of the Sixth International Conference on Principles of Knowledge Representation and Reasoning (KR 1998), pp. 488–498 (1998)
7. Matzkevich, I., Abramson, B.: The topological fusion of bayes nets. In: Dubois, D., Wellman, M.P., D'Ambrosio, B., Smets, P. (eds.) 8th Conf. on Uncertainty in Artificial Intelligence (1992)
8. Matzkevich, I., Abramson, B.: Some complexity considerations in the combination of belief networks. In: Heckerman, D., Mamdani, A. (eds.) Proc. of the 9th Conf. on Uncertainty in Artificial Intelligence (1993)

Basic Object Oriented Genetic Programming

Tony White, Jinfei Fan, and Franz Oppacher

School of Computer Science, Carleton University
1125 Colonel By Drive, Ottawa, Ontario, K1S 5B6 Canada
arpwhite@scs.carleton.ca, fanjinfei@gmail.com, oppacher@scs.carleton.ca

Abstract. This paper applies object-oriented concepts to genetic programming (GP) in order to improve the ability of GP to scale to larger problems. A technique called Basic Object-Oriented GP (Basic OOGP) is proposed that manipulates object instances incorporated in a computer program being represented as a linear array. Basic OOGP is applied to the even-parity problem and compared to GP, Liquid State GP and Traceless GP. The results indicate that OOGP can solve certain problems with smaller populations and fewer generations.

Keywords: Genetic Programming, Object-oriented Programming.

1 Introduction

Genetic Programming (GP) is an approach to automated computer program generation using genetic algorithms [1]. The original genetic programs used lisp-like expressions [2] for their representation. Since the original introduction of GP there have been several mechanisms proposed for program generation, such as strong-typed GP [3], grammar-based GP [3] and linear GP [3], all of which have attempted to improve the performance and broaden the scope of GPs applicability.

Genetic programming has been successfully applied in several engineering domains; however, there are limitations. The current GP algorithms cannot handle large-scale problems. For example, the generation of a modern computer operating system is far beyond the capability of any GP algorithm.

If we could find a way to drastically improve the performance of current GP algorithms/methods, there might be more GP applications. This paper is motivated to make a contribution towards this goal.

It is worth noting that there are implementations of traditional GP algorithms in object-oriented languages; this is not what is discussed in this paper.

Computer programming has developed from using machine code, to assembly language, to structured languages, to object-oriented languages, and now to model driven development. People are now studying a higher level of abstraction for computer software than simply using the notions of object and class; however, this too is of limited interest in this paper.

As the level of abstraction in software programming increases, as does the size of systems able to be developed and the ease with which they are maintained and upgraded.

Just as the GP paradigm already includes the idea of constraining the functions and types, it is still at the stage of using data and functions to solve the problem, which is analogous to the era of modular programming. Applying object-oriented programming should be a better idea according to human software engineering practice. There has been prior research in the area of object-oriented genetic programming (OOGP) [4], [5], but they are early stage research and this paper proposes a representation and genetic operators for OOGP.

This paper proposes extensions to the application of object-oriented concepts in genetic programming. The main contribution proposed is the method of operating on objects within the same executable as OOGP (hereafter called Basic OOGP), which is explained in detail in section 3. By experimental comparison of the performance of the Basic OOGP algorithm with 3 modern GP algorithms we demonstrate that this method has practical advantages. Experimental setup and results are contained in section 4. Conclusions and future work are described in section 5.

2 Related Work

Research in this area is limited. Many of the papers found containing the terms object oriented genetic programming are either not relevant to this area or reference the two papers described in the following two subsections. One of the earliest works in this area in due to Bruce [6] and built upon by the 2 papers described.

Although these two papers do include OOGP in their titles and they attempt to use the advantages of object oriented programming, they are different from what is proposed in this paper, in terms of applying OO concepts to software. The fundamental difference is in the nature of their representation; these two papers still represent the software program as a tree-based data structure in contrast to the linear representation proposed here. We should point out that linear representations are not new, linear GP having a well- developed, if recent, history [7] within mainstream GP.

2.1 An Initial Implementation of OOGP

The paper Initial Implementation of OOGP [4] discussed the basic idea of object-oriented GP, and provided an initial implementation, but there are many potentially useful aspects not yet implemented.

The chromosome is still represented as a tree. Traditional crossover and mutation operators apply to this algorithm with variables as terminals being the principal addition. This OOGP implementation also uses the Java language reflection feature to automatically find the classes and methods in an existing library.

Just as the title suggests, the technique utilize only a small number of object-oriented concepts. However, the potential of the idea is intriguing. Figure 1 shows an example of the tree representation and the program that it represents used in the initial implementation of OOGP.

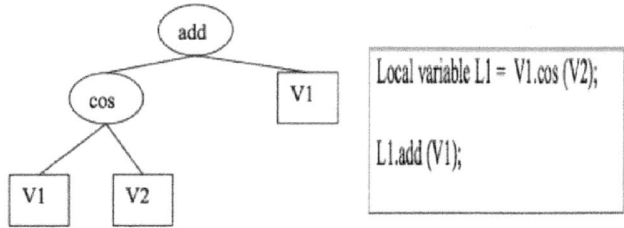

Fig. 1. Initial Implementation of OOGP

2.2 Exploiting Reflection in OOGP

The paper Exploiting Reflection in OOGP [5] listed the following reason (among others) for undertaking research into OOGP and is the inspiration for the research reported here:

Evolution is also possible at the object level, where the states of the objects evolve rather than the program code. [5]

Reflection is used in object oriented programming languages to discover what behaviour is available at run time. Using reflective techniques allows a program to dynamically construct calls to particular elements of a programming library at runtime.

In standard GP, we normally specify a limited number of functions and terminals. That is not the case in real world software engineering where programmers have access to a vast collection of existing libraries. Restricting the argument type and return type of the functions (or methods of objects), would help to generate more elegant programs. Reflection would certainly be a good technique in this regard. Given a set of APIs in a library that may have some overlapped functionality, those APIs that are never used by GP might be badly designed. It can also be considered as an evolutionary method in library design.

The research in [5] relies heavily on an existing library to generate GP programs. It tries to evolve objects instead of functions. In the examples given in the paper, GP explores graphical packages (library) in Java, and generates new programs that can draw lines, arcs, etc in a window. Despite this, the evaluation criteria are far from practical.

From an analysis of the 2 papers, we find that reflection is considered to be a useful technique for exploring an existing library using a modern programming environment, Although object oriented genetic programming has many advantages, it is hard to apply the idea in real domains and experiments. Problems related to syntax are difficult to resolve; however, grammar-based approaches are beginning to be investigated [8].

3 Basic OOGP

3.1 Conceptual Overview

In Basic OOGP, solution-building materials are objects. There are two types of object. Type A has data members and methods, and serves as a concrete placeholder. Type B is used as an abstraction of the operation/action (interface). Chromosomes are a linear sequence of genes. Each gene consists of two Type-A objects and one Type-B object as shown in Figure 2.

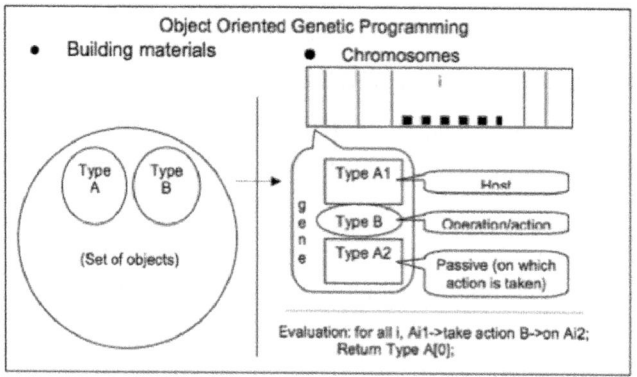

Fig. 2. Conceptual Basic OOGP Representation

From the abstraction shown in Figure 2, we observe the following several benefits of Basic OOGP. With reusable objects in different positions in Basic OOGP, there is no need to reproduce the same branch as in normal GP. Even in ADF, if two branches contain the same ADF function, these two branches still have to be generated. For the example in Figure 3, if f1 is an ADF, the three duplicated branches still have to be repeated. This is a disadvantage inherited from the tree-based representation. If such duplicates are common, Basic OOGP will be much faster because the search space is much smaller. No-ops (do nothing objects in Type B) are used to simulate the redundant genes that are observed in nature. With the introduction of a no-op, it is possible for Basic OOGP to automatically reduce the length of chromosome (when it finds there are too many no-ops in the chromosome).

No-op also serves another purpose. In normal GP, to solve a given problem, according to the complexity of the problem, an estimate has to be made to determine the parameters such as the maximum tree depth and the maximum number of tree nodes in order to get a reasonably sized search space. Otherwise, since it has the difficult job of generating many useless branches, even a

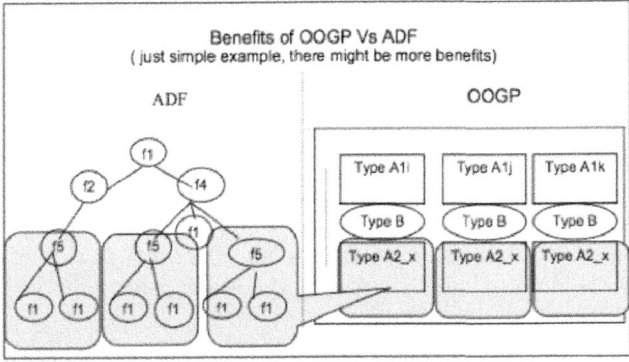

Fig. 3. The Benefits of Basic OOGP vs. GP

simple problem may not be solved in a large search space. With the help of no-op in Basic OOGP, the algorithm has the choice of filling no-ops into the chromosomes if the solution does not require all pieces of genes to be fully used. In Basic OOGP, the length of the chromosome is fixed and is determined for a particular problem domain. In tree-based GP, to increase the tree depth from N to N+1 for a binary tree, the total number of nodes has to be increased by 2N+1. Some sections of the chromosome might not be effective in the parent generation, but definitely could expose its functionality in the offspring; thus it successfully simulates a pseudo gene feature. This kind of section is saved and passed to the next generation. Because Basic OOGP is linear, there is no need to truncate the offspring tree. The offspring will remain the same chromosome length as that of its parents. In contrast, in normal tree-based GP, since the selected portion from two parents could have many nodes, the generated offspring might have more nodes than the pre-defined maximum number of nodes (there has to be a maximum value, otherwise the tree might grow to a size that is beyond the processing ability of the host computer). The truncation operation is always necessary in normal GP.

From the above discussion, we can see that Basic OOGP is somewhat different from traditional GP. It applies the concept of OO at the very core of the genetic algorithm. It is certainly far more than a simple OO implementation of conventional GP.

Because of the linear representation used by Basic OOGP, the search space is bounded by the length of the chromosome which will only increase when the number of nodes increases. For example, for a binary tree, if we set the maximum depth of the tree to (11+N), for each step K (12 to N), the number of tree nodes in the search space will increase by 2 to the power of K. Using Basic OOGP, if the original chromosome is (11+N), we only increase the number of searching nodes by 3*N. Hence, the rate of increase of our search space is substantially lower than for a tree-based GP. This will substantially reduce the computational search effort compared with normal GP.

3.2 Algorithm Description

Referring to Figure 2, building materials (referred to hereafter as a resource pool) comprise two types of base classes: A and B.

Type A (a concrete place holder, called Host in Figure 4). Classes hold the interim states and the concrete meaning of actual simulated objects. The interim state is similar to the branch in the tree representation. It holds the temporary results during the computation. The interim state can be considered to be similar to a register in a CPU, but its granularity is much larger than that of the register. This class has the data members as well as the implementations of the interfaces that are defined in Type B.

Type B (an abstraction of action, called Action in Figure 4). These are classes that define interfaces. The final implementation depends on the Type A classes but augmented with a null operation, or no-op.

The Resource Pool consists of a set of references of type A and B objects.

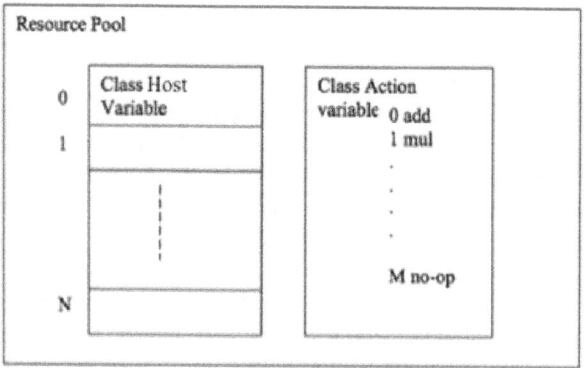

N: number of objects holding interim state
M: number of operations

Fig. 4. Resource Pool

The algorithm used for evolution is a standard genetic algorithm; however, the details of the chromosome used, crossover and mutation operators are significantly different. Specifically, each chromosome is a linear array of genes. Each gene in the array has three object pointers: one points to a type A object, called host; one points to a type B object called action and one points to a type A object called passive. When interpreted by the Basic OOGP engine it means that the object host will take action on the object passive.

While many forms of crossover operator could be envisaged, the crossover operator implemented for the experiments reported in this paper was: randomly pick a percentage of genes from one parent and exchange them with the equivalent positions in the second parent. This is uniform crossover implemented at the gene level.

The mutation operator implemented was: randomly pick two gene positions and generate new randomly generated genes. The Basic OOGP evaluation engine can then be specified as:

Algorithm 1.
  Evaluate(Gene *gene*)
  N = chromosome length
  for $i = 0$ to N **do**
    $j = gene[i]$.host
    $k = gene[i]$.action
    $l = gene[i]$.passive
    $host = resourcePool$.hosts[j]
    $action = resourcePool$.actions[k]
    $passive = resourcePool$.hosts[l]
    Invoke $host{\rightarrow}$actions($passive$)
  end for
  return $resourcePool$.hosts[0]

4 Experiments

The performance of Basic OOGP for an even-parity problem [2] was investigated and compared with 3 other techniques: conventional GP, Liquid State GP and Traceless GP.

Table 1. Configuration of Basic OOGP for the Even Parity Problem

Parameter	Value
Pop. Size	800
Function Set	AND, OR, NOT, XOR, RESET (reset to initial state), NO_OP
Crossover	Exchange 50% of genes based upon uniform distribution
Mutation	Mutation rate 0.5% of population size. Randomly generate the selected genes in the individual.
Selection	Keep the best 10% of individuals. Select one of the remaining individuals as parent A , randomly select another top 33.3% as parent B. Randomly select 50% of genes from parent B, and replace those in the same position in parent A., generate one offspring. Use this child in the next generation, replacing parent A.
Success	Predict all truth values in the truth table.

5 Results

The following tables provide results for Basic OOGP, which empirically demonstrate that Basic OOGP can indeed work. Results from two other types of GP are also provided for comparison.

Liquid State Genetic Programming (LSGP) [9] is a hybrid method combining both Liquid State Machine (LSM) and Genetic Programming techniques. It uses a dynamic memory for storing the inputs (the liquid). Table 2 shows the results from LSGP [9].

Table 2. Results for Liquid State GP

Problem	Population Size	Number of Generations	Success Rate GP(%)	Success Rate LSGP(%)
Even − 3	100	50	42	93
Even − 4	1000	50	9	82
Even − 5	5000	50	7	66
Even − 6	5000	500	4	54
Even − 7	5000	1000	−	14
Even − 8	10000	2000	−	12

Further experiments for the even parity problem of order 9 to 14 showed that Basic OOGP could solve the problem consistently; LSGP and GP could not. LSGP could solve the even parity problem for size 9 and above in less than 10% of all trials.

Traceless Genetic Programming (TGP) [10] is a hybrid method combining a technique for building the individuals and a technique for representing the individuals. TGP does not explicitly store the evolved computer programs. Table 3 shows the results for Traceless GP (TGP) [10].

Table 3. Results for Traceless GP

Problem	Population Size	Effort	Generation
Even − 3	50	33, 750	131
Even − 4	100	240, 000	480
Even − 5	500	2, 417, 500	967
Even − 6	1000	29, 136, 000	2428
Even − 7	2000	245, 900, 000	4918

The results of normal GP are not given because according to [10], [9], LSGP and TGP all out performed normal GP. Table 4 shows the average number of generations and standard deviation for Basic OOGP solving the even parity problem for orders 3 to 9. As can be observed, Basic OOGP solves the even parity problem in all runs (i.e., is 100% successful) and solves the problem in a fraction of the number of evaluations used in successful runs of the LSGP or TGP. It should be noted that for the 10 and 11 problem sizes a larger representation was used.

Table 4. Results for Basic OOGP

Problem	Avg Gen.	Std. Dev.
$Even-3$	1.70	131
$Even-4$	7.8	5.47
$Even-5$	45.90	18.08
$Even-6$	80.20	27.30
$Even-7$	190.20	70.71
$Even-8$	320.30	100.14
$Even-9$	420.10	71.88
$Even-10$	269.10	74.04
$Even-11$	274.80	278.78
$Even-12$	669.0	706.0

6 Conclusions

This paper proposes the application of OO concepts to GP in order to apply GP to large-scale problems. Through the introduction of OO programming, the computer program being evolved is represented as a linear array, which helps to reduce the search space. In the experiments reported in this paper, Basic OOGP can solve parity problems of up to order 12^1, while LSGP can only solve the even-parity problem up to order 8 with conventional GP failing to solve even smaller problems. This demonstration is promising and indicates that Basic OOGP is a feasible approach to program production. Furthermore, Basic OOGP has the potential to converge to the result faster than normal GP, even with a smaller population size and fewer generations, though the results may vary due to differing implementations of selection, mutation and crossover strategies. A more detailed examination of the roles of various operators in the two techniques is left as future research. In conclusion, the experiments show that incorporating OO concepts into GP is a promising direction to improve GP performance, though it is not clear at this point what is the ultimate complexity of problem that could be solved by GP. This remains an open question.

By applying Basic OOGP to problems in other domains we expect to be able to demonstrate that such an approach can improve the performance of GP in more complex software engineering domains.

References

1. Koza, J.R.: Genetic programming II: automatic discovery of reusable programs. MIT Press, Cambridge (1994)
2. Koza, J.R.: Genetic programming: on the programming of computers by means of natural selection. MIT Press, Cambridge (1992)

---

[1] Success with problems of up to size 14 has been achieved.

3. Poli, R., Langdon, W.B., McPhee, N.F.: A field guide to genetic programming (With contributions by J. R. Koza) (2008), Published via http://lulu.com and freely available at http://www.gp-field-guide.org.uk
http://www.gp-field-guide.org.uk
4. Abbott, R.: Object-oriented genetic programming an initial implementation. Future Gener. Comput. Syst. 16(9), 851–871 (2000)
5. Keijzer, M., O'Reilly, U.M., Lucas, S.M., Costa, E., Soule, T., Lucas, S.M.: Exploiting Reflection in Object Oriented Genetic Programming. In: Keijzer, M., O'Reilly, U.-M., Lucas, S., Costa, E., Soule, T. (eds.) EuroGP 2004. LNCS, vol. 3003, pp. 369–378. Springer, Heidelberg (2004)
6. Bruce, W.S.: Automatic generation of object-oriented programs using genetic programming. In: Proceedings of the First Annual Conference on Genetic Programming, GECCO 1996, pp. 267–272. MIT Press, Cambridge (1996)
7. Brameier, M.F., Banzhaf, W.: Linear Genetic Programming (Genetic and Evolutionary Computation). Springer, New York (2006)
8. Oppacher, Y., Oppacher, F., Deugo, D.: Evolving java objects using a grammar-based approach. In: GECCO, pp. 1891–1892 (2009)
9. Oltean, M.: Liquid State Genetic Programming. In: Beliczynski, B., Dzielinski, A., Iwanowski, M., Ribeiro, B. (eds.) ICANNGA 2007. LNCS, vol. 4431, pp. 220–229. Springer, Heidelberg (2007)
10. Oltean, M.: Solving even-parity problems using traceless genetic programming. In: Proceedings of the 2004 IEEE Congress on Evolutionary Computation, June 20-23, vol. 2, pp. 1813–1819. IEEE Press, Portland (2004)

Inferring Border Crossing Intentions with Hidden Markov Models

Gurmeet Singh[1], Kishan. G. Mehrotra[2], Chilukuri K. Mohan[2], and Thyagaraju Damarla[3]

[1] Indian Institute of Technology, Guwahati
[2] Department of EECS, Syracuse University
[3] Army Research Lab

Abstract. Law enforcement officials are confronted with the difficult task of monitoring large stretches of international borders in order to prevent illegal border crossings. Sensor technologies are currently in use to assist this task, and the availability of additional human intelligence reports can improve monitoring performance. This paper attempts to use human observations of subjects' behaviors (prior to border crossing) in order to make tentative inferences regarding their intentions to cross. We develop a Hidden Markov Model (HMM) approach to model border crossing intentions and their relation to physical observations, and show that HMM parameters can be learnt using location data obtained from samples of simulated physical paths of subjects. We use a probabilistic approach to fuse "soft" data (human observation reports) with "hard" (sensor) data. Additionally, HMM simulations are used to predict the probability with which crossings by these subjects might occur at different locations.

Keywords: Soft and Hard Data Fusion, Intention Modeling, Border Surveillance, Hidden Markov Models.

1 Introduction

Illegal border crossings constitute a problem addressed by governments at many international boundaries, and considerable efforts are devoted to surveillance, monitoring, and interdiction activities. This paper proposes to combine two kinds of data for this task: –*Soft Data,* viz., observations made by humans; and *Hard Data,* collected from sensors. In this paper, we use the word *subject* to indicate a group of people who are under observation, who may be intending to cross a border.

Decisions based on hard data alone, along with the fusion of data from multiple (hard) sensors, have been extensively studied and are fairly well understood [4]. In recent years, researchers have recognized the importance of soft data and the need to combine it effectively with hard data. For example, Laudy and Benedicte [7] has proposed various techniques to extract valuable information from soft data, and Hall *et al.* [3] have proposed a framework for dynamic hard and soft data fusion. But concrete methods for fusing the information collected from these two types of sources remain elusive. In this paper, we attempt to use multiple human observations to estimate the likelihood that a subject intends to cross the border, and to infer the location and time of such crossing.

Appropriate decisions can be taken to prevent the border crossing or catch the subjects upon such crossing. The final "belief" in the border crossing hypothesis is obtained by probabilistically combining three factors:

1. Observations from (hard) sensors;
2. Inferences of subjects' intentions from their partially observed paths (*via* human intelligence reports); and
3. Soft data from other human observations (e.g,, increased activity of unknown individuals).

Artificial Intelligence has been used to determining intentions from actions, in several scenarios. For example, the use of intention graphs is proposed in Youn and Oh [11], Hidden Markov Models (HMMs) [8] have been used to model sequenced behavior by Feldman and Balch [2] and sequenced intentions in Kelley et al. [6]. We explore the following questions: (1) If a partial path travelled by a subject is accessible to us, can we then predict whether the subject will cross the border or not?, (2) Can we assign levels of confidence with these decisions? (3) If it is determined that the subject will cross the border with high level of confidence then where and when will the crossing occur?

Section 2 presents the Hidden Markov Model approach as applied to representing and estimating intentions of subjects (to cross or not to cross). Section 3 discusses our probabilistic fusion approach. Simulation results are given in Section 4. Section 5 contains a description and example of the application of the HMM approach to predict subject paths. Concluding remarks are given in Section 6.

2 Intention Representation and Estimation

This section presents an HMM approach to representing subjects' intentions, and discusses how these HMM parameters may be learnt from available data and used for further predictions. An essential aspect of the problem is that simple "straight line paths" are not usually observed, making the intention inference task quite difficult. Subjects crossing the border are often in touch with their own information sources, who may inform the subject of perceived actions of the security personnel. In view of such information, the subject may change intent and exhibit behavior not consistent with the original intention.

We use one HMM for the case when a subject intends to cross the border, referred to HMM_C with intention *Cross,* and another for the case when a subject does not intend to cross the border, referred to HMM_{NC} with intention *Not-Cross*. Physical observations of a subject's movements correspond to states of the HMM, but HMM_C and HMM_{NC} differ with respect to the states in the models and the transition probabilities among the states. The subject with intention of crossing the border can be in one of the following states, as shown in Figure 1(a):

(1) approaching the border, (2) returning from the border, (3) Looking around in the region for a suitable location to cross, (4) waiting at a location for an appropriate time to proceed, and (5) crossing the border. The subject with no intention of crossing the border can also *approach* the border, try to *identify* a location by moving around, *wait* at a location in need for directions, *return* from the border and then again go back to one of the previous states, with transitions as shown in Figure 1(b).

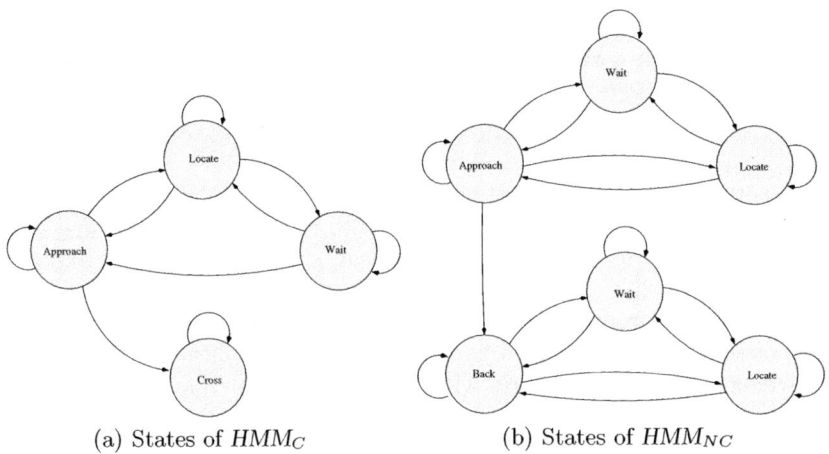

(a) States of HMM_C (b) States of HMM_{NC}

Fig. 1. States of Hidden Markov Models for Cross and Not-cross intentions

We now discusses how the HMM parameters are obtained. This is followed by a description of how the HMMs are used to infer intentions. We emphasize that the two steps are completely separate.

2.1 HMM Parameter Learning

Real data for border crossing scenarios is not available. Based on the models described above we simulate paths using a "Path Generator" algorithm. A set of paths is generated for the subject's intention to cross the border. Paths are generated by a Markov model for a given set of probability of moving in specified directions. Transition probabilities of changing intentions, such as from intention to cross to not-cross, are also needed.

Typical paths generated by these models are shown in Figure 2, one for each intention. A set of paths from Markov models for crossing intention is used to estimator the parameters of HMM_C. Similarly, another set is generated for the not-cross intention and parameters of HMM_{NC} are estimated. In this training stage the parameters of the HMM's are learnt using the standard Baum-Welch algorithm. These parameters mainly consist of the transition probabilities.

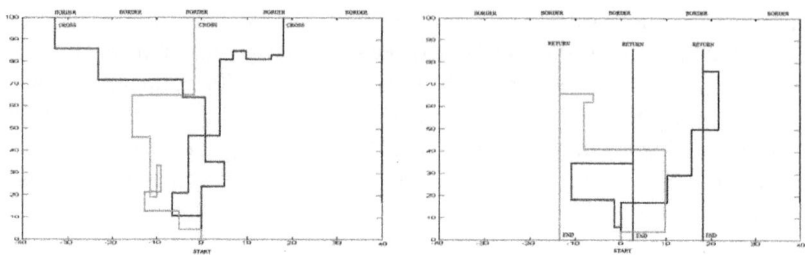

Fig. 2. Typical paths generated by Markov Models for cross and not-cross intentions. (a) Typical paths taken with intention to cross, (b) Typical paths taken with intention to not-cross.

The performance (of the system after training phase) is then tested on a different set (test set) of paths for intention to cross, not-cross, as well as for a third sets of paths where the subject changes its intention from cross to not-cross in the middle possibly due to (i) information received from its informants, (ii) other information from its surroundings such as being aware that someone is closely observing him/her, or (iii) spotting the patrol police van. That is, in the later scenario, the subject starts with the intention to cross the border but at some point in time changes intention (perhaps temporarily) to not-cross on the way to the border.

2.2 Intention Estimation

Our goal is to infer the subject's intention as the subject is in the middle of the path towards the border and not necessarily when he is very near the border. In addition, we allow the possibility that the subject may change its intention with time. For this reason, the intention recognition use of the HMM offers a good way to estimate intention of the subject using the information available till that time. Given a set of observations, by using the Forward Algorithm of the Baum-Welch algorithm [see [8]] we can estimate the probability that the subject is performing an activity corresponding to HMM_C or HMM_{NC}, and infer that the intention of the subject corresponds to the HMM with higher probability.

Algorithm <u>Infer Intention</u>:
Collect observations of subject's path;
Apply Baum-Welch Forward Algorithm to estimate p_C, the probability associated with HMM_C; and also estimate p_{NC}, the probability associated with HMM_{NC};
If $p_C > p_{NC}$, then infer that the intention is *Cross*,
else infer that the intention is Not-Cross.

Statistical inference about the subject's intention is based on the likelihood associated with the given sequence of observations, O. Let λ_c and λ_{nc} denote the vectors of parameters of the HMM's for cross and not-cross intentions. Then the Forward component of the Baum-Welch algorithm provides the probabilities $P(O|\lambda_c)$ and $P(O|\lambda_{nc})$. If the prior probabilities of the models $P(\lambda_c)$ and $P(\lambda_{nc})$ are known, then the posterior probabilities $P(\lambda_c|O)$ and $P(\lambda_{nc}|O)$ can be easily calculated. In the absence of any information, the priors are presumed equal for both models, but if inference from the soft data indicates that the subject is likely to cross the border, then by Bayesian rule these priors are adjusted accordingly:

$$P(\lambda_c|O) = \frac{P(O|\lambda_c) \times P(\lambda_c)}{P(O)}, \text{ and } P(\lambda_{nc}|O) = \frac{P(O|\lambda_{nc}) \times P(\lambda_{nc})}{P(O)}.$$

Given a set of observations O, the probability of intention to cross, $P(C)$, is estimated as:

$$P(C) = \frac{P(\lambda_c|O)}{P(\lambda_c|O) + P(\lambda_{nc}|O)} \quad (1)$$

and the probability of the intention of not-cross is assigned a value $P(N) = 1 - P(C)$ where λ_c and λ_{nc} are the estimated parameter vectors of HMM_C and HMM_{NC} respectively.

3 Probabilistic Fusion Approach

Strength in soft data is a combination of three factors:

1. Level of evidence e.g., the predefined probability of intention to cross;
2. Certainty of the soft sensor about the evidence, perhaps in the range [0-1], obtained using the knowledge elicitation techniques of Hall et al.[3]; and
3. Confidence in the credibility of the human intelligence provide, which may depends on the informer/observer's past performance.

The data fusion task involves updating the belief that the subject would cross, using a combination of human intelligence reports and observations of the physical path of the subject. This fusion may be performed in two ways:

1. Suppose we do not influence the prior probability of crossing by the soft information. In that case the prior probability to cross is $P(\lambda_c) = 0.5$. Given the evidence O, the probability of crossing is obtained as in Equation 1 and this probability is combined with the certainty associated with the soft data

$$P_{fused}(C) = 1 - \{1 - P(C)\} \times \{1 - P_{soft}(C)\} \quad (2)$$

where $P_{soft}(C)$ is the probability of crossing the border based on evidence provided by the soft sensor only and is calculated as discussed above.

2. Alternatively, $P_{soft}(C)$ can be used to adjust the prior $P(\lambda_c)$ and then calculate the posterior probability of the model. That is:

$$P_{fused}(\lambda_c|O) = P(O|\lambda_c) \times P_{fused}(\lambda_c); \ P_{fused}(\lambda_{nc}|O) = P(O|\lambda_{nc}) \times P_{fused}(\lambda_{nc})$$

where $P_{fused}(\lambda_c) = 1 - \{1 - P(\lambda_c)\} \times \{1 - P_{soft}(C)\}$ and $P_{fused}(\lambda_{nc}) = 1 - P_{fused}(\lambda_c)$. Finally,

$$P_{fused}(C) = \frac{P_{fused}(\lambda_c|O)}{P_{fused}(\lambda_c|O) + P_{fused}(\lambda_{nc}|O)} \quad (3)$$

On the other hand, if soft data suggests that the subject does not intend to cross, then $P(N)_{evidence}$ is fused with $P(N)$ to obtain $P(N)_{fused}$ and $P_{fused}(C) = 1 - P(N)_{fused}$. Again, the fusion can be done in two different ways as discussed above.

These two fusion methods have both been evaluated and compared, as discussed in the next section.

4 Simulation Results

In simulation experiments we use three levels of surveillance, viz. Low Surveillance during which around 20% of the path is expected to be seen, Medium Surveillance of around 40% and Alert Surveillance of around 60%. These results are compared with the ideal situation when there is 100% surveillance, i.e., the entire path followed by the subject is observed.

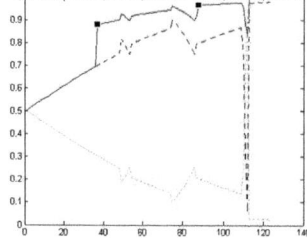
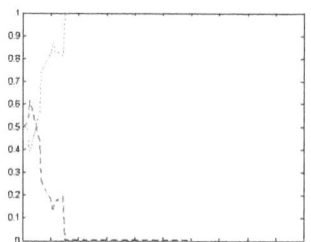

Fig. 3. Probabilities of cross and not-cross vs. time of observation. Probabilities of crossing (dashed line), not crossing (dotted line), and fused probability of crossing (solid line), (a) when the real intention is to cross (b) when the real intention is not-cross.

The HMMs for intentions to Cross and Not-Cross are learnt from the data set of 100 paths for each intention. The learnt HMM's are then tested on another set of 100 paths of each intention, and 100 paths of changing intentions.

In Figures 3 the red and blue curves denote the probability of intention to cross without fusion and with fusion respectively whereas the green curve is for not-cross.

The probability obtained from an analysis of soft sensor information clearly influences these values. For the purpose of simulation, we assumed that the probability of 'false alarm' associated with soft sensors information is 0.18; that is, the probability that soft sensors report an innocent person as suspicious is

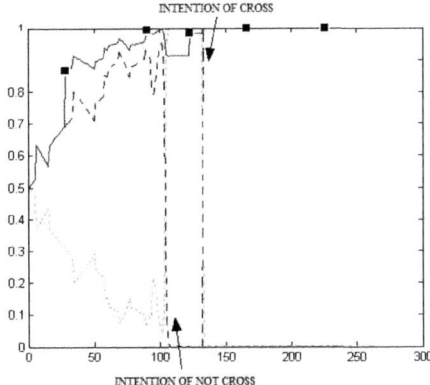

Fig. 4. Probabilities of crossing (dashed), not crossing (dotted) and fused probability (solid) of crossing (when the intention of subject is changing between cross or not to cross) vs. time of observation

0.18. Also, the probability that soft sensors wrongly conclude a subject to be innocent, is also assumed to be 0.18 (this probability is called the probability of *Miss*). Note that when the subject does not have intention to cross, the fusion comes into play when the human intelligence (soft data) results in a false alarm and informs that the subject will cross.

As mentioned earlier, in case of border surveillance scenarios it is infeasible to observe the entire path. Simulation was done on 20%, 40% and 60% of the complete path. We sampled the entire path for path observation followed by random period of non-observing part, depending on percentage of path to be seen. In each of these cases, 100 incomplete paths for each intention were generated 10 times. The performance of the learning algorithms is measured in terms of – how often the correct intention is predicted. In Tables 1, 2 and 3 results are reported for three case – when fusion is not implemented, when fusion with soft evidence was done, and when probability of false alarm and miss was considered for the intentions Cross, Not Cross and Changing intention respectively.

The above results are for fusion method given by Equation 2 out of the two methods discussed in Section on Fusion Aspect. Recall that in Equation 2 decisions of the HMMs and soft sensors are fused, whereas in fusion method given by Equation 3 soft sensor information is fused with the prior probabilities of the HMM models. In general, we have observed that the effect of soft sensor evidence is more prominent when Equation 2 is used instead of Equation 3 and hence it is a preferred approach. For example, if soft sensor report is contrary to HMMs results, then the overall probability changes by a more noticeable amount in Equation 2, often bring the probability below 0.5 from a value more than it, which is generally not the case with the use of Equation 2.

Table 1. Simulation results for the Intention to cross the border.[†] Accounting for soft sensor's probability for a 'Miss' in detection.

Percent Path Seen	Without Fusion	With Fusion	With SS 'Miss'[†]
20	89.20	95.09	94.66
40	88.66	95.68	95.25
60	88.54	95.74	95.13
100	86.58	95.79	95.17

Table 2. Simulation results for the Intention to not-to-cross the border. [‡] Accounting for soft sensor's probability for 'false alarm' In detection. This table has only three columns because there is no information from the soft sensor for the probability that a subject will change his/her intention.

Percent Path Seen	Without Fusion	With Fusion[‡]
20	70.97	69.98
40	76.03	74.85
60	76.68	75.73
100	77.60	76.59

Table 3. Simulation for Changing Cross Intention. [†] Accounting for soft sensor's probability for a 'Miss' In Detection.

Percent Path Seen	Without Fusion	With Fusion	With SS 'Miss'[†]
20	81.98	95.80	95.42
40	84.05	97.22	96.73
60	84.69	97.28	96.85
100	84.29	97.05	96.76

5 Path Prediction

Estimation of when and where a subject will cross the border is a significant component of border crossing problem. To accomplish this task the path observed so far is given as input to the Viterbi algorithm [8], which estimates the most probable state sequence which could produce the given sequence of observations. The last state in this state sequence, which corresponds to present state of the subject, becomes the state using which the future path prediction starts. We construct the remaining path, up to the border, using parameters of that HMM which has higher probability of generating the path observed so far.

However, the path so generated is a random path and represents one of the several possible paths that can be obtained by repeated use of HMM. By generating an ensemble of paths we can obtain a region where the subject is likely to cross with high level of confidence. Using the set of predicted data we can also find the average and minimum time required to reach each zone. Figure 5

Table 4. Predicted time of crossing for various zones of border

Time Instant	Zone-3	Zone-4	Zone-5	Zone-6	Zone-7
85	0	133.3	112.6	134	130
90	0	129.3	109	128.9	0
95	0	124.7	108.2	126.8	0
100	0	133	107	129	0
105	0	0	105	0	0

shows an example of Path Predictor after observing the subject for 60 minutes. It displays the probability of crossing border at each zone and the corresponding time of crossing.

Table 4 shows the predicted place and time of crossing for a subject having an intention to cross using Viterbi's algorithm for 1000 iterations. The place of crossing can be one of the several equal zones along the border. The table shows the predicted time of crossing at the different zones before the crossing takes place.

In realistic examples a subject is likely to follow a trail as opposed to walking through all possible areas within a region. In that case, the above approach will provide a short list of trails where the subject is likely to cross the border.

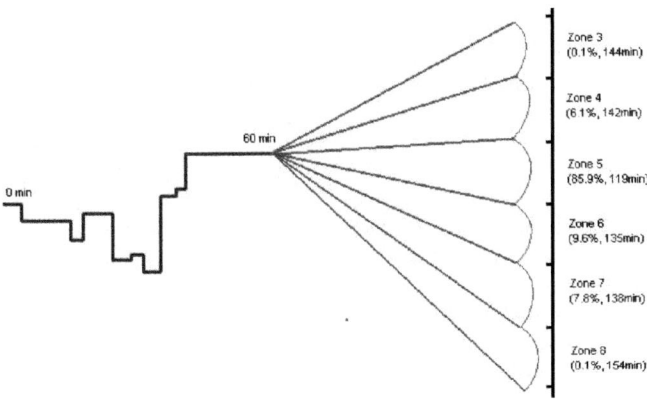

Fig. 5. Prediction of zone and time of border crossing given partially observed path

6 Conclusion

The main contribution of this paper is to demonstrate that given a partially observed path followed by a subject, it is possible to assess whether a subject intends to cross the border. Using the simulation it is also possible to estimate when and where the subject will cross the border. Soft sensor data can be very valuable in increasing or decreasing the amount of surveillance near the border and can also be used to better assessing the probability of intention to cross the border. In this work we do not consider geographical constraints nor do we consider the possibility that subjects path can be modeled using other method such as via a state equation. in our future work we plan to extend these results for more realistic scenarios and will consider sequential estimation approach for the case when soft and hard sensors data arrives at the fusion center at multiple, intermingled times.

Possible errors, imprecision, and uncertainty in the observations will also be considered in future work.

Acknowledgements

This research was partially supported by U.S. Army Research Contract W911NF-10-2-00505. The authors also thank Karthik Kuber for his assistance and helpful comments.

References

1. Brdiczka, O., Langet, M., Maisonnasse, J., Crowley, J.L.: Detecting Human Behavior Models From Multimodal Observation in a Smart Home. IEEE Transactions on Automation Science and Engineering
2. Feldman, A., Balch, T.: Representing Honey Bee Behavior for Recognition Using Human Trainable Models. Adapt. Behav. 12, 241–250 (2004)
3. Hall, D.L., Llinas, J., Neese, M.M., Mullen, T.: A Framework for Dynamic Hard/Soft Fusion. In: Proceedings of the 11th International Conferenec on Information Fusion (2008)
4. Brooks, R., Iyengar, S.: Multi-Sensor Fusion: Fundamentals and Applications with Software. Prentice Hall, Englewood Cliffs (1997)
5. Jones, R.E.T., Connors, E.S., Endsley, M.R.: Incorporating the Human Analyst into the Data Fusion Process by Modeling Situation Awareness Using Fuzzy Cognitive Maps. In: 12th International Conference on Information Fusion, Seattle, WA, USA, July 6-9 (2009)
6. Kelley, R., Tavakkoli, A., King, C., Nicolescu, M., Nicolescu, M., Bebis, G.: Understanding Human Intentions via Hidden Markov Models in Autonomous Mobile Robots. In: HRI 2008, Amsterdam, Netherlands (2008)
7. Laudy, C., Goujon, B.: Soft Data Analysis within a Decision Support System. In: 12th International Conference on Information Fusion, Seattle, WA, USA, July 6-9 (2009)
8. Rabiner, L.R.: A Tutorial on Hidden Markov Models and Selected Applications in Speech Recognition. Proceedings of the IEEE 77(2) (February 1989)
9. Rickard, J.T.: Level 2/3 Fusion in Conceptual Spaces. In: 10th International Conference on Information Fusion, Quebeck, Canada, July 9-12 (2007)
10. Sambhoos, K., Llinas, J., Little, E.: Graphical Methods for Real-Time Fusion and Estimation with Soft Message Data. In: 11th International Conference on Information Fusion, Cologne, Germany, June 30-July 3 (2008)
11. Youn, S.-J., Oh, K.-W.: Intention Recognition using a Graph Representation. World Academy of Science, Engineering and Technology 25 (2007)

A Framework for Autonomous Search in the Ecl^{i}ps^e Solver

Broderick Crawford[1,2], Ricardo Soto[1], Mauricio Montecinos[1], Carlos Castro[2], and Eric Monfroy[2,3]

[1] Pontificia Universidad Católica de Valparaíso, Chile
{broderick.crawford,ricardo.soto}@ucv.cl,
mauricio.montecinos@inf.ucv.cl
[2] Universidad Técnica Federico Santa María, Chile
carlos.castro@inf.utfsm.cl
[3] CNRS, LINA, Université de Nantes, France
eric.monfroy@inf.utfsm.cl

Abstract. Autonomous Search (AS) is a special feature allowing systems to improve their performance by self-adaptation. This approach has been recently adapted to Constraint Programming (CP) reporting promising results. However, as the research lies in a preliminary stage there is a lack of implementation frameworks and architectures. This hinders the research progress, which in particular, requires a considerable work in terms of experimentation. In this paper, we propose a new framework for implementing AS in CP. It allows a dynamic self-adaptation of the classic CP solving process and an easy update of its components, allowing developers to define their own AS-CP approaches. We believe this will help researchers to perform new AS experiments, and as a consequence to improve the current preliminary results.

1 Introduction

Constraint Programming (CP) is known to be an efficient software technology for modeling and solving constraint-based problems. Under this framework, problems are formulated as Constraint Satisfaction Problems (CSP). This formal problem representation mainly consists in a sequence of variables lying in a domain and a set of constraints. Solving the CSP implies to find a complete variable-instantiation that satisfies the whole set of constraints. The common approach to solve CSPs relies in creating a tree data structure holding the potential solutions by using a backtracking-based procedure. In general, two main phases are involved: an enumeration and a propagation phase. In the enumeration phase, a variable is instantiated to create a branch of the tree, while the propagation phase attempts to prune the tree by filtering from domains the values that do not lead to any solution. This is possible by using the so-called consistency properties [6].

In the enumeration phase, there is two important decisions to be made: the order in which the variables and values are selected. This selection refers to the variable and value ordering heuristics, and jointly constitutes the enumeration strategy. It turns out that this pair of decisions is crucial in the performance of the resolution process. For instance, if the right value is chosen on the first try for each variable, a solution can be found without performing backtracks. However, to decide a priori the correct heuristics is quite hard, as the effects of the strategy can be unpredictable.

A modern approach to handle this concern is called Autonomous Search (AS) [2]. AS is a special feature allowing systems to improve their performance by self-adaptation or supervised adaptation. Such an approach has been successfully applied in different solving and optimization techniques [3]. For instance, in evolutionary computing, excellent results have been obtained for parameter settings [5]. In this context, there exists a theoretical framework as well as different successful implementations. On the contrary, AS for CP is a more recent trend. Some few works have reported promising results based on a similar theoretical framework [1], but little work has been done in developing extensible frameworks and architectures for implementing AS in CP. This hinders the advances in this area, which in particular, requires a strong work in terms of experimentation.

The central idea of AS in CP is to perform a self-adaptation of the search process when it exhibits a poor performance. This roughly consists in performing an "on the fly" replacement of bad-performing strategies by another ones looking more promising. In this paper, we propose a new framework for implementing AS in CP. This new framework enables the required "on the fly" replacement by measuring the quality of strategies through a choice function. The choice function determines the performance of a given strategy in a given amount of time, and it is computed based upon a set of indicators and control parameters. Additionally, to guarantee the precision of the choice function, a genetic algorithm optimizes the control parameters. This framework has been implemented in the $Ecl^{i}ps^{e}$ Solver and it is supported by a 4-component architecture (described in Section 3). An important capability of this new framework is the possibility of easily update its components. This is useful for experimentation tasks. Developers are able to add new choice functions, new control parameter optimizers, and/or new ordering heuristics in order to test new AS approaches. We believe this framework is a useful support for experimentation tasks and can even be the basis for the definition of a general framework for AS in CP.

This paper is organized as follows. Section 2 presents the basic notions of CP and CSP solving. The architecture of the new framework is described in Section 3. The implementation details are presented in Section 4, followed by the conclusion and future work.

2 Constraint Programming

As previously mentioned, in CP problems are formulated as CSPs. Formally, a CSP $\mathcal{P}$ is defined by a triple $\mathcal{P} = \langle \mathcal{X}, \mathcal{D}, \mathcal{C} \rangle$ where:

- $\mathcal{X}$ is an n-tuple of variables $\mathcal{X} = \langle x_1, x_2, \ldots, x_n \rangle$.

- $\mathcal{D}$ is a corresponding n-tuple of domains $\mathcal{D} = \langle D_1, D_2, \ldots, D_n \rangle$ such that $x_i \in D_i$, and D_i is a set of values, for $i = 1, \ldots, n$.
- $\mathcal{C}$ is an m-tuple of constraints $\mathcal{C} = \langle C_1, C_2, \ldots, C_m \rangle$, and a constraint C_j is defined as a subset of the Cartesian product of domains $D_{j_1} \times \cdots \times D_{j_{n_j}}$, for $j = 1, \ldots, m$.

A solution to a CSP is an assignment $\{x_1 \to a_1, \ldots, x_n \to a_n\}$ such that $a_i \in D_i$ for $i = 1, \ldots, n$ and $(a_{j_1}, \ldots, a_{j_{n_j}}) \in C_j$, for $j = 1, \ldots, m$.

Algorithm 1. A general procedure for solving CSPs
1: $load\_CSP()$
2: **while not** $all\_variables\_fixed$ **or** $failure$ **do**
3:    $heuristic\_variable\_selection()$
4:    $heuristic\_value\_selection()$
5:    $propagate()$
6:    **if** $empty\_domain\_in\_future\_variable()$ **then**
7:      $shallow\_backtrack()$
8:    **end if**
9:    **if** $empty\_domain\_in\_current\_variable()$ **then**
10:     $backtrack()$
11:   **end if**
12: **end while**

Algorithm 1 represents a general procedure for solving CSPs. The goal is to iteratively generate partial solutions, backtracking when an inconsistency is detected, until a result is reached. The algorithm begins by loading the CSP model. Then, a while loop encloses a set of actions to be performed until fixing all the variables (i.e. assigning a consistent value) or a failure is detected (i.e. no solution is found). The first two enclosed actions correspond to the variable and value selection. The third action is a call to a propagation procedure, which is responsible for attempting to prune the tree. Finally two conditions are included to perform backtracks. A shallow backtrack corresponds to try the next value available from the domain of the current variable, and the backtracking returns to the most recently instantiated variable that has still values to reach a solution.

3 Architecture

Our framework is supported by the architecture proposed in [4]. This architecture consists in 4 components: SOLVE, OBSERVATION, ANALYSIS and UPDATE.

- The SOLVE component runs a generic CSP solving algorithm performing the aforementioned propagation and enumeration phases. The enumeration strategies used are taken from the quality rank, which is controlled by the UPDATE component.

- The OBSERVATION component aims at regarding and recording relevant information about the resolution process. These observations are called snapshots.
- The ANALYSIS component studies the snapshots taken by the OBSERVATION. It evaluates the different strategies, and provides indicators to the UPDATE component. Some indicators used are for instance, number of variables fixed by propagation, number of shallow backtracks, number of backtracks, current depth in the search tree, etc (a complete list of indicators used can be seen in [1]).
- The UPDATE component makes decisions using the choice function. The choice function determines the performance of a given strategy in a given amount of time. It is calculated based on the indicators given by the ANALYSIS component and a set of control parameters computed by an optimizer.

4 Framework Implementation

The framework has been designed to allow and easy modification of the UPDATE component. In fact, UPDATE is the most susceptible component to suffer modifications, since the most obvious experiment –in the context of AS – is to tune or replace the choice function or the optimizer.

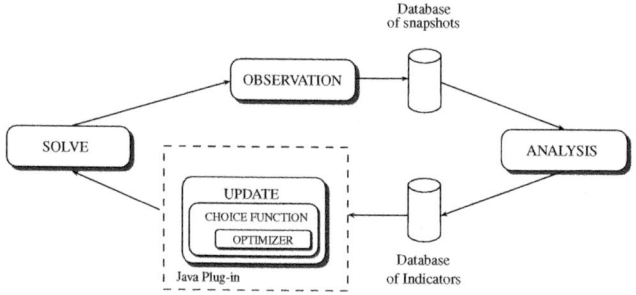

Fig. 1. Framework schema

Figure 1 depicts a general schema of the framework. The SOLVE, OBSERVATION, and ANALYSIS, component have been implemented in Ecl^ips^e. The UPDATE component as been designed as a plug-in for the framework. Indeed, we have implemented a Java version of the UPDATE component which computes the choice function and optimizes its control parameters through a genetic algorithm. Another version of the UPDATE component, which is currently under implementation, uses a swarm optimizer. The Ecl^ips^e-Java communication is very simple, it suffices to call the set_java_data/1 Ecl^ips^e predicate to send the indicators to the UPDATE component. UPDATE computes the choice function and selects the most promising strategy. This information is obtained from the SOLVE component by using the get_java_data/1 Ecl^ips^e predicate.

4.1 The UPDATE Plug-In

In the current UPDATE component, we use a choice function [7] that ranks and chooses between different enumeration strategies at each step (a step is every time the solver is invoked to fix a variable by enumeration). For any enumeration strategy S_j, the choice function f in step n for S_j is defined by equation 1, where l is the number of indicators considered and α is the control parameter (it manages the relevance of the indicator within the choice function). Such control parameters are computed by the genetic algorithm, which attempt to find the values for which the backtracks are minimized.

$$f_n(S_j) = \sum_{i=1}^{l} \alpha_i f_{in}(S_j) \qquad (1)$$

Additionally, to control the relevance of an indicator i for a strategy S_j in a period of time, we use a popular statistical technique for producing smoothed time series called exponential smoothing. The idea is to associate, for some indicators, greater importance to recent performance by exponentially decreasing weights to older observations. In this way, recent observations give relatively more weight that older ones. The exponential smoothing is applied to the computation of $f_{in}(S_j)$, which is defined by equations 2 and 3, where v_0 is the value of the indicator i for the strategy S_j in time 1, n is a given step of the process, β is the smoothing factor, and $0 < \beta < 1$.

$$f_{i1}(S_j) = v_0 \qquad (2)$$

$$f_{in}(S_j) = v_{n-1} + \beta_i f_{in-1}(S_j) \qquad (3)$$

Let us note that the speed at which the older observations are smoothed (dampened) depends on β. When β is close to 0, dampening is quick and when it is close to 1, dampening is slow.

The general solving procedure including AS can be seen in Algorithm 2. Three new function calls have been included: for calculating the indicators (line 11), the choice function (line 12), and for choosing promising strategies (line 13), that is, the ones with highest choice function[1]. They are called after constraint propagation to compute the real effects of the strategy (some indicators may be impacted by the propagation).

5 Conclusion and Future Work

In this work, we have presented a new framework for performing AS in CP. Based on a set of indicators, the framework measures the resolution process state to allow the replacement of strategies exhibiting poor performances. A main element of the framework is the UPDATE component, which can be seen as a plug-in of the architecture. This allows users to modify or replace the choice

---

[1] When strategies have the same score, one is selected randomically.

Algorithm 2. A procedure for solving CSPs including autonomous search
1: **while not** $all\_variables\_fixed$ **or** $failure$ **do**
2:     $heuristic\_variable\_selection()$
3:     $heuristic\_value\_selection()$
4:     $propagate()$
5:     **if** $empty\_domain\_in\_future\_variable()$ **then**
6:         $shallow\_backtrack()$
7:     **end if**
8:     **if** $empty\_domain\_in\_current\_variable()$ **then**
9:         $backtrack()$
10:    **end if**
11:    $calculate\_indicators()$
12:    $calculate\_choice\_function()$
13:    $enum\_strategy\_selection()$
14: **end while**

function and/or the optimizer in order to perform new AS-CP experiments. The framework has been tested with different instances of several CP-benchmarks (send+more=money, N-queens, N-linear equations, self referential quiz, magic squares, sudoku, knight tour problem, etc) by using the already presented UPDATE component, which has been plugged to the framework in a few hours by a master student with a limited Ecl^ips^e knowledge.

The framework introduced here is ongoing work, and it can certainly be extended by adding new UPDATE components. This may involve to implement new optimizers as well as the study of new statistical methods for improving the choice function.

References

1. Crawford, B., Montecinos, M., Castro, C., Monfroy, E.: A hyperheuristic approach to select enumeration strategies in constraint programming. In: Proceedings of ACT, pp. 265–267. IEEE Computer Society, Los Alamitos (2009)
2. Hamadi, Y., Monfroy, E., Saubion, F.: Special issue on autonomous search. Contraint Programming Letters 4 (2008)
3. Hamadi, Y., Monfroy, E., Saubion, F.: What is autonomous search? Technical Report MSR-TR-2008-80, Microsoft Research (2008)
4. Monfroy, E., Castro, C., Crawford, B.: Adaptive enumeration strategies and metabacktracks for constraint solving. In: Yakhno, T., Neuhold, E.J. (eds.) ADVIS 2006. LNCS, vol. 4243, pp. 354–363. Springer, Heidelberg (2006)
5. Robet, J., Lardeux, F., Saubion, F.: Autonomous control approach for local search. In: Stützle, T., Birattari, M., Hoos, H.H. (eds.) SLS 2009. LNCS, vol. 5752, pp. 130–134. Springer, Heidelberg (2009)
6. Rossi, F., van Beek, P., Walsh, T.: Handbook of Constraint Programming. Elsevier, Amsterdam (2006)
7. Soubeiga, E.: Development and Application of Hyperheuristics to Personnel Scheduling. PhD thesis, University of Nottingham School of Computer Science (2009)

Multimodal Representations, Indexing, Unexpectedness and Proteins

Eric Paquet[1,2] and Herna L. Viktor[2]

[1] National Research Council, 1200 Montreal Road, Ottawa,
Ontario, K1A 0R6, Canada
eric.paquet@nrc-cnrc.gc.ca
[2] School of Information Technology and Engineering, University of Ottawa,
800 King Edward, Ottawa, Ontario, K1N 6N5, Canada
hlviktor@site.uottawa.ca

Abstract. Complex systems, such as proteins, are inherently difficult to describe, analyse and interpret. A multimodal methodology which utilizes various diverse representations is needed to further our understanding of such intrinsically multifaceted systems. This paper presents a multimodal system designed to describe and interpret the content of the Protein Data Bank, a repository which contains tens of thousands of known proteins. We describe how complimentary modalities based on the amino acid sequence, the protein's backbone (or topology) and the envelope (the outer shape), are used when furthering our understanding of proteins' functionalities, behaviours and interactions. We illustrate our methodology against the human haemoglobin and show that the interplay between modalities allows our system to find unexpected, complimentary results with different domains of validity.

Keywords: Amino acid sequences, indexing, multimodal, protein, retrieval, shape, topology, unexpectedness, 3D.

1 Introduction

Multimodal systems have been successfully deployed across multiple domains, ranging from intelligent media browsers for teleconferencing to systems for analysing expressive gesture in music and dance performances [1, 2]. The success of multimodal approaches lies in the inherent assumption that each modality brings information that is absent from the others. That is, complementary perspectives on an otherwise partially known piece of information are provided. Usually, multimodal systems aim to combine the inputs from diverse media. For example, the combination of various images (visible, infrared and ultraviolet) in order to improve the contrast in air surveillance comes to mind [3]. However, a strength of multimodal analysis is that it may provide unexpected and complimentary, or even apparently conflicting, perspectives on a problem domain. This aspect is often overlooked, which may lead to unexplored knowledge being left undiscovered. This paper shows that such unexpected and complimentary modes exist when analyzing large protein repositories. We show how, when following a multimodal approach, potentially hidden results may be uncovered.

Specifically, we analyze the interplay in between various representations, or modalities, of proteins. We show how the use of various modalities improves the analysis and brings unexpected results. In Section 2, we present some useful notions and representations for proteins and we introduce our multimodal system. Section 3 explains how these multimodal representations are indexed. In Section 4, we apply our approach to the human haemoglobin. We show that a multimodal analysis greatly improves the overall analysis and helps to put in evidence unexpected results that would have been overlooked with a more superficial evaluation. Section 5 concludes the paper.

2 Three Modalities of Proteins

Proteomics researchers agree that proteins are complex systems that are difficult to interpret. This is especially evident when aiming to obtain a holistic view of their functionalities, to explain their interactions or to identify similarities and dissimilarities. A protein consists of one, or many, oriented chains of amino acids. The interaction between the amino acids and with their environment determines the shape of the protein [4]. For many years, researchers believed that the sequence of amino acid uniquely determined the shape of the protein. While this observation holds, up to a certain extent, researchers now agree that the complex reaction between the amino acids and their environment cannot be uniquely characterized by the sequence and that the "shape" should also be considered.

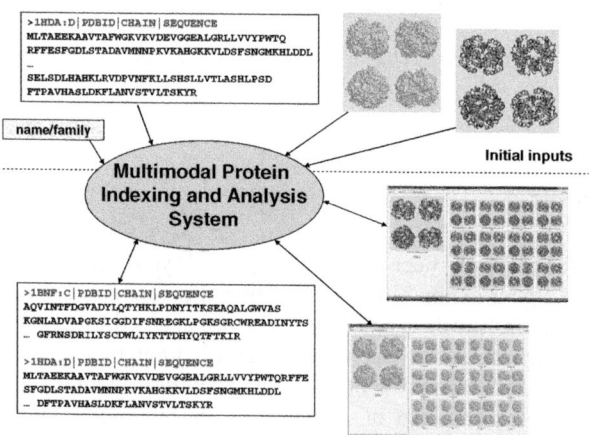

Fig. 1. Multimodal protein indexing and analysis system

A protein, however, does not have a shape "per se". It is ultimately constituted of atoms, which interact with one another according to quantum mechanics. However, it is possible to infer, from the position of the various atoms, various representations that characterized specific aspects of their "personality". The most common representation is the topology. What, then, does the topology refer to? We saw earlier that a protein is constituted of a sequence of amino acids. The position of the amino acid is determined by the position of the carbon atoms. These atoms thus determine the backbone, or the topology, of the protein. This topology is useful for classification, comparison and to understand the evolution of proteins.

Unfortunately, the topology tells us very little about the interactions of proteins. Understanding interactions is of great importance in order to better understand the mechanisms behind diseases and to design drugs in silico, i.e. when simulating the interaction of a drug and a protein entirely within a computer grid. Such information is provided in part by the envelope of the protein. The envelope refers to the outer surface bounding the protein and the part that is accessible from the environment. The envelope is thus the area through which most interactions take place.

Consider, for example, a situation where a molecular biologist wants to further explore a protein in order to potentially design a drug for a disease. Such a domain expert is interested in exploring the amino acid sequence, the topology as well as the envelope. A multimodal analysis is needed in order to aid him/her when exploring a protein's unique characteristics and interactions. Figure 1 provides an overview of our multimodal protein analysis system, designed to aid such a person. The figure shows that the user is able to input (a) the sequence, (b) the topology, (c) the envelope and/or (d) the protein name or the protein family it belongs too. The user also specifies the type(s) of output he/she requires, such as the sequence and topology. Our system then proceeds to find the most relevant proteins using a distance-based similarity search. It follows that, through interacting with the system, the user is able to change the representations required as he/she explores the protein repository and that this is an iterative process which is terminated by the user.

In the next sections we describe our algorithms and detail how we tested our multimodal system. We also show some unexpected results as derived from a multimodal analysis.

3 Algorithms and Experiments

All the data for our experiments are from the Protein Data Bank (PDB), which is a repository that contains 64.500 known proteins [5]. For each protein in the PDB, the amino acid sequence and the position of the constituent atoms is recorded. From this information, is it possible to determine, using the positions of the atoms, the topology and the envelopes of each protein with a molecular solver. In our case, we used the VMD molecular solver from the University of Illinois [6]. Thus, we have available, for each of the proteins in the PDB, the amino acid sequence, the topology and the outer shape or envelope. The topology and the envelope are represented as a triangular mesh, the 3D shape.

Recall that our objective is, for a given protein, to explore its amino acid sequence, its topology and its envelope. Next, we aim to compare the three representations to all the other proteins in the PDB. That is, given a particular protein, we are interested in determining the "closest" other proteins in the PDB, from three different perspectives. Here, we thus assess the proteins' modalities from three different and complementary perspectives. That is, their composition, their internal structure and the way they present themselves to the outside world, which shows the geometry involved in interaction. We further aim to determine the synergy and the unexpectedness of such a multimodal analysis. Indeed, as will be shown in Section 4, it is difficult to obtain a good understanding by simply considering one modality. Furthermore, the "a priori" knowledge associated with such a single modality may lead either to false, or over-generalized, conclusions. We will show that such erroneous conclusions may be avoided with a multimodal analysis.

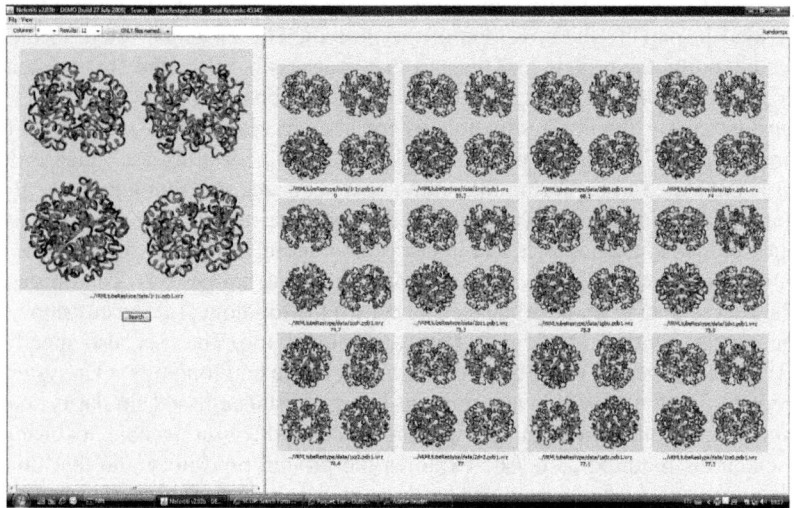

Fig. 2. First twelve results for 1r1y in terms of topology

In order to compare amino acid sequences and to determine their level of similarity, we used BLAST, which is a well established and publicly available tool from the National Center for Biotechnology Information. More details about BLAST and the underlying algorithms may be found in [7]. It follows that designing and implementing such comparison algorithms are not a trivial task. Similar sequences are often broken into a set of disconnected sub-sequences with unrelated inter-sequences of various lengths, which makes their comparison very difficult. For the purpose of this paper, it is sufficient to know that BLAST provides us, given a reference amino acid sequence, with a ranking of the most similar amino acid sequences. It is also important to realize that BLAST does not perform an exhaustive search, which would require a disproportionate amount of time. Rather, BLAST follows a stochastic approach that uses a heuristic algorithm to search and compare sequences, making it very efficient [7].

We developed our own algorithm to describe the three-dimensional shape associated with the topology and the envelope of each protein. Our algorithm accurately describes these intricate shapes and is able to process such a large amount of 3D data efficiently. The algorithm and its performance is detailed in [8] and [9], so it is only outlined here.

From the position of the atoms, the topologies and the envelopes were calculated with the University of Illinois molecular solver [6]. The 3D shape can be obtained from in one of two ways. Firstly, from the isosurface associated with the electronic density around the atoms forming the protein or, secondly, from the isosurface corresponding to the potential energy associated with their relative spatial configuration.

Our algorithm for 3D shape description then proceeds as follows. The 3D shape is represented by a triangular mesh. The tensor of inertia associated with the shape is calculated, from which a rotation invariant reference frame is defined on the Eigen vectors of the later. Then, the probability that a given triangle has a certain orientation and a certain distance with respect to the reference frame is calculated. The distributions associated with a given representation of two proteins may be compared either in terms of the Euclidian distance, which is a metric measure of similarity, or in terms of the Jensen-Shannon divergence. The Jensen-Shannon divergence is a probabilistic dissimilarity measure in between the two distributions. Given a 3D shape, our algorithm provides a ranking of the most similar shapes, which in our case are either associated with the topologies or the envelopes of the proteins.

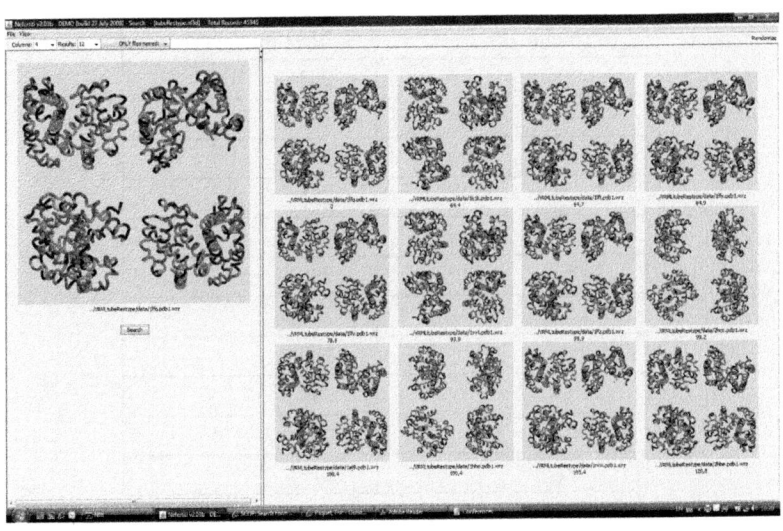

Fig. 3. First twelve results for 1lfq in terms of topology

In the next section we illustrate our results. We use, as a reference protein, the member 1r1y of the human haemoglobin; where 1r1y is the unique identifier of the protein in the PDB. However, our procedure may be followed for the multimodal analysis of any protein in the Protein Data Bank.

4 Unexpected Behaviour of the Human Haemoglobin

Bank. First, the topologies and the envelopes of all proteins were calculated with the University of Illinois molecular solver. Next, the three-dimensional shapes associated with these topologies and envelopes were indexed with our algorithm as described in Section 3. Recall that for the reference protein, we chose the `1r1y` member of the human haemoglobin. The amino acid sequence, the topology and the envelope of this protein were next compared to the entire PDB database. That is, for each one of the three representations, we conducted a similarity search against all the other proteins. Next, we discuss our results by dividing our analysis, in terms of amino acid sequence, topology and envelope.

4.1 Results in Terms of Amino Acid Sequence

The first eleven entries obtained when comparing the `1r1y` amino acid sequence with the other proteins as contained in the Protein Data Bank are shown in Table 1.

Table 1. Eleven highest Blast scores [6] for the comparison of the amino acid sequence of `1r1y` with the sequences of the other proteins in the Protein Data Bank

Protein	Blast max score
1r1x	283
1bz1	281
3ia3	281
1coh	281
1c7d	281
1aby	281
1lfq	281
1xzu	280
1aj9	280
1bzz	280
1bab	279

Our results show that the members of the human haemoglobin have a very similar amino acid sequence. This is to be expected. This sequence is the fruit of evolution and as such, it is very unlikely that for a given species, two distinct sequences would have produced the same functionality. Here, the aim is the transport of oxygen to the tissues. For many years, researchers believed this was "the end of the story". Even nowadays, it is still sometimes "a priori" assumed that most, if not all, useful information may be extracted from the amino acid sequence. As our multimodal analysis will show, this is clearly not the case.

4.2 Results in Terms of Topologies

The first twelve results for the topology similarity search are presented in Figure 2. It follows that 1lfq appears in the first position. For each protein, because of its intricate structure, four (4) views of the topology are shown. Recall that what is indexed is the actual 3D shape associated with the topology; the views are only utilised for visualization. The reference protein is shown to the left and the outcome of the search is shown to the right. As expected, the most similar structures all belong to the human haemoglobin.

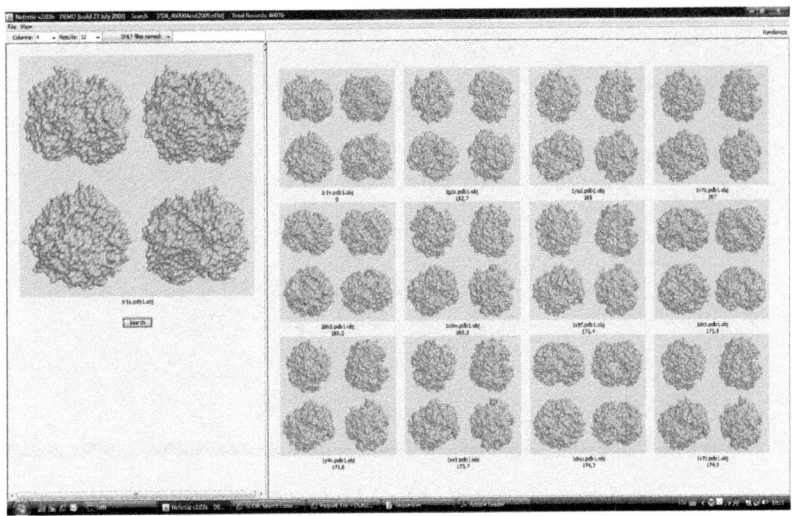

Fig. 4. First twelve results for 1r1y in terms of envelope

What is unexpected, however, is that an important fraction of the proteins detected during the amino acid sequence phase, are not within the results when searching for similarities based on the topology. For example, protein 1lfq which has a Blast max score of 281 (the max score depends on the level of sequence overlap and the amino acid scoring matrix and in this particular case, there is a good overlap) is not close in similarity to 1r1y, from the topological point of view (It is at distance 510, as opposed to 59.3 for the closest match. Typically, the threshold distance that divides similarity from dissimilarity, from the 3D shape point of view, is located around 100). This implies that there must be at least another topology for the haemoglobin to which 1lfq would belong. In order to find the new topology, the outcome of Section 4.1 is randomly sampled. A search is performed for each of the sampled proteins. From the results, one rapidly discovered that there are, in fact, two topologies. That is, the topology shown in Figures 2 versus the one depicted in Figure 3, which was obtained with 1lfq as a reference protein. This protein's shape is clearly distinct from the previous reference protein (1r1y). Such an unexpected result would have been totally hidden if the topology-based modality was not introduced.

One more unexpected result may be inferred by considering the first hundred (100) topology-based search results. The result at the 93rd position is not part of the human haemoglobin, but a member of the cow haemoglobin, entry 1hda in the Protein Data Bank. It shares a very close topology with its human counterpart. These two proteins have a similar topology since they both share the same function of carrying oxygen and it follows that such a function strongly constraints the possible shape of the protein.

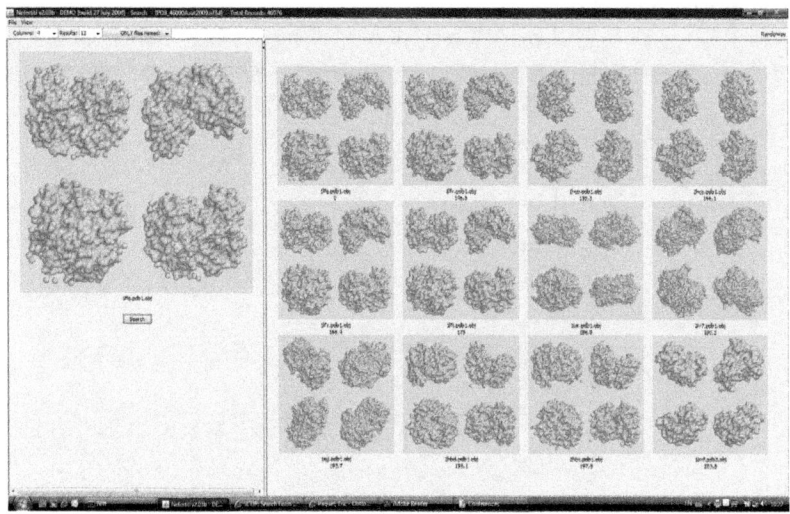

Fig. 5. First twelve results for 11fq in terms of envelope

A second unexpected behaviour is that the ranking based on the amino acid sequence is quite distinct from the one obtained with the topology. For example, 1vwt is the most similar protein to 1r1y from the topological point of view with a distance of 59.3. It has a Blast maximum score of 278 when considering the amino acid sequence. This means that there is a substantial level of dissimilarity, from the sequence point of view. On the other hand, 1r1x which is the most similar protein from the amino acid sequence point of view (see Table 1), has a distance of 510.1 from a topological perspective as opposed to the most similar one which has a distance of 59.3. Thus, highly similarity amino acid sequences do not necessarily imply a high similarity from the topological point of view, and vice versa.

4.3 Results in Terms of Envelopes

In the previous section, we explored the proteins in terms of their topology. As mentioned in Section 2, research indicates that the topology does not adequately describe protein-protein interactions, which are very important during drug design. The envelope is more suited for this task. In this section, our two previous reference proteins are compared against the entire PDB in terms of the 3D shape of their envelopes.

The results for the envelope may be divided in two (2) groups. The members of the first group have a 3D shape similar to the one of 1y1y as illustrated in Figure 4, while the members of the second group have a shape similar to the one of 1lfq as illustrated in Figure 5. Again, the similarity rankings differ from that of the amino acid sequences as presented in Section 4.1.

A few unexpected results appear. The first is that the ranking of the proteins in terms of their topologies and envelopes are not the same. In other words, some members of the human haemoglobin are more similar in terms of topology and others in terms of their envelopes. This implies, in general, that the topology is not a sufficient criterion to constraint the envelope. Importantly, the topology is thus not suitable for finding the modus operandi of the associated protein; recall that the envelope is related to interaction and of high interest for drug designers.

Perhaps even more surprising is the fact that, for a given rank, the envelope are much more dissimilar that their topological counterpart, i.e. that there is more variability in the envelope than in the topology.

Even more unexpected, non-members of the human haemoglobin appear much earlier in the envelope results, than with their topological counterpart. This is especially evident for the 1lfq subgroup. For instance, the result at the 7th position, 1uir, is not even a member of the globins family, but of the spermidine synthase family. This implies that low similarity amino acid sequences may have a very similar envelope. Although counterintuitive, this result may be explained by the fact that proteins are highly non-linear, complex molecular systems which are difficult to understand only in terms of amino acid sequences or topologies.

5 Conclusions

Complex systems, such as proteins, may be modelled using numerous complimentary representations. It is important to recognize that a true understanding of these systems may only be obtained through a multimodal analysis of such diverse representations. Due to the underlying complexity, it might be very difficult to capture all the relevant knowledge with a sole representation. We have shown that, in the case of proteins, a deeper understanding may be obtained by analyzing their similarity through three diverse modalities, i.e. in terms of amino acid sequences, topology and envelopes. It is also important to realize that it is the synergy, and not the combination, of the various representations that is important. As shown in Section 4, complimentary representations might lead to distinct results, which nevertheless hold in their respective domain of validity.

We are interested in further extending our initial system to include the ability for additional search modalities. To this end we are currently adding a text mining component, to enable users to enter free text, such as a disease description, as a search criterion. We also aim to add a component to provide users with additional links to relevant research articles from e.g. Medline.

Proteins are flexible structures, in the sense that the relative position of their constitutive atoms may vary over time. Our shape descriptor is mainly oriented toward rigid shapes, such as those that are routinely obtained from X-ray crystallography. However, shapes obtained by, for example, magnetic resonance imaging (MRI) under the form of a temporal sequence of 3D shapes, are representative of flexible or deformable 3D shapes. We plan to develop a descriptor that would be invariant to isometry and that would consequently allow for the description of such flexible shapes.

References

1. Tucker, S., Whittaker, S.: Accessing Multimodal Meeting Data: Systems, Problems and Possibilities. In: Bengio, S., Bourlard, H. (eds.) MLMI 2004. LNCS, vol. 3361, pp. 1–11. Springer, Heidelberg (2005)
2. Camurri, A., et al.: Communicating Expressiveness and Affect in Multimodal Interactive Systems. IEEE Multi Media 12(1), 43–53 (2005)
3. Jha, M.N., Levy, J., Gao, Y.: Advances in Remote Sensing for Oil Spill Disaster Management: State-of-the-Art Sensors Technology for Oil Spill Surveillance. Sensors 8, 236–255 (2008)
4. Zaki, M.J., Bystroff, C.: Protein Structure Prediction. Humana Press, Totowa (2007)
5. Berman, H.M., et al.: The Protein Data Bank. Nucleic Acids Research, 235–242 (2000)
6. Humphrey, W., et al.: VMD - Visual Molecular Dynamics. J. Molec. Graphics 14, 33–38 (1996)
7. Karlin, S., Altschul, S.F.: Applications and statistics for multiple high-scoring segments in molecular sequences. Proc. Natl. Acad. Sci. USA 90, 5873–5877 (1993)
8. Paquet, E., Viktor, H.L.: Addressing the Docking Problem: Finding Similar 3-D Protein Envelopes for Computer-aided Drug Design, Advances in Computational Biology. In: Advances in Experimental Medicine and Biology. Springer, Heidelberg (2010)
9. Paquet, E., Viktor, H.L.: CAPRI/MR: Exploring Protein Databases from a Structural and Physicochemical Point of View. In: 34th International Conference on Very Large Data Bases – VLDB, Auckland, New Zealand, August 24-30, pp. 1504–1507 (2008)

A Generic Approach for Mining Indirect Association Rules in Data Streams

Wen-Yang Lin[1], You-En Wei[1], and Chun-Hao Chen[2]

[1] Dept. of Computer Science and Information Engineering,
National University of Kaohsiung, Taiwan
wylin@nuk.edu.tw, waiewing@gmail.com
[2] Dept. of Computer Science and Information Engineering, Tamkang University, Taiwan
chchen@mail.tku.edu.tw

Abstract. An indirect association refers to an infrequent itempair, each item of which is highly co-occurring with a frequent itemset called "mediator". Although indirect associations have been recognized as powerful patterns in revealing interesting information hidden in many applications, such as recommendation ranking, substitute items or competitive items, and common web navigation path, etc., almost no work, to our knowledge, has investigated how to discover this type of patterns from streaming data. In this paper, the problem of mining indirect associations from data streams is considered. Unlike contemporary research work on stream data mining that investigates the problem individually from different types of streaming models, we treat the problem in a generic way. We propose a generic window model that can represent all classical streaming models and retain user flexibility in defining new models. In this context, a generic algorithm is developed, which guarantees no false positive rules and bounded support error as long as the window model is specifiable by the proposed generic model. Comprehensive experiments on both synthetic and real datasets have showed the effectiveness of the proposed approach as a generic way for finding indirect association rules over streaming data.

1 Introduction

Recently, the problem of mining interesting patterns or knowledge from large volumes of continuous, fast growing datasets over time, so-called data streams, has emerged as one of the most challenging issues to the data mining research community [1, 3]. Although over the past few years there is a large volume of literature on mining frequent patterns, such as itemsets, maximal itemsets, closed itemsets, etc., no work, to our knowledge, has endeavored to discover indirect associations, a recently coined new type of infrequent patterns. The term indirection association, first proposed by Tan *et al.* in 2000 [21], refers to an infrequent itempair, each item of which is highly co-occurring with a frequent itemset called "mediator". Indirect associations have been recognized as powerful patterns in revealing interesting information hidden in many applications, such as recommendation ranking [14], common web navigation path [20], and substitute items (or competitive items) [23], etc. For example, Coca-cola and Pepsi are competitive products and could be replaced by each other. So it is very likely that there is an indirect association rule revealing that consumers buy a kind of cookie tend to buy together with either Coca-cola or Pepsi but not both (Coca-cola, Pepsi | cookie).

In this paper, the problem of mining indirect associations from data streams is considered. Unlike contemporary research work on stream data mining that investigates the problem individually from different types of streaming models, we treat the problem in a unified way. A generic streaming window model that can encompass contemporary streaming window models and is endowed with user flexibility for defining specific models is proposed. In accordance with this model, we develop a generic algorithm for mining indirect associations over the generic streaming window model, which guarantees no false positive patterns and a bounded error on the quality of the discovered associations. We further demonstrate an efficient implementation of the generic algorithm. Comprehensive experiments on both synthetic and real datasets showed that the proposed algorithm is efficient and effectiveness in finding indirect association rules.

The remainder of this paper is organized as follows. Section 2 introduces contemporary stream window models and related work conducted based on these models. Our proposed generic window model, system framework and algorithm GIAMS for mining indirect association rules over streaming data are presented in Section 3. Some properties of the proposed algorithm are also discussed. The experimental results are presented in Section 4. Finally, in Section 5, conclusions and future work are described.

2 Related Work

Suppose that we have a data stream $S = (t_0, t_1, \ldots t_i, \ldots)$, where t_i denotes the transaction arriving at time i. Since data stream is a continuous and unlimited incoming data along with time, a window W is specified, representing the sequence of data arrived from t_i to t_j, denoted as $W[i, j] = (t_i, t_{i+1}, \ldots, t_j)$. In the literature [1], there are three main different types of window models for data stream mining, i.e., landmark window, time-fading window, and sliding window models.

- **Landmark model:** The landmark model monitors the entire history of stream data from a specific time point called landmark to the present time. For example, if window W_1 denotes the stream data from time t_i to t_j, then windows W_2 and W_3 will span stream data from t_i to t_{j+1} and t_i to t_{j+2}, respectively.
- **Time-fading model:** The time-fading model (also called damped model) assigns more weights to recently arrived transactions so that new transactions have higher weights than old ones. At every moment, based on a fixed decay rate d, a transaction processed n time steps ago is assigned a weight d^n, where $0 < d < 1$, and the occurrence of a pattern within that transaction is decreased accordingly.
- **Sliding window model:** A sliding window model keeps a window of size ω, monitoring the data within a fixed time [18] or a fixed number of transactions [8]. Only the data kept in the window is used for analysis; when a new transaction arrives, the oldest resident in the window is considered obsolete and deleted to make room for the new one.

The first work on mining frequent itemsets over data stream with landmark window model was proposed by Manku et al. [19]. They presented an algorithm, namely

Lossy Counting, for computing frequency counts exceeding a user-specified threshold over data streams. Although the derived frequent itemsets are approximate, the proposed algorithm guarantees that no false negative itemsets are generated. However, the performance of Lossy Counting is limited due to memory constraints. Since then, considerable work has been conducted for mining different frequent patterns over data streams, including frequent itemsets [5, 6, 12, 15, 17, 18, 25], maximal frequent itemsets [16], and closed frequent itemsets [8, 11]. Each method, however, is confined to a specific type of window model.

Existing researches on indirect association mining can be divided into two categories, either focusing on developing efficient algorithms [7, 13, 24] or extending the definition of indirect association for different applications [14, 20, 23].

The first indirect association mining approach was proposed by Tan et al. [21], called "INDIRECT". However, it is time-consuming for generating all frequent itemsets before mining indirect association. Wan and An [24] proposed an approach, called HI-mine, for improving the efficiency of the INDIRECT algorithm. Chen et al. [7] also proposed an indirect association mining approach that was similar to HI-mine.

3 A Generic Framework for Indirect Associations Mining

3.1 Proposed Generic Window Model

Definition 1. *Given a data stream* $S = (t_0, t_1, \ldots t_i, \ldots)$ *as defined before, a generic window model* Ψ *is represented as a four-tuple specification,* $\Psi(l, \omega, s, d)$*, where l denotes the timestamp at which the window starts,* ω *the window size, s the stride the window moves forward, and d is the decay rate.*

The stride notation s is introduced to allow the window moving forward in a batch of transactions, i.e., a block of size s. That is, if the current window under concern is $(t_{j-\omega+1}, t_{j-\omega+2}, \ldots, t_j)$, then the next window will be $(t_{j-\omega+s+1}, t_{j-\omega+s+2}, \ldots, t_{j+s})$, and the weight of a transaction within $(t_{j-s+1}, t_{j-s+2}, \ldots, t_j)$, say α, is decayed to αd, and the weight of a transaction within $(t_{j+1}, \ldots, t_{j+s})$ is 1.

Example 1. Let $\omega = 4$, $s = 2$, $l = t_1$, and $d = 0.9$. An illustration of the generic streaming window model is depicted in Figure 1. The first window $W_1 = W[1, 4] = (t_1, t_2, t_3, t_4)$ consists of two blocks, $B_1 = \{AH, AI\}$ and $B_2 = \{AH, AH\}$, for B_1 receiving weight 0.9 while B_2 receiving 1. Next, the window moves forward with stride 2. That means B_1 is outdated and a new block B_3 is added, resulting in a new window $W_2 = W[3, 6]$.

Below we show that this generic window model can be specified into any one of the contemporary models described in Section 2.

- Landmark model: $\Psi(l, \infty, 1, 1)$. Since $\omega = \infty$, there is no limitation on the window size and so the corresponding window at timestamp j is $(t_l, t_{l+1}, \ldots, t_j)$ and is $(t_l, t_{l+1}, \ldots, t_j, t_{j+1})$ at timestamp $j+1$.
- Time-fading model: $\Psi(l, \infty, 1, d)$. The parameter setting for this model is similar to landmark except that a decay rate less than 1 is specified.
- Sliding window model: $\Psi(l, \omega, 1, 1)$. Since the window size is limited to ω, the corresponding window at timestamp j is $(t_{j-\omega+1}, t_{j-\omega+2}, \ldots, t_j)$ and is $(t_{j-\omega+2}, \ldots, t_j, t_{j+1})$ at timestamp $j+1$.

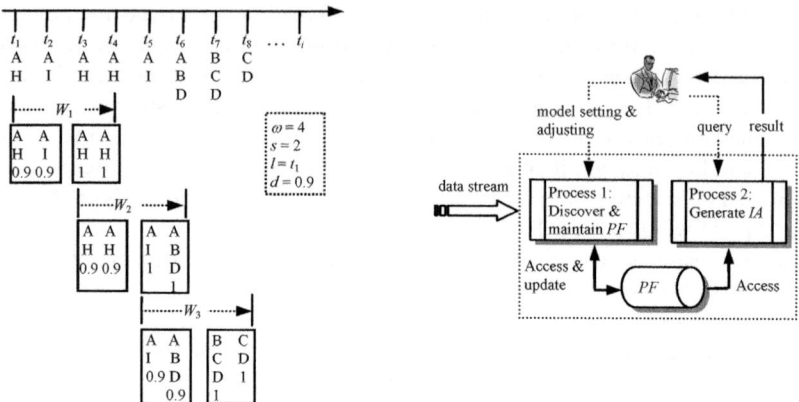

Fig. 1. An illustration of the generic window model   **Fig. 2.** A generic framework for indirect association mining

3.2 Generic Framework for Indirect Association Mining

Definition 2. *An itempair $\{a, b\}$ is indirectly associated via a mediator M, denoted as $\langle a, b|M \rangle$ if the following conditions hold:*

1. $sup(\{a, b\}) < \sigma_s$ *(Itempair support condition);*
2. $sup(\{a\} \cup M) \geq \sigma_f$ *and* $sup(\{b\} \cup M) \geq \sigma_f$ *(Mediator support condition);*
3. $dep(\{a\}, M) \geq \sigma_d$ *and* $dep(\{b\}, M) \geq \sigma_d$ *(Mediator dependence condition); where* $sup(A)$ *denotes the support of an itemset A, and* $dep(P, Q)$ *is a measure of the dependence between itemsets P and Q.*

In this paper, we follow the suggestion in [20, 21], adopting the well-known dependence function, *IS* measure $IS(P, Q)$ $(= sup(P, Q) / \text{sqrt}(sup(P) \times sup(Q)))$.

According to the paradigm in [21], the work of indirect association mining can be divided into two subtasks: First, discovers the set of frequent itemsets with support higher than σ_f, and then generates the set of qualified indirect associations from the frequent itemsets. Our framework adopts this paradigm, working in the following scenario: (1) The user sets the streaming window model to his need by specifying the parameters described previously; (2) The framework then executes the process for discovering and maintaining the set of potential frequent itemsets *PF* as the data continuously stream in; (3) At any moment once the user issues a query about the current indirect associations the second process for generating the qualified indirect associations is executed to generate from *PF* the set of indirect associations *IA*. Figure 2 depicts the generic streaming framework for indirect associations mining.

3.3 The Proposed Generic Algorithm GIAMS

Based on the framework in Figure 2, our proposed GIAMS (Generic Indirect Association Mining on Streams) algorithm consists of two concurrent processes: *PF-monitoring* and *IA-generation*. The first process is set off when the users specifies the window parameters to set the required window model, responsible for generating itemsets from the incoming block of transactions and inserting those potentially frequent itemsets into *PF*. The second process is activated when the user issues a query about the current indirect associations, responsible for generating the qualified patterns from the frequent itemsets maintained by process *PF-monitoring*. Algorithm 1 presents a sketch of GIAMS.

Algorithm 1. GIAMS

Input: Itempair support threshold σ_s, association support threshold σ_f, dependence threshold σ_d, stride s, decay rate d, window size ω, and support error ε.
Output: Indirect associations *IA*.

Initialization:
1:     Let N be the accumulated number of transactions, $N = 0$;
2:     Let η be the decayed accumulated number of transaction, $\eta = 0$;
3   Let *cbid* be the current block id, *cbid* = 0, *sbid* the starting block id of window, *sbid* = 1;
3:     **repeat**
4:         Process 1;
5:         Process 2;
6:     **until** terminate;

Process 1: *PF-monitoring*
1:     Reading the new coming block;
2:     $N = N + s$; $\eta = \eta \times d + s$;
3:     *cbid* = *cbid* + 1;
4:     **if** ($N > \omega$) **then**
5:         Block_delete(*sbid*, *PF*);   // Delete outdated block B_{sbid}
6:         *sbid* = *sbid* + 1;
7:         $N = N - s$;   // Decrease the transaction size in current window
8:         $\eta = \eta - s \times d^{cbid-sbid+1}$;   // Decrease the decayed transaction size in current window
9:     **endif**
10:    Insert(*PF*, σ_f, *cbid*, η); // Insert potential frequent itemsets in block *cbid* into *PF*
11:    Decay&Pruning(d, s, ε, *cbid*, *PF*); //Remove infrequent itemsets from *PF*

Process 2: *IA-generation*
1:     **if** user query request = true **then**
2:         IAgeneration(*PF*, σ_f, σ_d, σ_s, N); //Generate all indirect associations from *PF*

The most challenging and critical issue to the effectiveness of GIAMS is bounding the error. That is, under the constraints of only one pass of data scan and limited memory usage, how can we assure the error of the generated patterns is always bounded by a user specified range? Our approach to this end is, during the process of *PF-monitoring*, eliminating those itemsets with the least possibility to become frequent afterwards. More precisely, after processing the insertion of the new arriving block, we decay the accumulated count of each maintained itemset, and then prune any itemset X whose count is below a threshold indicated as follows:

$$X.count < \varepsilon \times s \times (d + d^2 + \ldots + d^{cbid-sbid+1}) = \varepsilon \times s \times \frac{1 - d^{cbid-sbid+1}}{1-d} \quad (1)$$

where *cbid* and *sbid* denote the identifiers of current block and the first block that X appears in *PF*, respectively, s is the stride, and ε is a user-specified tolerant error. Note that the term $s \times (d + d^2 + \ldots + d^{cbid-sbid+1})$ equals to the decayed amount of transactions between the first block that X appears and the current block within the current window. Therefore, we delete X when its count is far less than $\varepsilon \times s \times (d + d^2 + \ldots + d^{cbid - sbid + 1})$. Later, we will prove that this pruning condition guarantees that the error of the itemset support generated by our algorithms is always bounded by ε.

Another critical issue is the generation of indirect associations from frequent itemsets. A naïve approach is adopting the method used in the INDIRECT algorithm [21]. A novel and more efficient method is proposed based on the following theorem. Due to space limit, all of the proof is omitted.

Theorem 1. *The support of a mediator M should be no less than $\sigma_m = 2\sigma_f - \sigma_s$, i.e., $sup(M) \geq \sigma_m$.*

First, the set of frequent 1-itemsets and the set of 1-mediators $\mathcal{M}_1$ (with support larger than σ_m) are generated. Then, it generates 2-itemsets from those frequent 1-itemsets; simultaneously, those 2-itemsets generated with threshold less than σ_s form the indirect itempair set (*IIS*). The procedure then proceeds to generate all qualified indirect associations. First, a candidate indirect association rule is formed by simply combining an itempair from *IIS* and a 1-mediator from $\mathcal{M}_1$; the rule is output as a qualified association if it satisfies the dependence condition (σ_d) and mediator support threshold (σ_f). Next, the set of 2-mediators $\mathcal{M}_2$ is generated by performing apriori-gen on $\mathcal{M}_2$ and checking against σ_m. The above steps are repeated until no new mediator is generated.

3.4 Theoretical Analyses

Consider a generated frequent itemset X by GIAMS. Theorem 2 shows that GIAMS always guarantees a bound on the accuracy as long as the window model of concern can be specified by the proposed generic model.

Theorem 2. *Let the true support of itemset X, called Tsup(X), be the fraction of transactions so far containing X, and the estimated support of an itemset X, called Esup(X), is the fraction of transactions containing X accumulated by the proposed GIAMS algorithm. Then $Tsup(X) - Esup(X) \leq \varepsilon$.*

Recall that an indirect association is an itempair $\{a, b\}$ indirectly associated via a mediator M, denoted as $\langle a, b | M \rangle$, if it satisfies the three conditions. We show that if the mediator dependence threshold is set smaller than a specific value, then all indirect association patterns generated by our algorithms that satisfy the mediator support condition also satisfy the mediator dependence condition.

Lemma 1. *Let $\langle a, b | M \rangle$ be a candidate indirect association satisfies the mediator support condition. Then $\sigma_f - \varepsilon \leq dep(\{a\}, M), dep(\{b\}, M) \leq 1$.*

Corollary 1. *If the mediator dependence threshold is set as $\sigma_d \leq \sigma_f - \varepsilon$, then an indirect association satisfying the itempair support condition and mediator support condition also satisfies the mediator dependence condition. That is, it is a qualified indirect association.*

Although Corollary 1 suggests the appropriate setting of σ_d from the viewpoint of retaining all high mediator supported associations, it can be regarded as an alternative bound for pruning to take effect. That is, given σ_f and ε, we have to specify σ_d larger than $\sigma_f - \varepsilon$ to prune any candidate indirect associations with high mediator supports.

4 Experimental Results

A series of experiments were conducted to evaluate the efficiency and effectiveness of the GIAMS algorithm. Our purpose is to inspect: (1) As a generic algorithm, how GIAMS performs in various streaming models, especially the three classical models; (2) How is the influence to GIAMS of each parameter engaged in specifying the window model? Each evaluation was inspected from three aspects, including execution time, memory usage, and pattern accuracy.

All experiments were done on an AMD X3-425(2.7 GHz) PC with 3GB of main memory, running the Windows XP operating system. All programs were coded in Visual C++ 2008. A synthetic dataset, T5.I5.N0.1K.D1000K, generated by the program in [2] as well as a real web-news-click stream [4] extracted from msn.com for the entire day of September 28, 1999 were tested. Since similar phenomena on the synthetic data were observed, we only show the results on the real dataset.

Effect of Mediator Support Threshold: We first examine the effect of varying mediator support thresholds, which is ranging from 0.01 to 0.018. It can be seen from Figure 3(a) that in the case of landmark model, the smaller σ_f is, less execution time is spent due to smaller σ_f resulting in less number of itemsets. And, the overall trend is linearly proportional to the transaction size. The memory usage exhibits a similar phenomenon. The situation is different in the case of time-fading model. From Figure 3(d) we can observe both the execution time and memory usage are not affected by the mediator support threshold. This is because most of the time for time-fading model, compared with the other two models, is spent on the insertion of itemsets and the benefit of pruning fades away. The result for sliding model is omitted because it resembles that for landmark model.

Effect of Window Stride: The stride value is ranging from 10000 to 80000. Two noticeable phenomena are observed in the case of landmark model. First, the execution time decreases as the stride increases because larger strides encourage analogical transactions; more transactions can be merged together. Second, a larger stride also is helpful in saving the memory usage; as indicated in (1) the pruning threshold becomes stricter, so more itemsets will be pruned. Effect of larger strides, however, is on the contrary in the case of sliding model. The execution time increases in proportional to the stride because larger strides imply larger transaction block to be inserted and deleted in maintaining the *PF*. The memory usage also increases, but is not significant.

Effect of Decay Rate: Note that only the time-fading model depends on this factor. As exhibited in Figure 3(c), a smaller decay rate contributes to a longer execution time. This is because a smaller decay rate makes the support count decay more quickly, so the itemset lifetime becomes shorter, leading to more itemset insertion and deletion operations. This is also why the memory usage, though not significant, is smaller than that for larger decay rates.

Effect of Window Size: Not surprisingly the memory increases as the window size increases, as shown in Figure 3(f), while the performance gap is not significant. This is because the stride (block size) is the same, and most of the time for process *FP-monitoring* is spent on the block insertion and deletion.

Performance of Process IA-generation: We compare the performance of the proposed two methods for implementing process IA-generation. We only show the results for landmark model because the other models exhibit similar behavior. GIAMS-IND denotes the approach modified from algorithm INDIRECT while GIAMS-MED represents the more efficient method utilizing qualified mediator threshold. As illustrated in Figure 3(g), GIAMS-MED is faster than GIAMS-IND. The reason is that as shown in Figure 3(h), the number of candidate rules generated by GIAMS-IND is much more than that by GIAMS-MED. Both methods consume approximately the same amount of memory.

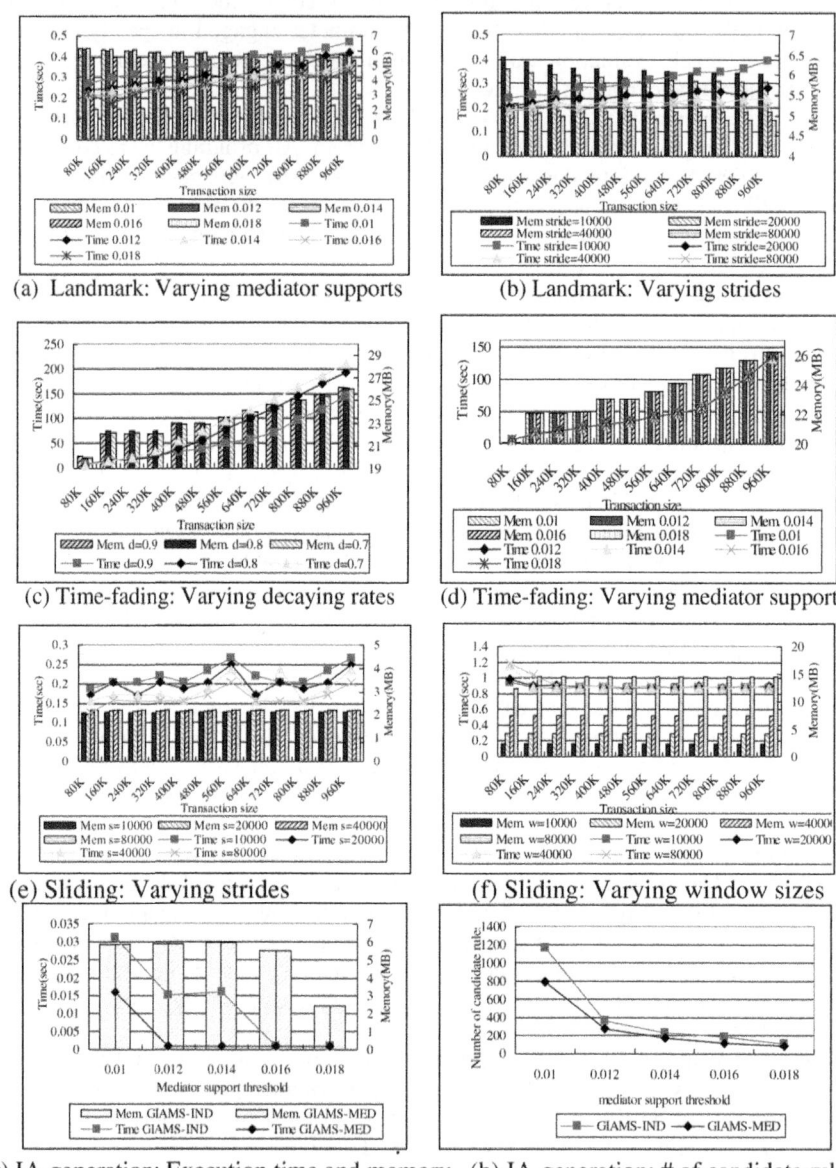

(a) Landmark: Varying mediator supports
(b) Landmark: Varying strides
(c) Time-fading: Varying decaying rates
(d) Time-fading: Varying mediator supports
(e) Sliding: Varying strides
(f) Sliding: Varying window sizes
(g) IA-generation: Execution time and memory
(h) IA-generation: # of candidate rules

Fig. 3. Experimental results on evaluating GIAMS over a real web-news-click stream

Accuracy: First, we check the difference between the true support and estimated support, measured by ASE (Average Support Error) = $\Sigma_{x \in F}(Tsup(x) - Esup(x))/|F|$, where F denotes the set of all frequent itemsets w.r.t. σ_f. The ASEs for all test cases with varying strides between 10000 and 80000 and σ_fs ranging from 0.01 to 0.018 were recorded. All of them are zero except the case for time-fading model with $d = 0.9$; all values are less than 3×10^{-7}. We also measured the accuracy of discovered indirect association rules by inspecting how many rules generated are correct patterns, i.e., recall. All the test cases exhibit 100% recalls.

5 Conclusions

In this paper, we have investigated the problem of indirect association mining from a generic viewpoint. We have proposed a generic stream window model that can encompass all classical streaming models and a generic mining algorithm that guarantees no false positive rules and bounded support error. An efficient implementation of GIAMS also was presented. Comprehensive experiments on both synthetic and real datasets have showed that the proposed generic algorithm is efficient and effectiveness in finding indirect association rules.

Recently, the design of adaptive data stream mining methods that can perform adaptively under constrained resources has emerged into an important and challenging research issue to the stream mining community [9, 22]. In the future, we will study how to apply or incorporate some adaptive technique such as load shedding [1] into our approach.

Acknowledgements

This work is partially supported by National Science Council of Taiwan under grant No. NSC97-2221-E-390-016-MY2.

References

1. Aggarwal, C.: Data Streams: Models and Algorithms. Springer, Heidelberg (2007)
2. Agrawal, R., Srikant, R.: Fast Algorithms for Mining Association Rules. In: 20th Int. Conf. Very Large Data Bases, pp. 487–499 (1994)
3. Babcock, B., Babu, S., Datar, M., Motwani, R., Widom, J.: Models and Issues in Data Stream Systems. In: 21st ACM Symp. Principles of Database Systems, pp. 1–16 (2002)
4. Cadez, I., Heckerman, D., Meek, C., Smyth, P., White, S.: Visualization of Navigation Patterns on a Web Site Using Model-based Clustering. In: 6th ACM Int. Conf. Knowledge Discovery and Data Mining, pp. 280–284 (2000)
5. Chang, J.H., Lee, W.S.: Finding Recent Frequent Itemsets Adaptively over Online Data Streams. In: 9th ACM Int. Conf. Knowledge Discovery and Data Mining, pp. 487–492 (2003)
6. Chang, J.H., Lee, W.S.: estWin: Adaptively Monitoring the Recent Change of Frequent Itemsets over Online Data Streams. In: 12th ACM Int. Conf. Information and Knowledge Management, pp. 536–539 (2003)

7. Chen, L., Bhowmick, S.S., Li, J.: Mining Temporal Indirect Associations. In: 10th Pacific-Asia Conf. Knowledge Discovery and Data Mining, pp. 425–434 (2006)
8. Chi, Y., Wung, H., Yu, P.S., Muntz, R.R.: Moment: Maintaining Closed Frequent Itemsets over a Stream Sliding Window. In: 4th IEEE Int. Conf. Data Mining, pp. 59–66 (2004)
9. Gaber, M.M., Zaslavsky, A., Krishnaswamy, S.: Towards an Adaptive Approach for Mining Data Streams in Resource Constrained Environments. In: 6th Int. Conf. Data Warehousing and Knowledge Discovery, pp. 189–198 (2004)
10. Hidber, C.: Online Association Rule Mining. ACM SIGMOD Record 28(2), 145–156 (1999)
11. Jiang, N., Gruenwald, L.: CFI-stream: Mining Closed Frequent Itemsets in Data Streams. In: Proc. 12th ACM Int. Conf. Knowledge Discovery and Data Mining, pp. 592–597 (2006)
12. Jin, R., Agrawal, G.: An Algorithm for In-core Frequent Itemset Mining on Streaming Data. In: 5th IEEE Int. Conf. Data Mining, pp. 210–217 (2005)
13. Kazienko, P.: IDRAM—Mining of Indirect Association Rules. In: Int. Conf. Intelligent Information Processing and Web Mining, pp. 77–86 (2005)
14. Kazienko, P., Kuzminska, K.: The Influence of Indirect Association Rules on Recommendation Ranking Lists. In: 5th Int. Conf. Intelligent Systems Design and Applications, pp. 482–487 (2005)
15. Koh, J.L., Shin, S.N.: An Approximate Approach for Mining Recently Frequent Itemsets from Data Streams. In: 8th Int. Conf. Data Warehousing and Knowledge Discovery, pp. 352–362 (2006)
16. Lee, D., Lee, W.: Finding Maximal Frequent Itemsets over Online Data Streams Adaptively. In: 5th IEEE Int. Conf. Data Mining, pp. 266–273 (2005)
17. Li, H.F., Lee, S.Y., Shan, M.K.: An Efficient Algorithm for Mining Frequent Itemsets over the Entire History of Data Streams. In: 1st Int. Workshop Knowledge Discovery in Data Streams, pp. 20–24 (2004)
18. Lin, C.H., Chiu, D.Y., Wu, Y.H., Chen, A.L.P.: Mining Frequent Itemsets from Data Streams with a Time-Sensitive Sliding Window. In: 5th SIAM Data Mining Conf., pp. 68–79 (2005)
19. Manku, G.S., Motwani, R.: Approximate Frequency Counts over Data Streams. In: 28th Int. Conf. Very Large Data Bases, pp. 346–357 (2002)
20. Tan, P.N., Kumar, V.: Mining Indirect Associations in Web Data. In: 3rd Int. Workshop Mining Web Log Data Across All Customers Touch Points, pp. 145–166 (2001)
21. Tan, P.N., Kumar, V., Srivastava, J.: Indirect Association: Mining Higher Order Dependencies in Data. In: 4th European Conf. Principles of Data Mining and Knowledge Discovery, pp. 632–637 (2000)
22. Teng, W.G., Chen, M.S., Yu, P.S.: Resource-Aware Mining with Variable Granularities in Data Streams. In: 4th SIAM Conf. Data Mining, pp. 527–531 (2004)
23. Teng, W.G., Hsieh, M.J., Chen, M.S.: On the Mining of Substitution Rules for Statistically Dependent Items. In: 2nd IEEE Int. Conf. Data Mining, pp. 442–449 (2002)
24. Wan, Q., An, A.: An Efficient Approach to Mining Indirect Associations. Journal of Intelligent Information System 27(2), 135–158 (2006)
25. Yu, J.X., Chong, Z., Lu, H., Zhou, A.: False Positive or False Negative: Mining Frequent Itemsets from High Speed Transactional Data Streams. In: 30th Int. Conf. Very Large Data Bases, pp. 204–215 (2004)

Status Quo Bias in Configuration Systems

Monika Mandl[1], Alexander Felfernig[1], Juha Tiihonen[2], and Klaus Isak[3]

[1] Institute for Software Technology, Graz University of Technology,
Inffeldgasse 16b, A-8010 Graz, Austria
{monika.mandl, alexander.felfernig}@ist.tugraz.at
[2] Computer Science and Engineering, Aalto University,
02015 TKK, Finland
juha.tiihonen@tkk.fi
[3] Institute for Applied Informatics, University of Klagenfurt,
Universitaetsstrasse 65-67, A-9020 Klagenfurt, Austria
kisak@configworks.com

Abstract. Product configuration systems are an important instrument to implement mass customization, a production paradigm that supports the manufacturing of highly-variant products under pricing conditions similar to mass production. A side-effect of the high diversity of products offered by a configurator is that the complexity of the alternatives may outstrip a user's capability to explore them and make a buying decision. A personalization of such systems through the calculation of feature recommendations (defaults) can support customers (users) in the specification of their requirements and thus can lead to a higher customer satisfaction. A major risk of defaults is that they can cause a status quo bias and therefore make users choose options that are, for example, not really needed to fulfill their requirements. In this paper we present the results of an empirical study that aimed to explore whether there exist status quo effects in product configuration scenarios.

Keywords: Configuration Systems, Interactive Selling, Consumer Buying Behavior, Consumer Decision Making.

1 Introduction

Following the paradigm of *mass customization*, the intelligent customizing of products and services is crucial for manufacturing companies to stay competitive. Configuration systems, which have a long tradition as a successful application area of Artificial Intelligence [1,2,3,4,5], have been recognized as ideal tools to assist customers in configuring complex products according to their requirements. Example domains where product configurators are applied are computers, cars, and financial services. An important task typically supported by configurators is to check the consistency of user requirements with the knowledge base, such that the amount of incorrect quotations and orders can be reduced.

One major problem of configuration systems is the high diversity of offered products. Users are often overwhelmed by the complexity of the alternatives, a phenomenon well known as mass confusion [9]. A possibility to help the user identifying meaningful alternatives that are compatible with their current preferences is to provide *defaults*. Defaults in the context of interactive configuration dialogs are preselected options used to express personalized feature recommendations. Felfernig et al. [7] conducted a study to investigate the impact of personalized feature recommendations in a knowledge-based recommendation process. Nearest neighbors and Naive Bayes voter algorithms have been applied for the calculation of default values. The results of this research indicate that supporting users with personalized defaults can lead to a higher satisfaction with the configuration process. In this paper we want to discuss further impacts of presenting such default values to users of configurator applications. We present the results of a case study conducted to figure out whether default values can have an impact on a user's selection behavior in product configuration sessions. The motivation for this empirical analysis is the existence of so-called status quo biases in human decision making [14].

The remainder of the paper is organized as follows. In the next section we discuss the concept of the status quo bias in human decision making. In Section 3 we introduce major functionalities of RecoMobile, an environment that supports the configuration of mobile phones and corresponding subscription features. In Section 4 we present the test design of our user study. In the following we discuss the results of our user study with the goal to point out to which extent a status quo effect exists in product configuration systems (Section 5). Finally, we discuss related work and conclude the paper.

2 Status Quo Bias in Decision Making

People have a strong tendency to accept preset values (representing the status quo) compared to other alternatives [11,13,14]. Samuelson and Zeckhauser [14] explored this effect, known as *status quo bias*, in a series of labor experiments. Their results implied that an alternative was significantly more often chosen when it was designated as the status quo. They also showed that the *status quo effect* increases with the number of alternatives. Kahnemann, Knetsch and Thaler [11] argue that the *status quo bias* can be explained by a notion of *loss aversion*. They explain that the status quo serves as a neutral reference point and users evaluate alternative options in terms of gains and losses relative to the reference point. Since individuals tend to regard losses as more important than gains in decision making under risk (i.e., alternatives with uncertain outcomes) [12] the possible disadvantages when changing the status quo appear larger than advantages.

A major risk of preset values is that they could be exploited for misleading users and making them choose options that are not really needed to fulfill their requirements. Bostrom and Ord defined the *status quo bias* as "a cognitive error, where one option is incorrectly judged to be better than another because it represents the status quo" [10]. Ritov and Barron [13] suggest counteracting the *status quo bias* by presenting the options in such a way that keeping as well as changing the status quo needs user input. They argue that "when both keeping and changing the status quo require action, people will be less inclined to err by favoring the status quo when it is worse" [13].

In this paper we want to focus on answering the question whether a *status quo bias* exists in the context of product configuration systems and whether it is possible to reduce this biasing effect by providing an interface supporting the interaction mechanisms introduced by Ritov and Barron [13].

3 The RecoMobile Prototype

RecoMobile is a knowledge-based configuration system for mobile phones and services enriched with recommendation functionalities to predict useful feature settings for the user [7]. Example pages of RecoMobile are depicted in Figure 1. After the specification of a few general attributes of the configuration domain (e.g., the *preferred phone style*) the system calculates personalized recommendations on the basis of user interactions of past configuration sessions. These recommendations are presented to the user as default proposals for the following questions regarding *mobile subscription details, privacy settings*, and the *phone*. After the specification of the user requirements, RecoMobile presents a phone selection page which enlists the set of phones that fulfill the given set of customer requirements. For each mobile phone the user can activate a fact sheet that is implemented as a direct link to the supplier's web page. Finally, it is possible to select the preferred mobile phone and to finish the session.

In the context of this paper we present a study where we used the RecoMobile prototype to explore whether there exist status quo effects in the context of product configuration sessions.

4 Study Design

Our experiment addressed two relevant questions. (1) *Are users of product configuration systems influenced by default settings even if these settings are uncommon?* (2) *Is it possible to counteract the status quo bias by providing a configuration interface where both keeping and changing the presented default settings needs user interaction?* To test the influence of uncommon defaults on the selection behavior of users we differentiate between three basic versions of RecoMobile (see Table 1). Users of RecoMobile Version A were not confronted with defaults, i.e., they had to specify each feature preference independent of any default proposals. Out of the resulting interaction log we selected for each feature (presented as questions within a configuration session) the alternative which was chosen least often and used it as default for Versions B and C. These two versions differ in the extent to which user interaction is required. In Version B user interaction is only required when the customer wants to change the recommended default setting (*low user involvement*). In Version C the acceptance as well as the changing of the default settings requires user interaction (*high user involvement* - see Figure 2). We conducted an online survey at the

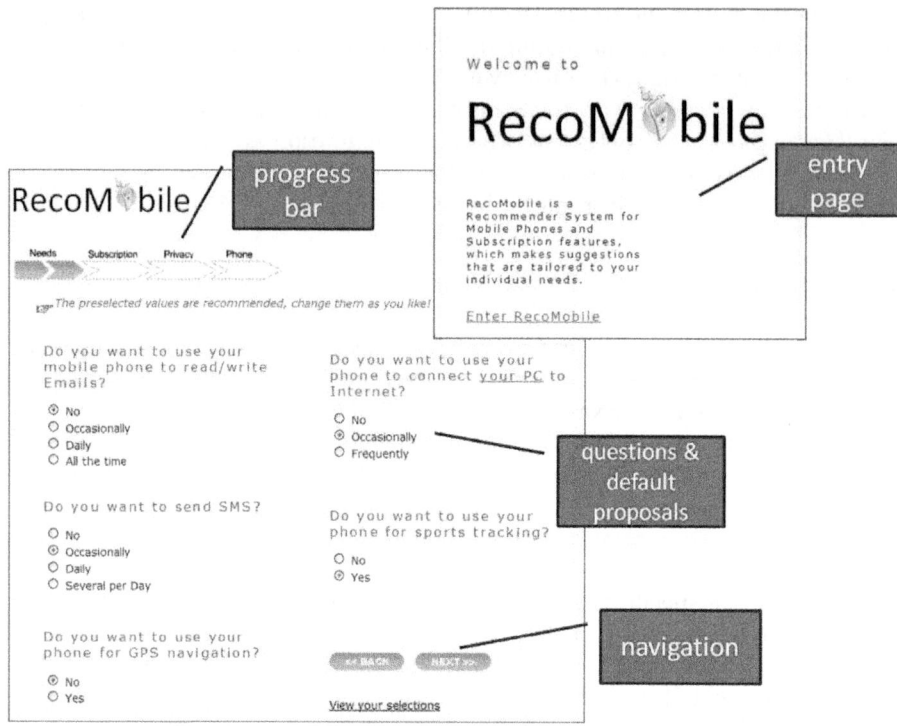

Fig. 1. RecoMobile user interface – in the case of default proposal acceptance no further user interfaction is needed (low involvement)

Table 1. RecoMobile – configurator versions in user study

Version	Default Type	Explanation
A	no defaults	no defaults were presented to the user
B	defaults without confirmation	unusual defaults were presented to the user – acceptance of defaults does not require additional interaction (low involvement version – see Figure 1)
C	defaults with confirmation	unusual defaults were presented to the user – acceptance of defaults requires additional interaction in terms of a confirmation (high involvement version – see Figure 2)

Graz University of Technology. N=143 subjects participated in the study. Each participant was assigned to one of the three configurator versions (see Table 1). The experiment was based on a scenario where the participants had to decide which mobile phone (including the corresponding services) they would select.

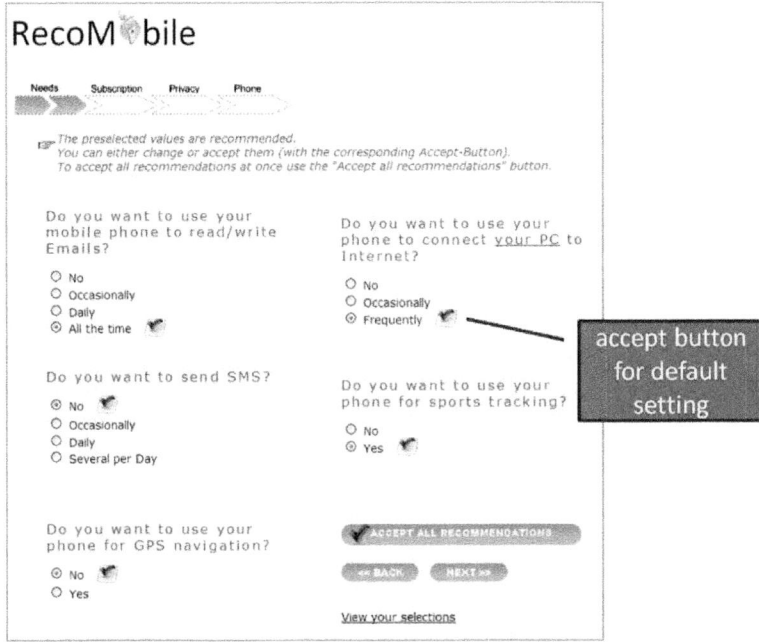

Fig. 2. Alternative version of the RecoMobile user interface – in the case of default proposal acceptance users have to explicitly confirm their selection (high involvement)

5 Results

In our evaluation we compared the data of the configurator version without default settings (Version A - see Table 1) with the data collected in Versions B and C. For each feature we conducted a chi-square test (the standard test procedure when dealing with data sets that express frequencies) to compare the selection behavior of the users. For many of the features we could observe significant differences in the selection distribution. A comparison of the selection behavior in the different configurator versions is given in Table 2.

For example, the evaluation results regarding the feature *Which charged services should be prohibited for SMS?* are depicted in Figure 3. For this feature the default in Versions B and C was set to alternative 3 - *Utility and Entertainment* - (that option which was chosen least often in Version A). In both versions the default setting obviously had a strong impact on the selection behavior of the users. Only 2 % of the users of Version A selected option 3 whereas in Version B 24 % chose this default option. The interesting result is that the version with high user involvement (Version C) did not counteract the status quo bias. 25.6 % of the users of Version C selected the default alternative. Contrary to the assumption of Ritov and Baron [13] people tend to stick to the status quo (the default option) even when user interaction is required to accept it.

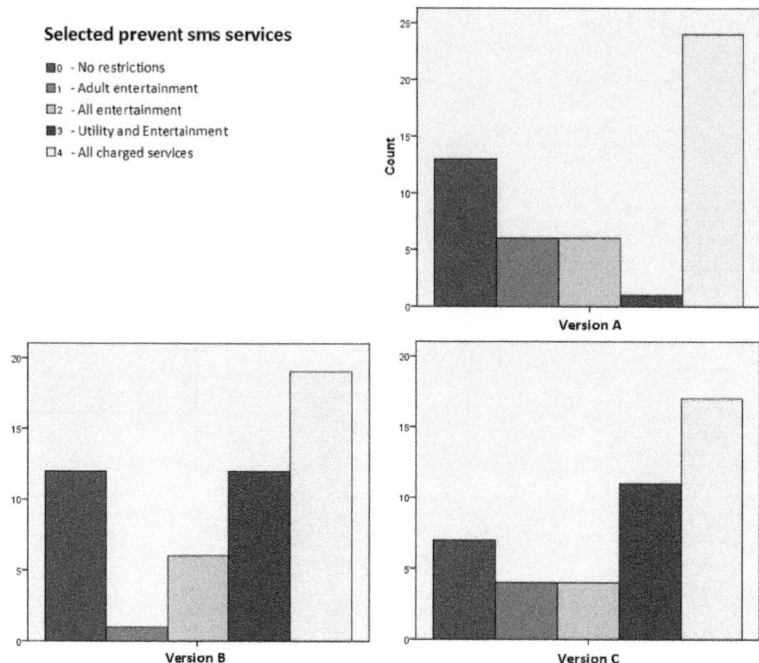

Fig. 3. Selections for *prohibit charged services for SMS* - the results of the conducted chi-square test show that the underlying distributions differ significantly (p=0.009 for Version A compared with Version B, p=0.020 for Version A compared with C)

In Figure 4 another example is shown for the feature *Which data package do you want?*. The default in Version B and C was set to option 5 (*2048 kbit/s (+ 29.90 euro)*), which was the most expensive alternative of this feature. In Version A 4 % of the users decided to choose this option - the mean value for the expenses for this attribute is 5.5 Euro (see Table 3). In Version B 16 % and in Version C 18.6 % of the users retained the status quo alternative. The mean value for the data package expenses in Version B is 12.8 Euro and in Version C 13.2 Euro. This example shows that exploiting the status quo effect can lead to selection of more expensive alternatives. Here again the status quo effect was not suppressed in Version C, where people had to confirm the default setting.

6 Related Work

Research in the field of human decision making has revealed that people have a strong tendency to keep the status quo when choosing among alternatives (see e.g. [10,11,13,14]). This decision bias has firstly been reported by Samuelson and Zeckhauser [14]. To our knowledge, such decision biases have not been analyzed in detail in the context of interactive configuration scenarios. The goal

Table 2. Comparison of value selection behavior in different configurator versions

Feature	Version A compared with Version B	Version A compared with Version C
use phone to read/write emails	p=0.079	p=0.133
use phone to connect PC to web	**p=0.025**	p=0.193
use phone to send SMS	p=0.302	p=0.395
use phone for sportstracking	p=0.211	p=0.825
use phone for GPS navigation	p=0.099	p=0.392
monthly minutes package	p=0.235	**p=0.014**
free sms messages included	p=0.323	**p=0.032**
selected data package	**p=0.001**	**p=0.004**
mobile antivirus service	**p=0.008**	**p=0.002**
mobile messenger	p=0.629	p=0.643
display number to receiver	p=0.100	p=0.090
publish phone number in phone book	**p=0.032**	p=0.497
prevent calls to foreign countries	p=0.260	p=0.107
prohibit charged services for calls	**p=0.014**	**p=0.011**
prohibit charged services for sms	**p=0.009**	**p=0.020**

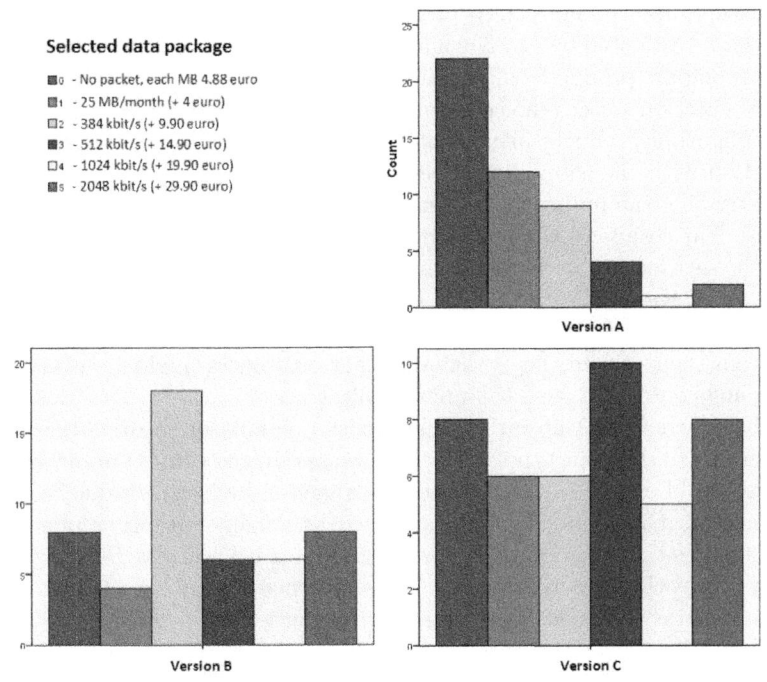

Fig. 4. Selections for *Monthly data package* - the results of the conducted chi-square test show that the underlying distributions differ significantly (p=0.001 for Version A compared with Version B, p=0.004 for Version A compared with Version C)

Table 3. Mean values for the monthly data package expenses

Version	Expenses (€) (mean value)
A	5.528
B	12.844
C	13.281

of our work was to investigate whether the status quo effect also exists in product configuration systems. Felfernig et al. [7] introduced an approach to integrate recommendation technologies with knowledge-based configuration. The results of this research indicate that supporting users with personalized feature recommendations (defaults) can lead to a higher satisfaction with the configuration process. The work presented in this paper is a logical contination of the work of [7] which extends the impact analysis of personalization concepts to the psychological phenomenon of decision biases.

Although product configuration systems support interactive decision processes with the goal to determine configurations that are useful for the customer, the integration of human decision psychology aspects has been ignored with only a few exceptions. Human choice processes within a product configuration task have been investigated by e.g. Kurniawan, So, and Tseng [16]. They conducted a study to compare product configuration tasks (choice of product attributes) with product selection tasks (choice of product alternatives). Their results suggest that configuring products instead of selecting products can increase customer satisfaction with the shopping process. The research of [17] and [9] was aimed at investigating the influences on consumer satisfaction in a configuration environment. The results of the research of Kamali and Loker [17] showed a higher consumer satisfaction with the website's navigation as involvement increased. Huffman and Kahn [9] explored the relationship between the number of choices during product configuration and user satisfaction with the configuration process. From their results they concluded that customers might be overwhelmed when being confronted with too many choices.

In the psychological literature there exist a couple of theories that explain the existence of different types of decision biases. In the context of our empirical study we could observe a status quo bias triggered by feature value recommendations, even if uncommon values are used as defaults. Another phenomenon that influences the selection behavior of consumers is known as *Decoy effect*. According to this theory consumers show a preference change between two options when a third asymmetrically dominating option is added to the consideration set. Decoy effects have been intensively investigated in different application contexts, see, for example [18,19,20,21,22,23]. The *Framing effect* describes the fact that presenting one and the same decision alternative in different variants can lead to choice reversals. Tversky and Kahnemann have shown that effect in a series of studies where they confronted participants with choice problems using variations in the framing of decision outcomes. They reported that "seemingly inconsequential changes in the formulation of choice problems caused significant shifts of preference" [24].

7 Conclusions

In this paper we have presented the results of an empirical study that had the goal to analyze the impact of the status quo bias in product configuration scenarios where defaults are presented as recommendations to users. The results of our study show that there exists a strong biasing effect even if uncommon values are presented as default values. Our findings show that, for example, status quo effects make users of a configuration system selecting more expensive solution alternatives. As a consequence of these results we have to increasingly turn our attention to ethical aspects when implementing product configurators since it is possible that users are mislead simply by the fact that some defaults are representing expensive solution alternatives (but are maybe not needed to fulfill the given requirements). Finally, we detected that providing the possibility of both keeping and changing the provided defaults (we called this the *high involvement user interface*) does not counteract the status quo bias. Our future work will include the investigation of additional decision phenomena in the context of knowledge-based configuration scenarios (e.g., framing or decoy effects).

Acknowledgments. The presented work has been developed within the scope of the research project XPLAIN-IT (funded by the Privatstiftung Kärnter Sparkasse).

References

1. Barker, V., O'Connor, D., Bachant, J., Soloway, E.: Expert systems for configuration at Digital: XCON and beyond. Communications of the ACM 32(3), 298–318 (1989)
2. Fleischanderl, G., Friedrich, G., Haselboeck, A., Schreiner, H., Stumptner, M.: Configuring Large Systems Using Generative Constraint Satisfaction. IEEE Intelligent Systems 13(4), 59–68 (1998)
3. Mittal, S., Frayman, F.: Towards a Generic Model of Configuration Tasks. In: 11th International Joint Conference on Artificial Intelligence, Detroit, MI, pp. 1395–1401 (1990)
4. Sabin, D., Weigel, R.: Product Configuration Frameworks - A Survey. IEEE Intelligent Systems 13(4), 42–49 (1998)
5. Stumptner, M.: An overview of knowledge-based configuration. AI Communications (AICOM) 10(2), 111–126 (1997)
6. Cöster, R., Gustavsson, A., Olsson, T., Rudström, A.: Enhancing web-based configuration with recommendations and cluster-based help. In: De Bra, P., Brusilovsky, P., Conejo, R. (eds.) AH 2002. LNCS, vol. 2347, Springer, Heidelberg (2002)
7. Felfernig, A., Mandl, M., Tiihonen, J., Schubert, M., Leitner, G.: Personalized User Interfaces for Product Configuration. In: International Conference on Intelligent User Interfaces (IUI 2010), pp. 317–320 (2010)

8. Mandl, M., Felfernig, A., Teppan, E., Schubert, M.: Consumer Decision Making in Knowledge-based Recommendation. Journal of Intelligent Information Systems (2010) (to appear)
9. Huffman, C., Kahn, B.: Variety for Sale: Mass Customization or Mass Confusion. Journal of Retailing 74, 491–513 (1998)
10. Bostrom, N., Ord, T.: The Reversal Test: Eliminating Status Quo Bias in Applied Ethics. Ethics (University of Chicago Press) 116(4), 656–679 (2006)
11. Kahneman, D., Knetsch, J., Thaler, R.: Anomalies: The Endowment Effect, Loss Aversion, and Status Quo Bias. The Journal of Economic Perspectives 5(1), 193–206 (1991)
12. Kahneman, D., Tversky, A.: Prospect theory: An analysis of decision under risk. Econometrica 47(2), 263–291 (1979)
13. Ritov, I., Baron, J.: Status-quo and omission biases. Journal of Risk and Uncertainty 5, 49–61 (1992)
14. Samuelson, W., Zeckhauser, R.: Status quo bias in decision making. Journal of Risk and Uncertainty 1(1), 7–59 (1988)
15. Mandl, M., Felfernig, A., Teppan, E., Schubert, M.: Consumer Decision Making in Knowledge-based Recommendation. Journal of Intelligent Information Systems (to appear)
16. Kurniawan, S.H., So, R., Tseng, M.: Consumer Decision Quality in Mass Customization. International Journal of Mass Customisation 1(2-3), 176–194 (2006)
17. Kamali, N., Loker, S.: Mass customization: On-line consumer involvement in product design. Journal of Computer-Mediated Communication 7(4) (2002)
18. Huber, J., Payne, W., Puto, C.: Adding Asymmetrically Dominated Alternatives: Violations of Regularity and the Similarity Hypothesis. Journal of Consumer Research 9(1), 90–98 (1982)
19. Simonson, I., Tversky, A.: Choice in context: Tradeoff contrast and extremeness aversion. Journal of Marketing Research 29(3), 281–295 (1992)
20. Yoon, S., Simonson, I.: Choice set configuration as a determinant of preference attribution and strength. Journal of Consumer Research 35(2), 324–336 (2008)
21. Teppan, E.C., Felfernig, A.: Calculating Decoy Items in Utility-Based Recommendation. In: Chien, B.-C., Hong, T.-P., Chen, S.-M., Ali, M. (eds.) IEA/AIE 2009. LNCS, vol. 5579, pp. 183–192. Springer, Heidelberg (2009)
22. Teppan, E., Felfernig, A.: Impacts of Decoy Elements on Result Set Evaluation in Knowledge-Based Recommendation. International Journal of Advanced Intelligence Paradigms 1(3), 358–373 (2009)
23. Felfernig, A., Gula, B., Leitner, G., Maier, M., Melcher, R., Schippel, S., Teppan, E.: A Dominance Model for the Calculation of Decoy Products in Recommendation Environments. In: AISB 2008 Symposium on Persuasive Technology, pp. 43–50 (2008)
24. Tversky, A., Kahneman, D.: The Framing of Decisions and the Psychology of Choice. Science, New Series 211, 453–458 (1981)

Improvement and Estimation of Prediction Accuracy of Soft Sensor Models Based on Time Difference

Hiromasa Kaneko and Kimito Funatsu

Department of Chemical System Engineering, Graduate School of Engineering,
The University of Tokyo, Hongo 7-3-1,
Bunkyo-ku, Tokyo, 113-8656, Japan
{hkaneko,funatsu}@chemsys.t.u-tokyo.ac.jp

Abstract. Soft sensors are widely used to estimate process variables that are difficult to measure online. However, their predictive accuracy gradually decreases with changes in the state of the plants. We have been constructing soft sensor models based on the time difference of an objective variable y and that of explanatory variables (time difference models) for reducing the effects of deterioration with age such as the drift and gradual changes in the state of plants without reconstruction of the models. In this paper, we have attempted to improve and estimate the prediction accuracy of time difference models, and proposed to handle multiple y values predicted from multiple intervals of time difference. An exponentially-weighted average is the final predicted value and the standard deviation is the index of its prediction accuracy. This method was applied to real industrial data and its usefulness was confirmed.

Keywords: Soft sensor, Time difference, Prediction error, Applicability domain, Ensemble prediction.

1 Introduction

Soft sensors are inferential models to estimate process variables that are difficult to measure online and have been widely used in industrial plant [1,2]. These models are constructed between those variables that are easy to measure online and those that are not, and an objective variable is then estimated using those model. Through the use of soft sensors, the values of objective variables can be estimated with a high degree of accuracy. Their use, however, involves some practical difficulties. One crucial difficulty is that their predictive accuracy gradually decreases due to changes in the state of chemical plants, catalyzing performance loss, sensor and process drift, and the like. In order to reduce the degradation of a soft sensor model, the updating of regression models [3] and Just-In-Time (JIT) modeling [4] have been proposed. Regression models are reconstructed with newest database in which new observed data is stored oline. While many excellent results have been reported based on the use of these methods, there remain some problems for the introduction of soft sensors into practice.

First of all, if soft sensor models are reconstructed with the inclusion of any abnormal data, their predictive ability can deteriorate [5]. Though such abnormal data must be detected with high accuracy in real time, under present circumstances it is difficult to accurately detect all of them. Second, reconstructed models have a high tendency to specialize in predictions over a narrow data range [6]. Subsequently, when rapid variations in the process variables occur, these models cannot predict the resulting variations in data with a high degree of accuracy. Third, if a soft sensor model is reconstructed, the parameters of the model, for example, the regression coefficients in linear regression modeling, are dramatically changed in some cases. Without the operators' understanding of a soft sensor model, the model cannot be practically applied. Whenever soft sensor models are reconstructed, operators check the parameters of the models so they will be safe for operation. This takes a lot of time and effort because it is not rare that tens of soft sensors are used in a plant [7]. Fourth, the data used to reconstruct soft sensor models are also affected by the drift. In the construction of the model, data must be selected from a database which includes both data affected by the drift and data after correction of the drift.

In order to solve these problems, it was proposed to construct soft sensor models based on the time difference of explanatory variables, X, and that of an objective variable, y, for reducing the effects of deterioration with age such as the drift and gradual changes in the state of plants without reconstruction of the models [8,9]. In other words, models which are not affected by these changes must be constructed using not the values of process variables, but the time difference in soft sensor modeling. A model whose construction is based on the time difference of X and that of y is referred to as a 'time difference model'. Time difference models can also have high predictive accuracy even after drift correction because the data is represented as the time difference that cannot be affected by the drift. We confirmed through the analysis of actual industrial data that the time difference model displayed high predictive accuracy for a period of three years, even when the model was never reconstructed [8,9]. However, its predictive accuracy was lower than that of the updating model.

On the one hand, we proposed to estimate the relationships between applicability domains (ADs) and the accuracy of prediction of soft sensor models quantitatively [6]. The larger the distances to models (DMs) are, the lower the estimated accuracy of prediction would be. We used the distances to the average of training data and to the nearest neighbor of training data as DMs, obtained the relationships between the DMs and prediction accuracy quantitatively, and then, false alarms could be prevented by estimating large prediction errors when the state was different from that of training data; further, actual y-analyzer faults could be detected with high accuracy [6].

Therefore in this paper, we have attempted to improve and estimate the prediction accuracy of time difference models and focused attention on an interval of time difference. In terms of prediction by using a time difference model, when the interval is small, a rapid variation in process variables could be accounted for, but after the variation, predictive accuracy could be low because a state of a plant before the variation and a state after the variation are different and the time difference model could not accounted for the difference if an interval of time difference is small. On the other hand, when the interval is large, the difference between before and after a variation in process variables could be accounted for, but a rapid variation could not.

Thus, we have proposed a ensemble prediction method handling multiple y values predicted by using multiple intervals of time difference of X with a time difference model. A final predicted value is an exponentially-weighted average of the multiple predicted values, that is, the larger a interval is, the exponentially less a weight is, because time difference from the old value would have less and less influence as it recedes into the past. Besides, it is expected that variance of final prediction errors could be small from the point of view of ensemble prediction.

In addition, predictive accuracy of a final predicted value would be high if variation in the multiple predicted values is small and vice versa. Therefore, we use the standard deviation of the multiple predicted values (SD) as a DM, that is, a index of prediction accuracy of a final predicted value. By using the proposed method, we can estimate a final predicted value of y and its predictive accuracy.

2 Method

We explain the proposed ensemble prediction method handling multiple y values predicted by using multiple intervals of time difference of X. Before that, we briefly introduce the time difference modeling method and DMs.

2.1 Time Difference Modeling Method

In a traditional procedure, modeling relationship between explanatory variables, X(t), and an objective variable, y(t), is done by regression methods after preparing data, X(t) and y(t), related to time t. In terms of prediction, the constructed model predicts the value of y(t') with the new data x(t').

In time difference modeling, time difference of X and that of y, $\Delta X(t)$ and $\Delta y(t)$, are first calculated between the present values, X(t) and y(t), and those in some time i before the target time X(t-i) and y(t-i).

$$\Delta X(t) = X(t) - X(t-i)$$
$$\Delta y(t) = y(t) - y(t-i) \quad (1)$$

Then, relationship between $\Delta X(t)$ and $\Delta y(t)$ is modeled by regression methods.

$$\Delta y(t) = f(\Delta X(t)) + e \quad (2)$$

where f is a regression model and e is a vector of calculation errors. In terms of prediction, the constructed model, f, predicts the time difference of y(t'), $\Delta y(t')$, using the time difference of the new data, $\Delta x(t')$, calculated as follows:

$$\Delta x(t') = x(t') - x(t'-i)$$
$$\Delta y_{pred}(t') = f(\Delta x(t')) \quad (3)$$

$y_{pred}(t')$ can be calculated as follows:

$$y_{pred}(t') = \Delta y_{pred}(t') + y(t'-i) \quad (4)$$

because y(t'-i) is given previously. This method can be easily expanded to a case that a interval i is not constant.

By constructing time difference models, the effects of deterioration with age such as the drift and gradual changes in the state of plants can be accounted for, because data is represented as time difference that cannot be affected by these factors.

2.2 Distance to Model

Previously, we proposed a method to estimate the relationships between DMs and the accuracy of prediction of soft sensor models quantitatively [6]. For example, the Euclidean distance to the average of training data (ED) is used as a DM. The ED of explanatory variables of data, x, is defined as follows:

$$ED = \sqrt{(x-\mu)^T (x-\mu)} \tag{5}$$

where μ is a vector of the average of training data. When there is correlation among the variables, the Mahalanobis distance is often used as the distance. The MD of x is defined as follows:

$$MD = \sqrt{(x-\mu)^T \Sigma^{-1} (x-\mu)} \tag{6}$$

where Σ is the variance-covariance matrix of training data. The absolute prediction errors will increase with the DMs, and their distributions will become wider. By quantifying the relationships between these distances and a index of prediction accuracy, for example, the standard deviation of prediction errors [6], we can estimate standard deviations of the prediction errors for test data.

2.3 Proposed Ensemble Prediction Method with Time Difference Model

Fig. 1 shows the basic concept of the proposed method. In prediction of a y-value of time k, the time difference of y from time k-j, k-2j, ..., k-nj is predicted by using the time difference model, f, obtained in Eq. (2) as follows:

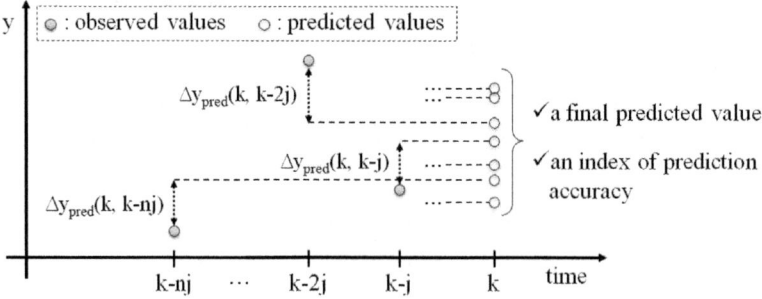

Fig. 1. The basic concept of the proposed method

$$y_{pred}(k, k-j) = f(\Delta x(k, k-j)) + y(k, k-j)$$
$$y_{pred}(k, k-2j) = f(\Delta x(k, k-2j)) + y(k, k-2j)$$
$$\vdots \qquad (7)$$
$$y_{pred}(k, k-nj) = f(\Delta x(k, k-nj)) + y(k, k-nj)$$

where j is unit time; n is a positive integer; and Δ (p,q) represents time difference between time p and q. A final predicted value, $y_{pred}(k)$, is an exponentially-weighted average of the multiple predicted values above, which is calculated as follows:

$$y_{pred}(k) = \alpha \begin{pmatrix} y_{pred}(k, k-j) + (1-\alpha)y_{pred}(k, k-2j) + \\ \cdots + (1-\alpha)^{r-1} y_{pred}(k, k-rj) + \cdots \end{pmatrix} \qquad (8)$$

where α is a constant smoothing factor between 0 and 1. This predicted value would account for both a rapid variation in process variables and the difference between before and after the variation.

After that, we use the SD as a DM that is defined as follows:

$$SD = \sqrt{\frac{1}{n-1} \sum_{r=1}^{n} (y_{pred}(k, k-rj) - \mu')^2} \qquad (9)$$

where μ' is an average of multiple predicted y values. This is the proposed index of prediction accuracy of $y_{pred}(k)$. If variation in multiple predicted y values is small, the SD is small and prediction errors of $y_{pred}(k)$ are estimated as small and vice versa. By quantifying the relationship between the SD and the calculation errors of training and/or validation data, we can estimate the prediction errors for test data.

3 Results and Discussion

We analyzed the data obtained from the operation of a distillation column at Mizushima works, Mitsubishi Chemical Corporation [4,6,8,9]. An y variable is the concentration of the bottom product having a lower boiling point, and X variables are 19 variables such as temperature and pressure. The measurement interval of y is 30 minutes and that of X is 1 minute. We collected data from monitoring that took place from 2002 to 2006, and used data from January to March 2003 for training data because plant tests took place. A disturbance and a plant inspection took place in 2002 and basically a plant inspection did every year. Data that reflects variations caused by y-analyzer fault were eliminated in advance.

3.1 Prediction Accuracy

First, we verified the predictive accuracy of the proposed method. In order to incorporate the dynamics of process variables into soft sensor models, X included each explanatory variable that was delayed for durations ranging from 0 minute to 60 minutes in steps of 10 minutes. The three methods listed below were applied.

A: Update a model constructed with the values of X and those of y.
B: Do not update a model constructed with the time difference of X and that of y.
C: Do not update a model constructed with the time difference of X and that of y. A final predicted value is an exponentially-weighted average of multiple predicted values (Proposed method)

For method B, The time difference was calculated between the present values and those that were 30 minutes before the present time. For method C, the intervals of the time difference were from 30 minutes to 1440 minutes in steps of 30 minutes. The α value optimized by using the training data was 0.56. The partial least squares (PLS) method was used to construct each regression model because the support vector regression (SVR) model had almost the same predictive accuracy as that of PLS for this distillation column [6]. The results from distance-based JIT models are not presented here, but they were almost identical to those of the updating models.

Fig. 2 shows the RMSE values per month for each method from April 2003 to December 2006. RMSE (root mean square error) is defined as follows:

$$\text{RMSE} = \sqrt{\frac{\sum (y_{obs} - y_{pred})^2}{N}} \tag{10}$$

where y_{obs} is the measured y value, y_{pred} is the predicted y value, and N is the number of data.

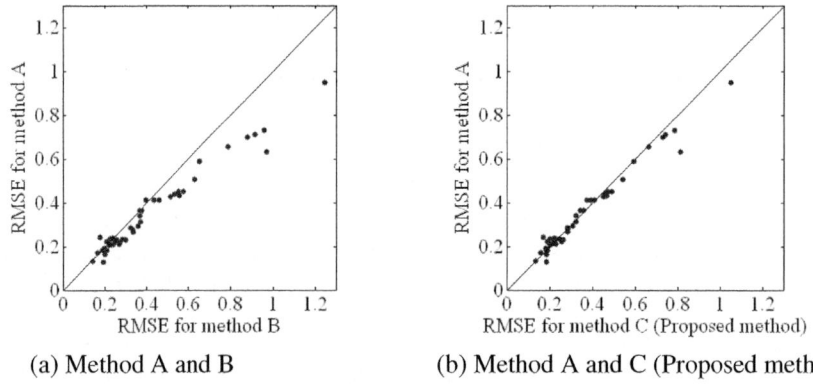

(a) Method A and B   (b) Method A and C (Proposed method)

Fig. 2. The RMSE values per month for each method

As shown in Fig. 2(a), the predictive accuracy of A was higher than that of B totally. On the other hand, the RMSE values for C were almost the same as those for A from Fig. 1(b). The predictive accuracy improved by using the exponentially-weighted average of multiple predicted values. It is important that the time difference model in B and C was constructed using only data from January to March 2003 and never reconstructed. It is possible for a predictive model to be constructed without updating by using the time difference and multiple predicted values. Additionally, when the RMSE values were small, those of C tended to be smaller than those of A. It is very practical because high prediction accuracy in the state of the plant is desired.

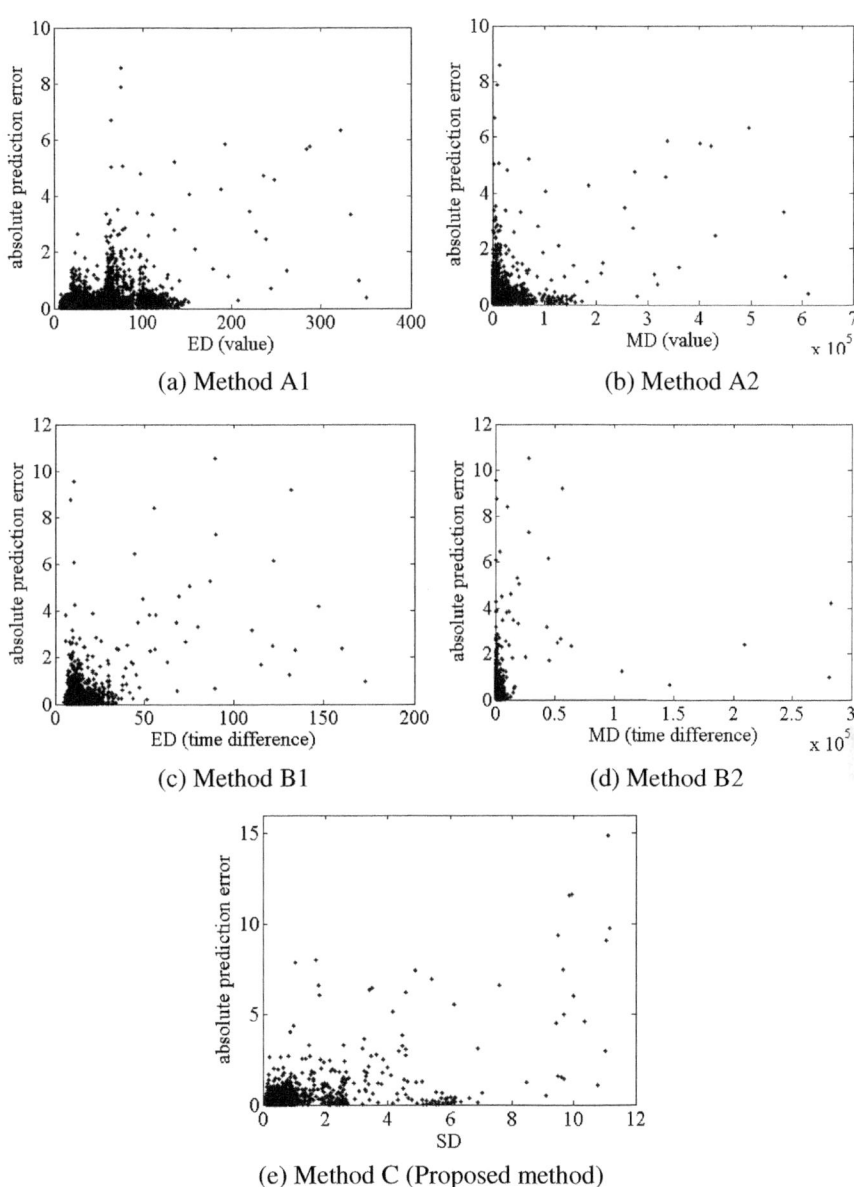

Fig. 3. The relationships between DMs and the absolute prediction errors in 2002

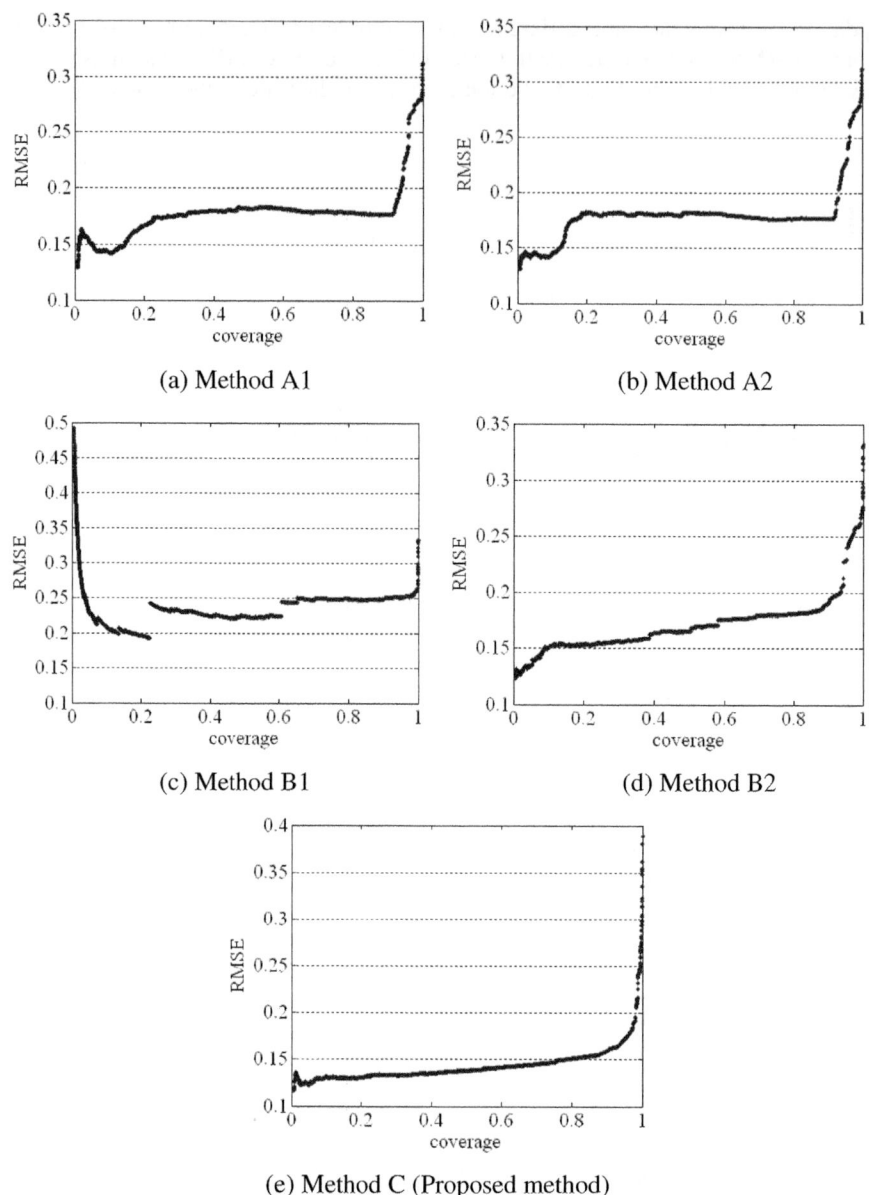

Fig. 4. The relationships between the coverage and the absolute prediction errors

3.2 Estimation of Prediction Accuracy

Next, we estimated the prediction accuracy. The five DMs listed below were applied.
A1: The ED of the values of X
A2: The MD of the values of X

B1: The ED of the time difference of X
B2: The MD of the time difference of X
C: The SD (Proposed method)

The regression models of (A1 and A2), (B1 and B2), and C are corresponding to those of A, B, and C in the "3.1 Presiction Accuracy" section, respectively. The results from the distances to the nearest neighbor of training data are not presented here, but they were almost identical to those of the average of training data.

Fig. 3 shows the calculation results of the relationships between DMs and the absolute prediction errors in 2002. For A1 and A2, the absolute prediction errors increased with the DMs, and their distributions became wider, but for B1 and B2, this tendency was not shown so much. This could be because the DMs were calculated by using not the values but the time difference, and then the DMs were small in the unsteady-state of the plant. The proposed method showed the clear tendency that the absolute prediction errors increased with the DM, and its distribution became wider.

Subsequently, the data were sorted in ascending order of each DM and we calculated the coverage that is the rate of the number of the data inside each AD to the total number of the data N_{all}. The coverage of the mth data is defined as follow:

$$\text{coverage}_m = N_{in, m} / N_{all} \tag{11}$$

where $N_{in, m}$ is the number of the data whose DMs are smaller than those of the mth data. The relationships between the coverage and the absolute prediction errors are shown in Fig. 4. The mth RMSE value is calculated with the $N_{in, m}$-data. It is desired for ADs that the smaller the values of the coverage are, the smaller the RMSE values are and vice versa. This tendency was shown in the figures except (c) and the line of the proposed method was smoother than the others. In addition, by comparing the values of the coverage where the RMSE values were 0.15 and 0.2, for example, those of the proposed method were larger than those of the other methods, respectively. This means that the proposed model could predict more number of data with higher predictive accuracy.

Fig. 5 shows the relationships between the coverage and the absolute prediction errors in each year using the proposed method. The tendency was identical in all years. Therefore, we confirmed the high performance of the proposed index.

4 Conclusion

In this study, for improving and estimating the accuracy of the prediction of time difference models, we proposed the ensemble prediction method with time difference. A final predicted value is an exponentially-weighted average of the multiple predicted values and we use the standard deviation of them as a index of prediction accuracy of the final predicted value. The proposed method was applied to the actual industrial data obtained from the operation of a distillation column. The proposed model achieved high predictive accuracy and could predict more number of data with higher predictive accuracy than traditional models. We therefore confirmed the usefulness of the proposed method without reconstruction of the soft sensor model. If the relationship between the SD and the standard deviation of the prediction errors is modeled as is done in [6], we can estimate not only the value of y but also the prediction accuracy of new data. The proposed methods can be combined with other kinds of regression methods, and thus can be used in various fields of the soft sensor. It is expected that the problems of maintenance of soft sensor models are reduced by using our methods.

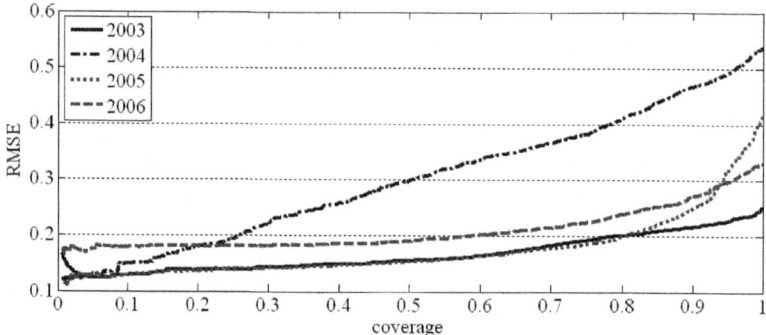

Fig. 5. The relationships between the coverage and the absolute prediction errors in each year using the proposed method

Acknowledgments. The authors acknowledge the support of Mizushima works, Mitsubishi Chemical and the financial support of Japan Society for the Promotion of Science.

References

1. Kano, M., Nakagawa, Y.: Data-Based Process Monitoring, Process Control, and Quality Improvement: Recent Developments and Applications in Steel Industry. Comput. Chem. Eng. 32, 12–24 (2008)
2. Kadlec, P., Gabrys, B., Strandt, S.: Data-Driven Soft Sensors in the Process Industry. Comput. Chem. Eng. 33, 795–814 (2009)
3. Qin, S.J.: Recursive PLS Algorithms for Adaptive Data Modelling. Comput. Chem. Eng. 22, 503–514 (1998)
4. Cheng, C., Chiu, M.S.: A New Data-based Methodology for Nonlinear Process Modeling. Chem. Eng. Sci. 59, 2801–2810 (2004)
5. Kaneko, H., Arakawa, M., Funatsu, K.: Development of a New Soft Sensor Method Using Independent Component Analysis and Partial Least Squares. AIChE J. 55, 87–98 (2009)
6. Kaneko, H., Arakawa, M., Funatsu K.: Applicability Domains and Accuracy of Prediction of Soft Sensor Models. AIChE J. (2010) (in press)
7. Ookita, K.: Operation and quality control for chemical plants by soft sensors. CICSJ Bull. 24, 31–33 (2006) (in Japanese)
8. Kaneko, H., Arakawa, M., Funatsu, K.: Approaches to Deterioration of Predictive Accuracy for Practical Soft Sensors. In: Proceedings of PSE ASIA 2010, P054(USB) (2010)
9. Kaneko, H., Funatsu K.: Maintenance-Free Soft Sensor Models with Time Difference of Process Variables. Chemom. Intell. Lab. Syst. (accepted)

Network Defense Strategies for Maximization of Network Survivability

Frank Yeong-Sung Lin[1], Hong-Hsu Yen[2], Pei-Yu Chen[1,3,4,*], and Ya-Fang Wen[1]

[1] Department of Information Management, National Taiwan University
[2] Department of Information Management, Shih Hsin University
[3] Information and Communication Security Technology Center
[4] Institute Information Industry
Taipei, Taiwan, R.O.C.
yslin@im.ntu.edu.tw, hhyen@cc.shu.edu.tw,
d96006@im.ntu.edu.tw, r94048@im.ntu.edu.tw

Abstract. The Internet has brought about several threats of information security to individuals and cooperates. It is difficult to keep a network completely safe because cyber attackers can launch attacks through networks without limitations of time and space. As a result, it is an important and critical issue be able to efficiently evaluate network survivability. In this paper, an innovative metric called the Degree of Disconnectivity (DOD) is proposed, which is used to evaluate the damage level of the network. A network attack-defense scenario is also considered in this problem, in which the attack and defense actions are composed by many rounds with each round containing two stages. In the first stage, defenders deploy limited resources on the nodes resulting in attackers needing to increase attack costs to compromise the nodes. In the second stage, the attacker uses his limited budget to launch attacks, trying to maximize the damage of the network. The Lagrangean Relaxation Method is applied to obtain optimal solutions for the problem.

Keywords: Information System Survivability, Degree of Disconnectivity, Lagrangean Relaxation, Mathematical Programming, Optimization, Network Attack and Defense, Resource Allocation.

1 Introduction

With growth of internet use, the number of experienced computer security breaches has increased exponentially in recent years, especially impacting on businesses that are increasingly dependent on being connected to the Internet. The computer networks of these businesses are more vulnerable to access from outside hackers or cyber attackers, who could launch attacks without the constraints of time and space. In addition to the rapid growth in rate of cyber attack threats, another factor that influences overall network security is the network protection of the network defender [1]. Although it is impossible to keep a network completely safe, the problem of the network security thus gradually shifts to the issue of survivability [2]. As knowing

---
* Corresponding author.

how to evaluate the survivability of the Internet is a critical issue, more and more researchers are focusing on the definitions of the survivability and evaluation of network survivability.

However, to enhance or reduce the network survivability, both network defender and cyber attacker usually need to invest a fixed number of resources in the network. The interaction between cyber attackers and network defenders is like information warfare, and how to efficiently allocate scarce resources to the network for both cyber attacker and network defender is a significant issue. Hence, the attack-defense situation can be formulated as a min-max or max-min problem. As a result, researchers can solve this kind of attack-defense problem of network security by mathematical programming approaches, such as game theory [3], Simulated Annealing [4], Lagrangean Relaxation Method [5]. In [5], the authors propose a novel metric called Degree of Disconnectivity (DOD) that is used to measure the damage level, or survivability, of a network. The DOD value is calculated by (1), in which a larger DOD value represents a greater damage level, which also implies lower network survivability. An attacker's objective is to minimize the total attack cost, whereas a defender's objective is to maximize the minimized total attack cost. The DOD is used as a threshold to determine whether a network has been compromised.

Considering the defense and attacker scenarios in the real world, it is more reasonable that the attacker fully utilizes his budget to cause maximal impact to the network rather than simply minimize his total attack cost. In this paper, the attack and defense actions are composed by rounds, where each round contains two stages. In the first stage, a defender deploys defense resources to the nodes in the network, whereas in the second stage, an attacker launches attacks to compromise nodes in the network in order to cause maximal impact to the network. Survivability and damage of the network is measured in terms of the DOD value.

$$\frac{\sum (number\ of\ broken\ nodes\ on\ shortest\ path\ of\ each\ OD\ pair)}{number\ of\ all\ OD\ pairs\ of\ a\ network} \quad (1)$$

2 Problem Formulation and Notations

In this paper, the two roles of attackers and defenders are described in this optimization model. An attacker wants to compromise nodes that cause maximal impact to the network, while a defender tries to minimize the damage by deploying defense resources to nodes. Both the attacker and the defender have perfect information about the network topology. Damage and impact of the network is measured in terms of the Degree of Disconnectivity (DOD) value here. If a node is compromised by the attacker, it is dysfunctional and thus cannot be used for transmitting information. Meanwhile, both sides have a budget limitation, and the total attack costs and defense resources are under their entire expenditure. The cost of compromising a node is related to the defense budget allocated to it. The result is that attackers have to allocate more costs on a node once more defense budget is allocated to the corresponding node. Note that in this context, the terms "resource" and "budget" are used interchangeably to describe the effort spent on nodes by attackers or defenders.

In the real world, network attack and defense are continuous processes. For modeling purposes, the problem is defined as an attack-defense problem, with each round contains the two stages of defense and attack. First, the defender tries to minimize network damage, i.e., DOD value, by deploying defense resources to the nodes in the network. The allocation is restricted by the defender's defense budget. Next, attackers start to launch attacks that compromise nodes, trying to maximize the damage, i.e. DOD value, of the network. Since attackers only have a limited budget and thus have to make good use of their budget, they need to decide which nodes to attack in order to cause the greatest impact to network operation. The given parameters and decision variables of the problem are shown in Table 1.

Table 1. Given Parameters and Decision Variables

Given parameter Notation	Description
V	Index set of nodes
W	Index set of OD pair
P_w	Set of all candidate paths of an OD pair w, where $w \in W$
M	Large enough number of processing cost that indicates a node has been compromised
ε	Small enough number of processing cost that indicates a node is functional
δ_{pi}	Indicator function, 1 if node i is on path p, 0 otherwise, where $i \in V$ and $p \in P_w$
d_i	Existing defense resources on node i, used for condition which has more than 1 round, where $i \in V$
$a_i(b_i+d_i)$	Attack cost of node i, which is a function of b_i+d_i, where $i \in V$
q_i	State of node i before this round. 1 if node i is inoperable, 0 otherwise, used for a condition which has more than 1 round, where $i \in V$
A	Attacker's total budget in this round
B	Defender's defense budget in this round

Decision variable Notation	Description
x_p	1 if path p is chosen, 0 otherwise, where $p \in P_w$
y_i	1 if node i is compromised by attacker, 0 otherwise (where $i \in V$)
t_{wi}	1 if node i is used by OD pair w, 0 otherwise, where $i \in V$ and $w \in W$
c_i	Processing cost of node i, which is ε if i is functional, M if i is compromised by attacker, where $i \in V$
b_i	Defense budget allocated to node i, where $i \in V$

The problem is then formulated as the following min-max problem:
Objective function:

$$\min_{b_i} \max_{y_i} \frac{\sum_{w \in W} \sum_{i \in V} t_{wi} c_i}{|W| \times M} \qquad \text{(IP 1)}$$

Subject to:

$$c_i = (y_i + q_i)M + [1-(y_i+q_i)]\varepsilon \qquad \forall i \in V \quad \text{(IP 1.1)}$$

$$\sum_{i \in V} t_{wi} c_i \leq \sum_{i \in V} \delta_{pi} c_i \qquad \forall p \in P_w, w \in W \quad \text{(IP 1.2)}$$

$$\sum_{p \in P_w} x_p \delta_{pi} = t_{wi} \qquad \forall i \in V, w \in W \quad \text{(IP 1.3)}$$

$$\sum_{i \in V} y_i a_i (b_i + d_i) \leq A \qquad \text{(IP 1.4)}$$

$$\sum_{i \in V} b_i \leq B \qquad \text{(IP 1.5)}$$

$$0 \leq b_i \leq B \qquad \forall i \in V \quad \text{(IP 1.6)}$$

$$y_i + q_i \leq 1 \qquad \forall i \in V \quad \text{(IP 1.7)}$$

$$\sum_{p \in P_w} x_p = 1 \qquad \forall w \in W \quad \text{(IP 1.8)}$$

$$x_p = 0 \text{ or } 1 \qquad \forall p \in P_w, w \in W \quad \text{(IP 1.9)}$$

$$y_i = 0 \text{ or } 1 \qquad \forall i \in V \quad \text{(IP 1.10)}$$

$$t_{wi} = 0 \text{ or } 1 \qquad \forall i \in V, w \in W. \quad \text{(IP 1.11)}$$

Objective function (IP 1) is to minimize the maximized damage of the network. That is, the attacker tries to maximize the DOD value by deciding which nodes to attack (denoted by y_i), while the defender tries to minimize the DOD value by deciding to which nodes defense resources should be allocated (denoted by b_i). Constraint (IP 1.1) describes the definition of processing cost c_i, which is ε if i is functional, M if i is compromised. Constraint (IP 1.2) requires the selected path for an OD pair w should have the minimal cost. Constraint (IP 1.3) represents the relationship between $x_p \delta_{pi}$ and t_{wi}. Constraint (IP 1.4) restricts the total attack cost spent and should not exceed attacker's budget A. Constraint (IP 1.5) specifies that the total defense budget allocated to nodes should not exceed defense budget B. Constraint (IP 1.6) indicates that variable b_i is continuous and bounded by 0 and defense budget B. Constraint (IP 1.7) specifies the attacker cannot attack a node that is already dysfunctional. Constraints (IP 1.8) and (IP 1.9) jointly limit the possibility that only one of the candidate paths of an OD pair w can be selected. Lastly, constraints (IP 1.9) to (IP 1.11) impose binary restrictions on decision variables.

3 Solution Approach

3.1 Solution Approach for Solving the Inner Problem of (IP 1)

3.1.1 Lagrangean Relaxation

In order to solve the (IP 1), the problem is decomposed into an inner problem and outer problem. The inner problem with constraints (IP 1.1), (IP 1.3) is reformulated and added as a redundant constraint (IP 1.12), as shown below. Note that the optimal solution is not affected if an equation is relaxed into an inequality version.

$$c_i \leq (y_i + q_i) M + [1 - (y_i + q_i)] \varepsilon \qquad \forall i \in V \quad \text{(IP 1.1')}$$

$$\sum_{p \in P_w} x_p \delta_{pi} \leq t_{wi} \qquad \forall i \in V, w \in W \quad \text{(IP 1.3')}$$

$$c_i = \varepsilon \text{ or } M \qquad \forall i \in V. \quad \text{(IP 1.12)}$$

By applying the Lagrangean Relaxation Method [6], the inner problem of (IP 1) is then transformed into the following Lagrangean Relaxation problem (LR 1), where constraints (IP 1.1'), (IP 1.2), (IP 1.3') and (IP 1.4) are relaxed. With a vector of Lagrangean multipliers, the inner problem of (IP 1) is transformed. (LR 1) is decomposed into three independent and easily solvable subproblems, as the following shows in more detail.

Subproblem 1.1 (related to decision variable x_p):

$$Z_{\text{Sub 1.1}}(\mu^3) = \min \sum_{w \in W} \sum_{i \in V} \sum_{p \in P_w} \mu_{wi}^3 \delta_{pi} x_p \qquad \text{(Sub 1.1)}$$

Subject to: (IP 1.8), (IP 1.9).

Dijkstra's shortest path algorithm can be applied to (Sub 1.1) since the node weight μ_{wi}^3 is non-negative. The time complexity of this problem is $O(|W| \times |V|^2)$, where $|W|$ is the number of OD pairs.

Subproblem 1.2 (related to decision variable y_i):

$$Z_{\text{Sub 1.2}}(\mu^1, \mu^4) = \min \sum_{i \in V} \left[\mu_i^1(\varepsilon - M) + \mu^4 a_i(b_i + d_i) \right] y_i \qquad \text{(Sub 1.2)}$$

Subject to: (IP 1.7), (IP 1.10).

(Sub 1.2) can be simply and optimally solved by examining the coefficient of y_i, for each node i, if the coefficient $\left[\mu_i^1(\varepsilon - M) + \mu^4 a_i(b_i + d_i) \right]$ is positive or the value of q_i is one, the value of y_i is set to zero; conversely, if $\left[\mu_i^1(\varepsilon - M) + \mu^4 a_i(b_i + d_i) \right]$ is non-positive and the value of the value of q_i is equal to zero, y_i is set to one. The time complexity of (Sub 1.2) is $O(|V|)$.

Subproblem 1.3 (related to decision variable t_{wi} and c_i):

$$Z_{\text{Sub 1.3}}(\mu^1, \mu^2, \mu^3) = \min \sum_{i \in V} \left\{ \left[\mu_i^1 - \sum_{w \in W} \sum_{p \in P_w} \mu_{wp}^2 \delta_{pi} + \sum_{w \in W} \left(\frac{-1}{|W| \times M} + \sum_{p \in P_w} \mu_{wp}^2 \right) t_{wi} \right] c_i - \sum_{w \in W} \mu_{wi}^3 t_{wi} \right\} \qquad \text{(Sub 1.3)}$$

Subject to:

$$t_{wi} = 0 \text{ or } 1 \qquad \forall i \in V, w \in W \quad \text{(Sub 1.3.1)}$$

$$c_i = \varepsilon \text{ or } M \qquad \forall i \in V. \quad \text{(Sub 1.3.2)}$$

In (Sub 1.3), both decision variable t_{wi} and c_i have two options. As a result, the value of t_{wi} and c_i can be determined by applying an exhaustive search to obtain the minimal value. The time complexity here is $O(|W| \times |V|)$.

The Lagrangean Relaxation problem (LR 1) can be solved optimally if all the above subproblems are solved optimally. By the weak duality theorem [7], for any set of multipliers, the solution to the dual problem is a lower bound on the primal problem (IP 1). To acquire the tightest lower bound, the value of Lagrangean multipliers needs to be adjusted to maximize the optimal value of the dual problem.

The dual problem can be solved in many ways; here the subgradient method is adopted to solve the dual problem.

3.1.2 Getting Primal Feasible Solutions

By applying the Lagrangean Relaxation Method, a theoretical lower bound on the primal objective function can be found. This approach provides some suggestions for obtaining feasible solutions for the primal problem. However, the result of the dual problem may be invalid when compared to the original problem since some important and complex constraints are relaxed. Therefore, a heuristic is needed here to make infeasible solutions feasible. In order to obtain primal feasible solutions and an upper bound on the inner problem of (IP 1), the outcome of (LR 1) and Lagrangean multipliers are used as hints for deriving solutions. The concept of the proposed heuristic is described below.

Recall that subproblem 1.1 is related to decision variable x_p, which determines the path to be used for an OD pair. By using this hint provided by subproblem 1.1, for each OD pair w, the chosen path is used to traverse from source to destination and calculate the number of times a node is used by all OD pairs, called node popularity (NP). A node with a larger NP value is likely to be an attack target, since more paths use this node. Therefore, attacking this node may result in larger DOD. An additional important issue for the attacker is the attack cost (AC) of the nodes in the network, because the attacker has a budget limitation. Hence, both factors are considered in the heuristic for getting primal feasible solutions. Once the popularity of nodes is determined; all nodes according to the ratio of NP to AC are sorted in descending order. For each of the sorted vertices i, the total attack cost and attack budget A is checked. If node i can be compromised without exceeding attack budget A, node i is selected as the attack target; else go to next node until all nodes are examined.

3.2 Solution Approach for Solving (IP 1)

The result of the inner problem represents the attack strategy under a certain initial defense budget allocation policy. The objective (IP 1) is to minimize the maximized damage of the network under intentional attacks. Therefore, the outcome of the inner problem can be used as the input of the outer problem for developing a better budget allocation policy. From the current attack strategy, the defender can adjust the budget allocated to nodes in the network according to certain reallocation policies. After budget adjustment, the inner problem is solved again to derive an attack strategy under the new defense budget allocation policy. This procedure is repeated a number of times until an equilibrium is achieved.

The concept used to adjust the defense budget allocation policy is similar to the subgradient method, in which the budget allocated to each node is redistributed according to current step size. This subgradient-like method is described as follows. Initially, the status of each node after attack is checked. If the node is uncompromised, this suggests that the defense resources (budget) allotted to this node is inadequate (more than needed) or it is unworthy for an attacker to attack this node as the node has too great a defense budget. Therefore, we can extract a fraction of the defense resources from the nodes unaffected by attacks, and allocate it to compromised nodes. The amount extracted is related to the step size coefficient, and it is halved if the optimal solution of (IP 1) does not improve in a given iteration limit.

Another factor that is related to the deducted defense resources is the importance of a node. In general, the greater the number of times a node is used by all OD paths implies higher importance. When a node with a larger number of times used by all OD paths is compromised, it provides a higher contribution to the growth of the objective function value (i.e., DOD value), compared with a node with a smaller number of times. As a result, only a small amount of defense resources should be extracted from nodes with higher usage. In the proposed subgradient-like method, the importance factor to measure the importance of a node is used, which is calculated by t_i / t_{total}, where t_i is the average number of times node i used by all OD paths, and t_{total} is the summation of t_i ($\forall i \in V$). An uncompromised node with greater importance factor will have a lower amount of defense resources extracted.

4 Computational Experiments

4.1 Experiment Environment

For the proposed Lagrangean Relaxation algorithm, two simple algorithms were implemented using C++ language, and the program was executed on a PC with AMD 3.6 GHz quad-core CPU. Here three types of network topology acted as attack targets: the grid network, random network and scale-free network. To determine which budget allocation policy is more effective under different cases, two initial budget allocation policies were designed uniform and degree based. The former distributed the defense budget evenly to all nodes in the network, while the latter allocated budget to each node according to the percentage of a node's degree.

4.2 Computational Experiment of (IP 1)

To prove the effectiveness of the Lagrangean Relaxation algorithm and the proposed heuristic, two simple algorithms were developed, introduced in Table 2 and Table 3 respectively for comparison purposes.

Table 2. SA_1 Algorithm

```
//initialization
total_attack_cost = 0;
sort all nodes by their attack cost in ascending order;
for each node i { //already sorted
        if ( total_attack_cost + attack_cost _i <= TOTAL_ATTACK_BUDGET
        AND (node i is not compromised OR compromised but repaired)){
                compromise node i;
                total_attack_cost += attack_cost_i;
        }
}
calculate DOD;
return DOD;
```

Table 3. SA_2 Algorithm

```
//initialization
total_attack_cost = 0;
sort all nodes by their node degree in descending order;
for each node i { //already sorted
        if ( total_attack_cost + attack_cost _i <= TOTAL_ATTACK_BUDGET
        AND (node i is not compromised OR compromised but repaired)){
                compromise node i;
                total_attack_cost += attack_cost_i;
        }
}
calculate DOD;
return DOD;
```

4.3 Experiment Results of the Inner Problem of (IP 1)

Fig. 1 compares the performance between the proposed Lagrangean Relaxation algorithm and two simple algorithms (SA_1 and SA_2); also, the gap between LR (UB) and LB is presented. Each point on the chart represents the DOD value of different node numbers and topologies under degree-based initial budget allocation strategy. The proposed LR algorithm has a better performance than the two competitive algorithms: it always causes the lowest network survivability (highest DOD value) in all network topologies and sizes. Although the gaps between LR and LB are generally small (about 12% on average), when the network size grows, the gap becomes larger, especially in the grid network case. The performance of the simple algorithm 2 (SA_2) is better than that of simple algorithm 1 (SA_1). The result again shows the importance of allocating more defense resources on vital nodes.

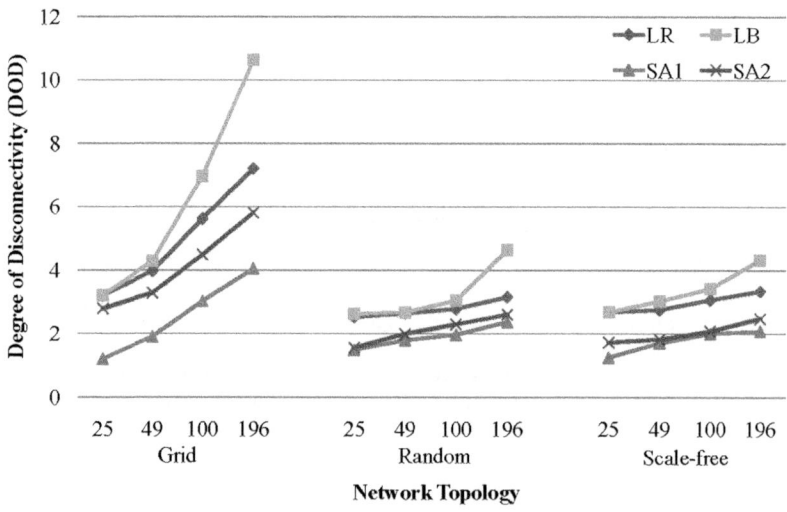

Fig. 1. Survivability of Different Network Sizes and Topologies

4.4 Experiment Result of (IP 1)

Fig. 2 illustrates the survivability (i.e., DOD value) of the network under different topologies, node numbers and initial budget allocation policies. Networks with a degree-based initial budget allocation strategy are more robust compared to those with a uniform strategy, and the damage of the network was less under the same attack and defense budgets. This finding suggests that the defense resources should be allotted according to the importance of each node in the network. The survivability of grid networks is lower than others using the DOD metric, since grid networks are more regular and more connected. As a result, some nodes are used more often by OD pairs. If these nodes are compromised by the attacker, the DOD value increases considerably. Random networks have higher survivability compared with scale-free networks because nodes in random networks are arbitrarily connected to each other rather than attached to nodes with a higher degree and therefore have lower damage when encounter intentional attacks.

The survivability of different network topologies, node numbers, and budget reallocation strategies is demonstrated in Fig. 2. From the above bar chart, the subgradient-like heuristic and both budget reallocation strategies perform well in all conditions (approximately 20% improvement on average), and the degree-based budget reallocation strategy is better than the uniform reallocation. However, the difference between two budget allocation strategies is not significant. One possible reason is that in the proposed heuristic, the extracted defense resources are allotted to compromised nodes according to certain strategies. These nodes selected by the attacker already indicate which nodes are more important, although node degree is related to the importance of nodes. As a result, the gap between the two strategies is small.

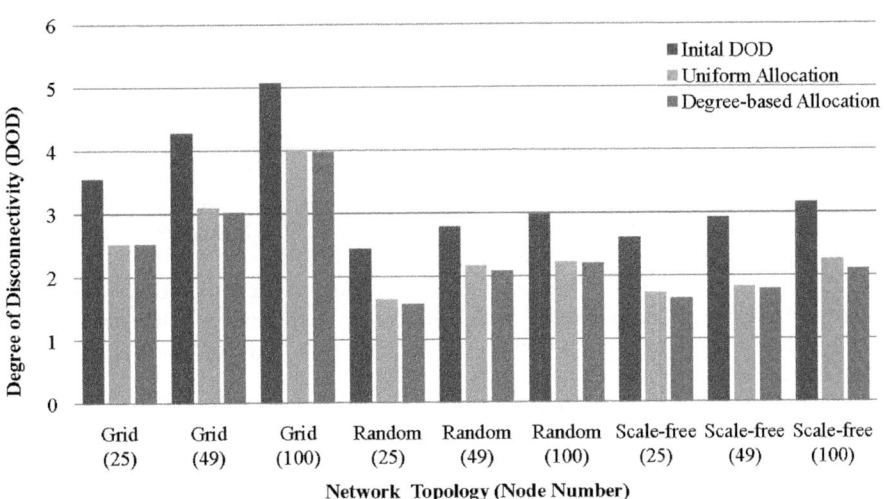

Fig. 2. Survivability of Different Networks and Reallocation Strategies

5 Conclusion

In this paper, a generic mathematical programming model that can be used for solving network attack-defense problem is proposed by simulating the role of the defender and the attacker. The attacker tries to maximize network damage by compromising nodes in the network, whereas the defender's goal is to minimize the impact by deploying defense resources to nodes, and thus enhancing their defense capability. In this context, the Degree of Disconnectivity (DOD) is used to measure the damage of the network. The inner problem is solved by a Lagrangean Relaxation-based algorithm, and the solution to min-max problem is obtained from the subgradient-like heuristic and budget adjustment algorithm.

The main contribution of this research is the generic mathematical model for solving the network attack-defense problem. The proposed Lagrangean Relaxation-based algorithm and subgradient-like heuristic have been proved to be effective and can be applied to real-world networks, such as grid, random and scale-free networks. Also, the survivability of networks with different topologies, sizes, and budget allocation policies has been examined. Their survivability can be significantly improved by adjusting the defense budget allocation. From the outcomes of the experiments, the defense resources should be allocated according to the importance of nodes. Moreover, the attack and defense scenarios take rounds to the end. As a result, the number of rounds, e.g., from 1 to N, could be further expanded.

Acknowledgments. This research was supported by the National Science Council of Taiwan, Republic of China, under grant NSC-99-2221-E-002-132.

References

1. Symantec.: Symantec Global Internet Security Threat Report Trends for 2009, Symantec Corporation, vol. XV (April 2010)
2. Ellison, R.J., Fisher, D.A., Linger, R.C., Lipson, H.F., Longstaff, T., Mead, N.R.: Survivable Network Systems: An Emerging Discipline, Technical Report CMU/SEI-97-TR-013 (November 1997)
3. Jiang, W., Fang, B.X., Zhang, H.L., Tian, Z.H.: A Game Theoretic Method for Decision and Analysis of the Optimal Active Defense Strategy. In: The International Conference on Computational Intelligence and Security, pp. 819–823 (2007)
4. Lin, F.Y.S., Tsang, P.H., Chen, P.Y., Chen, H.T.: Maximization of Network Robustness Considering the Effect of Escalation and Accumulated Experience of Intelligent Attackers. In: The 15th World Multi-Conference on Systemics, Cybernetics and Informatics (July 2009)
5. Lin, F.Y.S., Yen, H.H., Chen, P.Y., Wen, Y.F.: An Evaluation of Network Survivability Considering Degree of Disconnectivity. In: The 6th International Conference on Hybrid Artificial Intelligence Systems (May 2011)
6. Fisher, M.L.: The Lagrangian Relaxation Method for Solving Integer Programming Problems. Management Science 27(1), 1–18 (1981)
7. Geoffrion, M.: Lagrangean Relaxation and its Use in Integer Programming. Mathematical Programming Study 2, 82–114 (1974)

PryGuard: A Secure Distributed Authentication Protocol for Pervasive Computing Environment

Chowdhury Hasan, Mohammad Adibuzzaman, Ferdaus Kawsar, Munirul Haque, and Sheikh Iqbal Ahamed

Marquette University, Milwaukee, Wisconsin, USA
{chowdhury.hasan,mohammad.adibuzzaman,ferdaus.kawsar,
md.haque,sheikh.ahamed}@marquette.edu

Abstract. Handheld devices have become so commonplace nowadays that they are an integral part of our everyday life. Proliferation of these mobile handheld devices equipped with wide range of capabilities has bolstered widespread popularity of pervasive computing applications. In such applications many devices interact with each other by forming ad hoc wireless networks. The necessity of such unavoidable inter-device dependency along with volatile nature of connectivity and the lack of a fixed infrastructure for authentication and authorization, devices are susceptible and vulnerable to malicious active and passive snoopers. If a device registers a malicious device as its valid neighbor, the security and privacy of entire system might be jeopardized. Such sensitivity to malevolent activity necessitates the need for a robust mechanism to maintain a list of valid devices that will help to prevent malicious devices from authenticating successfully. In this paper, we present the feasibility of using a decentralized protocol in order to prevent malicious devices from participating illicitly into the ad hoc networks.

Keywords: Computer network security, Device discovery, Handheld devices, Pervasive computing, ambassador node, sequence generator.

1 Introduction

Pervasive computing is becoming an inseparable part of our daily life due to the development of technology for low-cost pervasive devices and inexpensive but powerful wireless communication. Pervasive computing which was the brain child of Weiser [3] is now proving its viability in industry, education, hospitals, healthcare, battlefields etc. The computational power and memory capacity of pervasive computing devices have significantly grown in the last few years. However, it is still not comparable to the storage capacity of other memory devices employed in distributed computing environment having a fixed infrastructure scenario. Low battery power places another restriction on pervasive computing devices, as mentioned in [23]. Due to such constraints, the dependency of one device on another is an important aspect of the pervasive computing environment.

Pervasive computing environment is characterized as a volatile one because each device can join and leave arbitrarily. So, the knowledge of current, valid neighbors is important for each device. If a device registers a malicious device as its valid neighbor, the security and privacy of the whole system might fall down. For example in a hospital premise doctors, nurses and other medical staff need to make contact with one another through a wireless network. A hacker with a malicious intent can easily enter the hospital premise as a guest or visitor. He could then communicate with others in the guise of a doctor and access sensitive and crucial patient records. In order to defend such destructive activity a valid device discovery protocol is of utmost importance for pervasive computing applications. This discovery protocol is responsible for updating devices with the latest information regarding current, valid neighbors. Thus it prevents any valid device in the network from communicating with a malicious user. This paper presents the architecture and development of such a protocol as a middleware service for pervasive computing applications.

Several papers [4-6] presented resource discovery protocols for dynamically changing environments. But the impact of the protocol on issues like scalability, performance, and battery power optimization was not addressed by them. The insecure service discovery protocol which allows each device in the network to access the services offered by other devices has been followed by some researchers for the purpose of service discovery. Intentional naming system (INS)/Twine [7], INS [8], and service discovery in DEAPspace [9] are instances of such sort of discovery protocol. Service location protocol (SLP) [10] and salutation consortium's protocol [11] follow a typical access control approach to conduct secure service discovery. Secure service discovery service (SSDS) [12] makes use of a trusted central server to provide the required security features. Although device discovery is a component of service discovery, the major goal of these service discovery protocols is to make available services accessible while maintaining security and privacy. However, several active and passive attack scenarios have been ignored by them.

The concern of our paper is different as it considers advanced issues beyond the usual ones managed by these service discovery protocols. Popovski et al. [13] proposes a protocol for device discovery in a pervasive computing environment in which the unique addresses of individual devices are not identifiable. In [14-16] Bluetooth wireless technology has been described. However, these protocols are not suitable as middleware service because they operate in the media access control (MAC) layer. In [17] the adaptability issue has been partially addressed. The central focus of universal plug and play (UPnP) [18, 19], Jini [20, 21] and open services gateway initiative (OSGI) [22] protocols is to defeat the impediment of heterogeneity while discovering devices. Our objective, however, is towards devising a protocol for maintaining a list of valid devices and building a model for discovering valid devices that is impregnable to active and passive malicious attacks.

Hopper and Blum (HB) have exhibited the use of HB protocol in [1,2] for achieving a secure human–computer (HC) authentication, where the user communicates using a dumb terminal in the presence of active and passive eavesdroppers. In [27] a variant of this technique is used for device authentication. In our proposed model, we adapted these protocols for device discovery, using an ad hoc scenario, in such a way so that the device discovery mechanism works in a distributed mode by ensuring participation of each network device.

The incorporation of security in each step of authentication process from a design point of view and the capability to safeguard major security attacks are the exclusive features of our model. It is also lightweight because it does not require a fixed infrastructure and all the computations involve only binary AND, XOR and rotate operations, which have very low computational overhead. In addition to providing safeguard against the common active and passive attacks, our proposed scheme is capable of protecting a novel type of attack introduced in this paper.

The rest of this paper is organized as follows: The related works in literature are reviewed in Section 2. Section 3 introduces the novel security threat for distributed authentication protocols designed based on the LPN (learning parity with noise) technique. In Section 4 we mention the novelty of our approach by listing its exclusive features and comparing those with couple of existing approaches. Section 5 presents a concise overview of our complete approach. We present in detail the proposed scheme in Section 6 along with the assumptions and protocol descriptions. Here we show how PryGuard withstands the active and passive security attacks in the context of an ad hoc network environment. Finally, we conclude the paper in Section 7.

2 Related Works

Implementations of several device discovery protocols for ad hoc network environment are presented in [13, 14, 15, 17]. These protocols are compatible with different wireless communication protocols like Bluetooth, 802.11, etc. but none of them have the capability to serve as middleware service. These protocols do not provide any application interface required for application developers to use a middleware service rather they work in the MAC and network layers. In [32] the security of Web service has been talked about. However, the focus of this software system is to enable the interoperability of applications over the Internet. Martin and Hung presented a security policy [33] to manage different security-related concerns including confidentiality, integrity and availability for VoIP. However, our main focus is on secure device discovery in pervasive ad hoc networks. Aleksy et al. proposed a three-tier approach to successfully handle the heterogeneity and interoperability issues [34].

In PryGuard, we have focused on making a validity decision based on behavior. Popovski et al. [13] have proposed a randomized distributed algorithm for device discovery with collision avoidance using a stack algorithm. Unfortunately, it fails to adapt itself with the dynamically changing environment of an ad hoc network.

In UPnP protocol [18, 19], a controlled device broadcasts its presence and services it is willing to provide. Jini [20, 21] constructs a community where each device can use the services provided by other devices belonging to the same community. A framework is provided in OSGI [22] for a home network that allows the user to communicate through devices with heterogeneous communication protocols. However, none of them considered the issue of maintaining a valid device list and corresponding security threats. Research studies in [14, 15, 16] discussed Bluetooth wireless communication. However, this protocol does not address power optimization and fault tolerance. In addition, it does not include the impact of malicious attacks which we have discussed in our findings.

In [24, 25] the authors have shown that the Hopper–Blum protocol can be applied to strengthen the security features of RFID. In our approach, the task of sending challenges is distributed to all valid nodes. Thus, the ambassador only combines the replies of other devices. Due to the limitation of battery power, it is not feasible for a specific node to manage all the challenges and calculations as/when the ad hoc network grows in size. In [27] a similar distributed protocol called ILDD is proposed where recommendations from the valid devices are returned to the ambassador in plain format, like 'true' or 'false'. Unfortunately, this spoils the purpose of noise inclusion since by observing the recommendations a passive intruder could determine whether a response was correct or noisy. We have explained this loophole in ILDD in the following section where we demonstrate that the authentication protocol presented in ILDD is not fully secure. In fact, any distributed authentication protocol based on the LPN technique is vulnerable to a special type of security flaw which we have termed as the "Noise Recognition Attack" as it enables the attacker to distinguish noisy responses from valid ones. Our proposed distributed authentication protocol, PryGuard resolves this security attack.

3 The Noise Recognition Attack

The purpose of inserting noise into the returned answers is to thwart the attacker from learning the correct ones. The knowledge of a specific number of valid challenge-response pairs may result into re-generating the actual secret by applying the Gaussian Elimination method. The existing distributed authentication protocols based on the LPN technique suffer from a severe security threat as they might allow the attacker to differentiate which of the responses are noisy. We have termed this security flaw as the Noise Recognition Attack. It basically stems from the opportunity offered to the attacker to distinguish valid responses from noises by using the correlations among captured data. This attack scenario is depicted in figure 1.

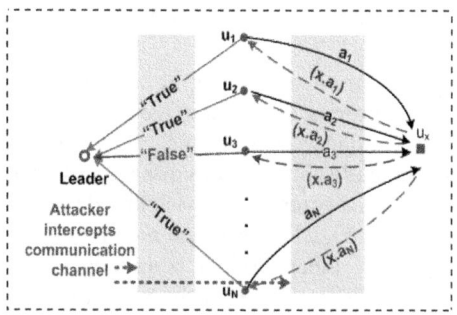

Fig. 1. Overview of existing distributed device discovery protocol and security loophole

Basic architecture of the protocol is described here. When a new member, u_x attempts to join the network, sends request to the leader node. Subsequently the leader distributes the authentication process by forwarding the request to other valid members, $u_1, u_2, ..., u_N$. Afterwards pairs of challenge-response messages are exchanged between each valid member and the new member. Although this step is similar to the basic operation of HB protocol, it distributes the entire process. Later, recommendations from the valid members are forwarded to the leader. As shown in the figure, all these transactions are in threat of being exposed to any potential attacker. If the attacker listens to the communication channel, according to the format of recommendations returned to the leader, he would be able to figure out which of the challenge-response pairs were valid in the earlier communications. It is assumed that the attacker captures messages exchanged between each valid member and the new member but due to the inclusion of noise he cannot apply the Gaussian Elimination method. However, the attacker's knowledge of figuring out which of the recommendations were "false" enables him to apply the Gaussian elimination method. For example, in the scenario considered above, the attacker knows that the challenge-response pair exchanged between u_3 and u_x was not valid whereas the other pairs were valid. Such capability of predicting the wrong equations goes opposite to the LPN concept. Most importantly, it ultimately enables the attacker to correctly apply the Gaussian Elimination method on the valid challenge-response pairs to obtain the secret.

4 Characteristics

PryGuard is intended to offer a feasible device authentication protocol for ad hoc network environment in a distributed fashion. It is featured with several novel and exclusive characteristics in order to accomplish that purpose. Our model is different from others due to its capability of being used as a middleware service. Some other exclusive features of PryGuard along with comparison with competitive approaches are mentioned below.

1) The HB protocol is for human-computer authentication in which a user communicates with the computer through an insecure channel. This protocol was not designed for an ad hoc environment. In an ad hoc network any member can join and leave at any time. Our protocol is adaptive to deal with the constraints of such dynamic environment.
2) HB protocol is based on a single server–client scenario. However, an ad hoc network is formed by a numerous devices joining and leaving arbitrarily. Instead of providing a challenge from a single device/server, we modified the method so that each valid device sends only one challenge to every other node present in the network. Such distributed approach obviously makes it more scalable.
3) We have used the concept of an ambassador node which is chosen based on trust level. For the trust portion we used the trust model proposed in [26]. This model updates the trust level of each device dynamically, depending on its interaction and behavior with other devices. Remaining battery power of a device can be another significant criterion for selecting the ambassador node.
4) In our model, we consider the impact of noise inclusion. Due to the fact that a device is only required to make a certain number of correct responses, there is possibility of a user being successfully authenticated without having the secret.

These users will impact the authentication processes later as they will also actively involve. A new term β has been incorporated; β denotes the expected number of malicious devices present in the valid neighbor list. The number of valid responses required to get access to the network is adjusted based on this parameter.

5) We have introduced the use of a specific function to enable the process of bit position generation and distribution. This function is a bijection and shared among the valid devices only. The ambassador selects particular bit position for each member and uses the bijective function to encrypt it.

5 Overview of PryGuard

Major steps of PryGuard are described here. After the execution of PryGuard each device in the network is updated with a list of valid neighbors. This is suitable for large corporate building, college campus, etc. where numerous devices communicate with each other using wireless connection. In our model, each device maintains a list known as a valid neighbor list that contains the nodes that have been declared valid by the ambassador node. The terms, leader node and ambassador node, are used interchangeably in this paper. Ambassador node initiates the process and finally computes validity of each device upon getting replies from existing valid devices. Later this updated list is broadcast within the network. We are using the term valid device to denote those devices that are already in the valid neighbor list of others. This also indicates that these devices have passed the challenge–response based device update mechanism in the previous phase. If a malicious device can generate the required number of correct responses to pass the challenge–response phase, it will also be updated as a valid device in the valid neighbor list of other devices.

First of all, one of the nodes is selected as the ambassador node. We assume that selection of ambassador node is based on mutual trust ratings among the network members. The ambassador node initiates the device authentication process. Whenever a new member intends to join the network this process is triggered. Again, the ambassador triggers the process periodically to update its valid neighbor list. It is assumed that a common secret is possessed by a valid member. The authentication process runs on this assumption. Main task is to verify whether a new member or an existing member possesses the correct version of current secret as the secret is also changed periodically. The verification process is distributed over all the network members and it is administered by the ambassador. The ambassador uses a sequence generator function to assign a bit position of the current secret to individual members which are already in the valid neighbor list. These devices then send challenges either to the new member or to one another. Afterwards each of them replies with the corresponding response. Based on the correctness of the response the challengers forward their replies to the ambassador. Finally the ambassador decides which are the valid devices based on the number of correct responses made by each device. At the end, this updated list is broadcast within the network. An overview of the architecture of PryGuard is depicted in the figure below. Elaborate description on the details of the protocol follows next.

6 Details of Our Solution

The effectiveness of our model is based on several realistic assumptions. *Assumptions*

1) The overall authentication process is centered at verifying the possession of a common secret. We assume that this secret is exchanged prior to any attempt of communication. Possession of this secret characterizes a device as valid. All the devices which have passed the authentication phase know this specific secret x of length n. Malicious devices present in the network do not know the secret x.
2) The secret is changed periodically over the period of time. A special function is used to derive new secret from the old one, i.e. f(x) becomes the new secret. All the valid devices know this function. Even if a malicious intruder gains the secret x at some stage, he will not be able to generate the new secret as f is not known. There is no chance of f being accessed by a malicious user because it is never communicated through the transmission medium. In this way the secrecy of x is ensured.
3) In order to formulate a distributed protocol, our model disseminates the entire task over all the valid network members. The ambassador uses a sequence generator function g to determine bit sequence for each valid device and assigns this bit sequence to it. It is assumed that g is a special bijective function and g and its inverse are possessed by the valid devices only. Like f, it is never transmitted through the medium as well. Hence, malicious attackers cannot access this.
4) The authentication process is initiated each time a device joins the network. However, the system is triggered periodically if there is a long idle period. This initiation starts with the challenges sent by the ambassador node.

Authentication Protocol

PryGuard device authentication comprises several sequential steps. These steps are discussed in detail below. Following tasks are executed either when a new device requests to join the network or after the passage of a specific amount of time.

Step 1) The ambassador node sends a challenge to every other node in the network. After receiving that challenge every device generates the new secret f(x). This new secret is used in responding to all further challenges.

Step 2) The ambassador node determines a bit position taken from [1,n] for each individual device present in the current valid neighbor list and this bit position is assigned to the corresponding device. In essence, this is done to make each valid member responsible for only one bit of the current secret x. The bit position is exchanged using the secret function g.

Step 3) Each valid device in the network sends all other devices in the network a random challenge a of length n. So, each sender S will send arbitrary challenges $a_1, a_2, a_3, \ldots, a_{N-1}$ to $u_1, u_2, u_3, \ldots, u_{N-1}$ respectively, where u_i denotes ith member and N being the total number of devices in the network.

Step 4) Each member $u_1, u_2, u_3, \ldots, u_{N-1}$ calculates the binary inner product of secret x and the received challenge a i.e. $(a \cdot x)$ modulo 2 and sends their responses to S. S calculates $(a \cdot x)$ modulo 2 as well for all the challenges $a_1, a_2, a_3, \ldots, a_{N-1}$ and compares the results with corresponding responses.

Step 5) Now suppose, sender S was actually the kth network member u_k and was assigned the bit position b_k by the ambassador in step 2. Then, for all the matched results, S will send the value of b_kth bit of x i.e. $x[b_k]$ to the ambassador. This will be counted as a true recommendation to the ambassador for the corresponding member to which the response belongs. Conversely, for the devices which fail to correctly compute the binary inner product, S will send the negated value of b_kth bit of x i.e. $\overline{x[b_k]}$ to the ambassador. This will be counted as a false recommendation to the ambassador for the corresponding member to which the response belongs. So, from the point of view of the ambassador node, a true recommendation (exact value of the corresponding bit of x) indicates that the receiver node calculated the correct response while a false recommendation (negated value of the corresponding bit of x) denotes an incorrect response.

Step 6) steps 2–5 are continued for each of the valid devices $u_1, u_2, u_3, \ldots, u_\Delta$ present in the network where Δ denotes the total number of valid devices.

Step 7) After receiving all of the recommendations, the ambassador calculates the validity of each device including itself then broadcasts the valid result. Finally, every other device updates its valid neighbor list according to the ambassador's broadcasted results.

All these tasks are executed sequentially among the different participating entities. The ambassador/leader initiates the authentication process. It generates random challenges $a_1, a_2, \ldots, a_N$ and mutually exclusive sequence numbers $b_1, b_2, \ldots, b_n$ for all the current valid members and sends those to them individually. Upon receiving the challenge and the sequence number each valid member updates the old secret and forwards the challenge to the new member. The new member also updates its version of the current secret and uses it to compute the binary inner product. Then, it sends the resultant parity bits p'_k to corresponding network members. The valid member itself computes the parity bit p_k using its version of the secret key. Then based on the comparison between p_k and p'_k, it returns its decision, in terms of the value of $x[b_k]$ to the ambassador. After getting decisions from all the existing members, the ambassador picks the predefined n recommendations which corresponds to n bits of x and forms the value of the secret possessed by the new member. If this secret matches exactly with the original x, the authentication process succeeds.

The calculation of determining the validity of a node based on recommendations from its peers needs to be explained since we intend to consider the impact of noise inclusion on such decision. A device that is already in the valid neighbor list will be updated as a valid device if $K \geq \text{ceil}\left((1 - \eta) \times l\right) - \beta \ldots (1)$, where K is the number of true recommendations from other devices to the ambassador about the device being validated, η is the maximum allowable percentage of noise (i.e. intentional incorrect answer), l is the total number of challenges received by one valid device. As each valid device receives challenges from every other valid device except itself, $l = \Delta - 1$, β is the expected number of malicious devices present in the network that have updated themselves in the valid neighbor list, and Δ is the total number of valid devices in the network. So, (1) can be rewritten as $K \geq \text{ceil}\left((1 - \eta) \times (\Delta - 1)\right) - \beta \ldots (2)$. A new device that has just joined the network but is not in the valid neighbor list of other devices will receive a challenge from every valid device in the network. Thus, it will actually receive Δ challenges. As a result, l in (1) has been replaced by Δ. So, a new device will be updated as a valid device if $K \geq \text{ceil}((1 - \eta) \times \Delta) - \beta \ldots$ (3). As the output of the term $((1 - \eta) \times \Delta)$ can be a float value we are considering the ceiling of the term. For example, if the output of $((1 - \eta) \times \Delta)$ is 3.8, we take it as 4.

We need to clarify some points. A device cannot send challenge until it is proven to be a valid device. It can only answer to challenges sent to it. Any challenge made by such a device is discarded since the ambassador knows the current valid devices. Likewise, it cannot send a recommendation about any device to the ambassador. Even if this device sends challenges or recommendations, those will be ignored because the trusted nodes will not accept any challenge or recommendation from any node that is not in their trusted neighbor list. When an ambassador node leaves the network, it selects another node in the network as the ambassador based on criteria like trust rating, remaining battery power etc. and broadcasts a message about the new ambassador. The optimal selection criteria for choosing an ambassador are of great concern and we plan to delve into that issue in future.

7 Conclusions and Future Work

In this paper, we have presented the design and development of PryGuard, a secure device authentication protocol. We have also shown a novel type of security attack. PryGuard adopted the well-known Hopper–Blum algorithm in an ad hoc network environment. The unique characteristics of our proposed model have been specified clearly. The lightweight distributed protocol is able to handle the scalability issue of dynamic environment like mobile ad hoc network or pervasive computing environment. In future, we plan to focus on several issues like challenges to a selective number of devices, loss of challenge or response due to collision, replay attack, spoofing etc.

References

1. Hopper, N., Blum, M.: A secure human computer authentication scheme., Carnegie Mellon Univ., Pittsburgh, PA, Tech. Rep. CMU-CS-00-139 (2000)
2. Hopper, N.J., Blum, M.: Secure human identification protocols. In: Boyd, C. (ed.) ASIACRYPT 2001. LNCS, vol. 2248, pp. 52–66. Springer, Heidelberg (2001)
3. Weiser, M.: Some computer problems in ubiquitous computing. Communications of the ACM 36(7), 75–84 (1993)
4. Eronen, P., Nikander, P.: Decentralized Jini security. In: Proceedings of the Network. Distributed. System Security Symposium, San Diego, CA (February 2001)
5. Hewlett Packard CoolTown (2008), http://cooltown.hp.com
6. UC Berkeley. The Ninja Project: Enabling internet scale services from arbitrarily small devices (2008), http://ninja.cs.berkeley.edu
7. Balazinska, M., Balakrishnan, H., Karger, D.: INS/Twine: A scalable peer-to-peer architecture for intentional resource discovery. In: Proceedings of the International Conference on Pervasive Computing, Zurich, Switzerland (2002)
8. Adjie-Winoto, W., Schwartz, E., Balakrishnan, H., Lilley, J.: The design and implementation of an intentional naming system. In: Proceedings of the 17th ACM Symposium on Operating Systems Principles (SOSP 1999), Kiawah Island, SC (1999)

9. Nidd, M.: Service discovery in DEAPspace. IEEE Pers. Communications 8(4), 39–45 (2001)
10. Guttman, E., Perkins, C., Veizades, J.: Service location protocol. Version 2, http://www.ietf.org/rfc/rfc2608.txt
11. The Salutation Consortium, Inc. Salutation architecture specification (1999), http://ftp.salutation.org/salute/sa20e1a21.ps
12. Czerwinski, S., Zhao, B.Y., Hodes, T., Joseph, A., Katz, R.: An architecture for a secure service discovery service. In: Procedings of the 5th Annual International Conference on Mobile Computing Networks (MobiCom 1999), Seattle, WA (1999)
13. Popovski, P., Kozlova, T., Gavrilovska, L., Prasad, R.: Device discovery in short-range wireless ad hoc networks. IEEE Networks 3, 1361–1365 (2002)
14. Zaruba, G.V., Gupta, V.: Simplified Bluetooth device discovery— Analysis and simulation. In: Proceedings of the 37th Hawaii International Conference on Systems Sciences, pp. 307–315 (January 2004)
15. Ferraguto, F., Mambrini, G., Panconesi, A., Petrioli, C.: A newapproach to device discovery and scatternet formation in Bluetooth networks. In: Proceedings of the 18th International Parallel Distributed Process. Symposium, pp. 221–228 (April 2004)
16. Zaruba, G.V., Chlamtac, I.: Accelerating Bluetooth inquiry for personal area networks. In: Proceedings of IEEE Global Telecommunication Conference, vol. 2, pp. 702–706 (December 2003)
17. Sohrabi, K., Gao, J., Ailawadhi, V., Pottie, G.J.: Protocols for selforganization of a wireless sensor network. Proceedings of IEEE Pers. Communication 7(5), 16–27 (2000)
18. Universal Plug and Play Forum. About universal plug and playtechnology (2008), http://www.upnp.org/about/default.asp#technology
19. Universal Plug and Play. Understanding universal plug and play: A white paper (June 2000), http://upnp.org/resources/whitepapers.asp
20. Sun Microsystems. Jini network technology (2008), http://www.sun.com/jini
21. Sun Microsystems. The community resource for Jini technology (2008), http://www.jini.org
22. Dobrev, P., Famolari, D., Kurzke, C., Miller, B.: Device and service discovery in home networks with OSGI. Proceedings of IEEE Communications Magazine 40(8), 86–92 (2002)
23. Satyanarayanan, M.: Fundamental challenges in mobile computing. In: Proceedings of 15th ACM Symposium on Principles of Distributed Computing, pp. 1–7 (May 1996)
24. Weis, S.A.: Security parallels between people and pervasive devices. In: Proceedings of 3rd IEEE International Conference on Pervasive Computing Communications Workshops, pp. 105–109 (2005)
25. Juels, A., Weis, S.A.: Authenticating pervasive devices with human protocols. In: Shoup, V. (ed.) CRYPTO 2005. LNCS, vol. 3621, pp. 293–308. Springer, Heidelberg (2005)
26. Sharmin, M., Ahmed, S., Ahamed, S.I.: An adaptive lightweight trust reliant secure resource discovery for pervasive computing environments. In: Proc. of PerCom 2006, Pisa, Italy, pp. 258–263 (2006)
27. Haque, M., Ahamed, S.I.: An Impregnable Lightweight Device Discovery (ILDD) Model for the Pervasive Computing Environment of Enterprise Applications. IEEE Transactions on Systems, Man, and Cybernetics 38(3), 334–346 (2008)
28. Sharmin, M., Ahmed, S., Ahamed, S.I.: MARKS (middleware adaptability for resource discovery, knowledge usability and self-healing) in pervasive computing environments. In: Proc. of 3rd Int. Conf. Inf. Technol.: New Gen, pp. 306–313 (April 2006)

29. Ahmed, S., Sharmin, M., Ahamed, S.I.: Knowledge usability and its characteristics for pervasive computing. In: Proc. 2005 Int. Conf. pervasive Syst. Computing (PSC 2005), Las Vegas, NV, pp. 206–209 (2005)
30. Sharmin, M., Ahmed, S., Ahamed, S.I.: SAFE-RD (Secure, adaptive, fault tolerant, and efficient resource discovery). in pervasive computing environments. In: Proc. IEEE Int. Conf. Inf. Technol (ITCC 2005), Las Vegas, NV, pp. 271–276 (2005)
31. Ahmed, S., Sharmin, M., Ahamed, S.I.: GETS (Generic, efficient, transparent and secured) self-healing service for pervasive computing application. Proceedings of International Journal of Network Security 4(3), 271–281 (2007)
32. Carminati, B., Ferrari, E., Hung, P.C.K.: Web services composition: A security perspective. In: Proceedings of the 21st Int. Conference on Data Engineering (ICDE 2005), Japan, April 8–9 (2005)
33. Martin, M.V., Hung, P.C.K.: Toward a security policy for VoIP applications. In: Proceedings of the 18th Annual Can. Conf. Electr. Comput. Eng (CCECE 2005), Saskatoon, SK, Canada (May 2005)
34. Aleksy, M., Schader, M., Tapper, C.: Interoperability and interchangeability of middleware components in a three-tier CORBA-environmentstate of the art. In: Proc. 3rd Int. Conf. Enterprise Distrib. Object Comput (EDOC 1999), pp. 204–213 (1999)

A Global Unsupervised Data Discretization Algorithm Based on Collective Correlation Coefficient

An Zeng[1,2], Qi-Gang Gao[2], and Dan Pan[3]

[1] Guangdong University of Technology
[2] Dalhousie University
[3] Saint Mary's University

Abstract. Data discretization is an important task for certain types of data mining algorithms such as association rule discovery and Bayesian learning. For those algorithms, proper discretization not only can significantly improve the quality and understandability of discovered knowledge, but also can reduce the running time. We present a Global Unsupervised Discretization Algorithm based on Collective Correlation Coefficient (GUDA-CCC) that provides the following attractive merits. 1) It does not require class labels from training data. 2) It preserves the ranks of attribute importance in a data set and meanwhile minimizes the information loss measured by mean square error. The attribute importance is calibrated by the CCC derived from principal component analysis (PCA). The idea behind GUDA-CCC is that to stick closely to an original data set might be the best policy, especially when other available information is not reliable enough to be leveraged in the discretization. Experiments on benchmark data sets illustrate the effectiveness of the GUDA-CCC algorithm.

Keywords: Data discretization, Data mining, Principal component analysis, Collective correlation coefficient.

1 Introduction

Since the Internet applications and other information systems emerged rapidly, the amount of data becoming available has grown exponentially. Meanwhile, handling large amount and different types of data can be challenging because the current data analysis algorithms (e.g. machine learning and statistics learning) are not effective enough to discover knowledge from a wide spectrum of data analysis tasks. In fact, some algorithms, such as Associate rule mining, Bayesian networks and Rough Set theory, can only directly deal with discrete data rather than continuous numerical data.

As a data-preprocessing step, discretization is to partition continuous numerical values of an attribute in a data set into a finite number of intervals as discrete values. A great number of research findings have indicated that discretization results exert a great impact on the performance of data mining algorithms. A good discretization algorithm not only can simplify continuous attributes to aid people in more easily comprehending data and results, but also can make subsequent procedures such as classification more effective and efficient.

In this paper, we propose a Global Unsupervised Discretization Algorithm based on Collective Correlation Coefficient algorithm (GUDA-CCC). Without requiring the class labels in a data set, the algorithm is able to simultaneously optimize the numbers of data intervals corresponding to all continuous attributes in a data set.

In the GUDA-CCC algorithm, we attempt to stick closely to an original data set as far as possible by two ways when determining the number of discrete intervals and positioning these intervals (or cut-points). One way is to preserve the rank of attribute importance in an attribute set by applying an optimization approach when determining the number of data intervals. In the method, the attribute importance is measured as the Collective Correlation Coefficient (CCC), which is in essence an index quantifying each continuous attribute's contribution to the state spaces comprised of continuous attributes in a data set as a whole [1]. In addition, we can consider discretization as a quantization procedure since quantization is the process of subdividing the range of a signal into non-overlapping regions (i.e. data intervals) and assigning a numerical value to represent each region. Hence, the other way is to minimize the discretization errors using the Lloyd-Max Scalar Quantization [2], when positioning the data intervals according to the given numbers of data intervals.

The rest of the paper is organized as follows. In section 2, we introduce related work. Section 3 articulates the algorithm and relevant thoughts. Key components used in the GUDA-CCC algorithm are detailed in section 4. A set of empirical results and corresponding comparisons with other discretization algorithms are shown in section 5. Section 6 concludes the paper.

2 Related Works

So far, there have been a wide range of discretization methods. In paper [3], six separated dimensions were utilized to classify these discretization methods: unsupervised versus supervised, static versus dynamic, local versus global, splitting (top-down) versus merging (bottom-up), direct versus incremental, and univariate versus multivariate.

Due to failing to take into consideration the interdependency among the attributes to be discretized, univariate discretization algorithms seem difficult to obtain much better discretization scheme for a multi-attribute data set. The characteristics of dynamic methods constrain the generality of these discretization algorithms. Local discretization approaches don't fully explore the whole instance space so that there is still some room for improvement. Such an algorithm as merging methods is computation-consuming and probably unfeasible for high-dimensional and large-scale data sets. Supervised approaches require much more information, i.e. class labels, in comparison with unsupervised ones. Whereas for many applications such as information retrieval unlabeled instances are readily available while labeled ones are too expensive to achieve since labeling the instances requires substantial human involvements. In addition, in supervised approaches, leveraging the class labels to obtain the discretization scheme is likely to result in the overfitting or oversimplifying of the acquired knowledge.

Recently, many supervised discretization algorithms have been advocated. CAIM discretization algorithm is one of them [4]. The aim of the CAIM algorithm is to generate a possibly minimal number of data intervals and at the same time, to maximize the class-attribute interdependence. However, CAIM's overemphasis on the minimal number of achieved intervals and the class with the most instances might negatively impact the quality of the resulting discretization scheme in some data sets. As an enhanced version of CAIM, class-attribute contingency coefficient discretization algorithm (CACC) overcomes the above-mentioned drawbacks of CAIM to some extent [5].

In fact, as the paper [6] indicated, some simple unsupervised discretization methods, such as proportional discretization and fixed frequency discretization, can outperform a commonly-used supervised method (entropy minimization discretization) in the context of naïve-Bayes learning. Additionally, in order to better handle continuous attributes, some researchers employed fuzzy sets to represent interval events in the domains of continuous attributes, i.e. permitting continuous attributes to be located on the interval boundaries to partially belong to multiple intervals [7]. Recently, a new semi-supervised discretization method requiring very low informative prior on data was proposed to discretize the numerical domain of a continuous input variable, while keeping the information relative to the prediction of classes [8].

An unsupervised data discretization algorithm based on Principal Component Analysis (PCA) was presented in paper [9]. In order to discover the hidden/important relationships among attributes and ensure that the latent correlations were preserved, the algorithm first discretize principal components in orthogonal space with simple distance-based clustering, then map the cut-points in a principal component back to the corresponding attribute in the original data set with either K-NN method or direct projection method.

However, most current discretization algorithms do not take into consideration how to preserve the rank of attribute importance during discretization. As a matter of fact, attributes in a data set are not equally important. We can imagine what the final results will be when the most important attribute becomes negligible after it is discretized. Although what is important is likely to depend on the task at hand, intuitively, the closer the importance rank of the discretized attributes is to that of the original ones, the better the discretization results are, especially when other available information is not reliable enough to be leveraged in the discretization.

Thus, based on relative importance of different attributes in the attribute set measured by a PCA-based quantitative index, i.e., CCC [1], we proposed the GUDA-CCC algorithm with few predefined parameters to acquire the discretization scheme under the circumstances of preserving the rank of inherent attribute importance embedded in a data set and minimizing the discretization errors measured by the mean square errors. Furthermore, the algorithm requires no class label and deals with all attributes simultaneously, rather than one attribute at a time.

Although both the GUDA-CCC method proposed by this paper and the work in paper [9] use the same well-known PCA algorithm to capture the interdependences among all attributes of a data set, our algorithm does not need to discretize the principal components in Eigenspace and accordingly, is no need to reproject Eigen cut-points to the original dimensions. Rather, GUDA-CCC leverages the CCC values derived from PCA to define the attribute importance and keeps the rank of the attribute importance unchanged when directly discretizing all the continuous numerical attributes simultaneously.

3 Methodology

Before introducing the GUDA-CCC algorithm, we first define the formal terminology on discretization.

3.1 Formal Terminology Definition

In a data set comprising T instances, each instance comprises M attributes and a corresponding class label. Suppose all M attributes are numerical and continuous, for a continuous numerical attribute A_m ($m \in [1, 2, ..., M]$), $A_m = [a_{1m}, a_{2m}, ..., a_{km}, ..., a_{Tm}]$. A discretization scheme D is composed of a set of data intervals:

$$D = \{(c_1, c_2], (c_2, c_3], \cdots, (c_{n-1}, c_n], (c_n, c_{n+1})\} \qquad (1)$$

The definition of discretization is demonstrated in Fig. 1.

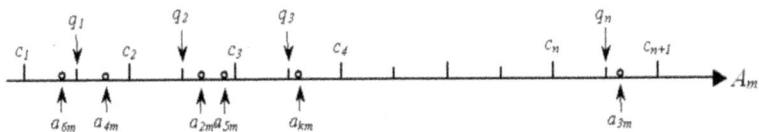

Fig. 1. A discretization illustration

In Figure 1, there are n data intervals (i.e. quantization levels), and $c_1, c_2, ..., c_{n+1}$ are the cut-points to position the corresponding intervals. Moreover, there is a numerical value (i.e. quantum) representing every interval. For example, q_2 and q_n represent $(c_2, c_3]$ and $(c_n, c_{n+1}]$, respectively. So, the discretized A_m (i.e. A_m^D) consists of $a_{km}^D \in \{q_1, q_2, \cdots q_n\}$ ($k=1, 2, ..., T$), i.e.,

$$a_{km}^D = q_i, \text{if } a_{km} \in (c_i, c_{i+1}], \ k=1, 2, ..., T; \ i=1, 2, ..., n \qquad (2)$$

Thus, after the discretization, $a_{6m}, a_{4m}, a_{2m}, a_{5m}$ and a_{3m} are corresponding to q_1, q_1, q_2, q_2 and q_n, respectively.

Note that, some discretization algorithms don't calculate the quanta: $q_1, q_2, \cdots q_n$, but merely utilize the nominal values to label and distinguish different data intervals. However, in our algorithm, the numerical values must be calculated to facilitate the subsequent processes.

In fact, the discretization process above can be considered as the quantization procedure. Quantization is the process of subdividing the range of a signal into non-overlapping regions (i.e. data intervals). An output level (i.e. quantum) is then assigned to represent a region. If the A_m can be seen as a signal, the quantization is to find a set of quanta $a_{km}^D \in \{q_1, q_2, \cdots q_n\}$ and a set of corresponding partition endpoints (i.e. cut-points) $\{c_1, c_2, \ldots, c_{n+1}\}$ that satisfy a certain optimization criterion, such as the minimal Mean-Square Error (MSE) distortion.

3.2 The Descriptions of GUDA-CCC Algorithm

In the GUDA-CCC algorithm, we emphasize to preserve the rank of the attribute importance in a data set during discretization. We now look at the issue how to define the attribute importance in an attribute set. Our previous work [1] offered a PCA-based quantitative index (i.e., CCC) method to measure relative importance of different attributes in a data set. The CCC is the weighted mean of the correlation between an original attribute and all principal components.

Intuitively, the fewer the discretization errors are, the better the discretization results are when all other conditions are the same. Paper [2] showed Lloyd-Max Scalar Quantization was able to approximate the continuous set of values with a finite set of discrete values that minimizes the mean square error. Hence, we attempt to capitalize on the Lloyd-Max Scalar Quantization to determine the values of cut points and quanta corresponding to each continuous attribute, given its probability density function and number of data intervals (i.e. quantization levels). In real-life applications, we usually assume or estimate the probability density function of continuous attributes in a data set with parametric estimation or non-parametric estimation methods. Here, the quanta are employed to calculate the CCC values of the discretized attributes.

Then, the next question is how to decide the number of data intervals under the circumstances of preserving the rank of the attribute importance in an attribute set during the discretization. We perceive that an optimization algorithm might be a good solution. In paper [10], Artificial Fish-Swarm Algorithm (AFSA) exhibits similar features to Genetic Algorithm, such as stochastic global optimization in high dimension data and requires no gradient information regarding the objective function. Moreover, in comparison with Genetic Algorithm, it has higher convergence speed, fewer parameters to be tuned, whereupon can be easily implemented. That's why the GUDA-CCC algorithm capitalizes on AFSA to globally optimize the numbers of data intervals for all continuous attributes in a data set. The optimization goal is to keep the rank of CCC values after discretization as close as possible to that before discretization.

The GUDA-CCC algorithm is structured as follows.

1. Before the discretization, calculate the CCCbefore values [1] of each numeric attribute in the original data set. Then, RANKbefore is obtained by decreasingly sorting attributes according to their corresponding CCCbefore values.
2. Search for the vector *L*, which consists of the number of quantization levels (i.e. data intervals) for each continuous numeric attribute in the attribute set, with the help of AFSA [10] to satisfy the condition that the *Rankafter* is close to the RANK*before*. Here, *Rankafter* is the rank of each discretized attribute by decreasingly sorting attributes based on their *CCCafter* values.
    a) Generate the candidates of the optimal *L* with the help of the AFSA[10];

b) For each individual numeric attribute, upon the number of quantization levels (the number of data intervals) is determined, the Lloyd-Max Quantization [2] can be invoked to decide the values of cut points and the quanta to satisfy the requirement that the discretization (quantization) errors are minimal, given the probability density function of each attribute A_m: $p(x)$. In general, attributes are assumed to have normal distribution.
c) With the obtained discretization scheme D on the data set (i.e. the number of data intervals, the values of cut points and the quanta corresponding to all attributes in the data set), a discretized data set can be produced.
d) Calculate the *CCCafter* values of each attribute in the discretized data set. *RANKafter* is obtained by sorting the attributes according to their *CCCafter* values accordingly.
e) Compute the fitness value with the fitness function $f(D)$ by comparing the *RANKafter* with the *RANKbefore*. Owing to limited space, detailed explanations on the fitness function has to be omitted.
f) Check whether or not the stop criteria are satisfied. Here, the criteria can be that maximal iteration times are reached or the obtained fitness value is bigger than a preset threshold or the variations of fitness values obtained in successive n steps are less than a preset threshold.
  i. If not, go back to step a);
  ii. Otherwise, output the current discretized data set and the current discretization scheme D. The end.

4 Key Components in GUDA-CCC

The sub-algorithms employed in GUDA-CCC are based on the computation method for generating CCC values [1], Artificial Fish-Swarm Algorithm (AFSA) [10] and Lloyd-Max Scalar Quantization [2]. In essence, the CCC value based on PCA is the weighted mean of the correlation between an original variable and all principal components. Paper [1] provides a detailed description of CCC.

4.1 Artificial Fish-Swarm Algorithm (AFSA) [10]

To build a simulated fish school, fish's behaviors were designed through three rules in paper [11]: 1) Avoid collisions with nearby flockmates; 2) Endeavor to match velocity with nearby flockmates so as to keep the same direction; 3) Attempt to stay close to nearby flockmates. Based on the three rules, the paper [10] designed AFSA to simulate synchronized group behavior of fish school, such as prey, follow and swarm motion. It shows similar features to Genetic Algorithm, including stochastic global optimization in high dimension problems, needless of gradient information regarding the objective function, higher convergence speed, less parameters to be tuned and simpler implementation. It is worth to mention that AFSA doesn't provide crossover and mutation operators.

The steps of AFSA detail as follows.

1) Initialize n artificial fish.
2) Set up a bulletin board to record the best fish and corresponding target value, i.e. fitness value;

3) Catch a fish, check whether or not the stop condition of the algorithm is satisfied. If so, stop; or else, do the following steps;
   a) Check whether or not the 'Follow' action can be executed. If yes, execute it; or else, 'Prey'. And then, compare the current results with the Bulletin Board to see which one is better. If the current results are better, update the Bulletin Board with them;
   b) Based on the current state, judge whether or not the 'Swarm' action can be executed. If so, execute it; or else, 'Prey'. And then, compare the current results with the Bulletin Board to see which one is better. If the current results are better, update the Bulletin Board with them;
4) Do step 3) until all n artificial fish have been visited;

The detailed steps of prey, follow and swarm action were shown in paper [11].

4.2 Lloyd-Max Scalar Quantization

Lloyd-Max Scalar Quantization [2] is to approximate the continuous set of values with a finite set of discrete values under the circumstances of minimizing the mean square error between the original continuous values and the discrete values. The input of a quantizer is the original data and the number of quantization levels (i.e. the data intervals), and the output is the locations of cut points and corresponding quanta.

Let a n-level quantizer be A^D, and a set of quanta be $\{q_1, q_2, ..., q_n\}$. When the sample x (input) falls in the range of c_i to c_{i+1}, the output level q_i (quantum) can be denoted as:

$$q_i = A^D(x) = A^D(c_i < x \leq c_{i+1}), \quad i = 1, 2, \cdots n \quad (3)$$

Intuitively, the quantization can be described as follows:

$$-\infty < c_1 < q_1 \leq c_2 < q_2 \leq \cdots \leq c_n < q_n < c_{n+1} < +\infty \quad (4)$$

Quantization error e can be defined as:

$$e = x - q_i = x - A^D(x) \quad (5)$$

Assume the probability density function of an attribute be $p(x)$, the mean square error σ_e^2 can be denoted as:

$$\sigma_e^2 = \sum_{i=1}^{n} \int_{c_i}^{c_{i+1}} (x - q_i)^2 p(x) dx \quad (6)$$

To get a minimal mean square error quantizer, we can get

$$\frac{\partial}{\partial c_i} [\int_{c_{i-1}}^{c_i} (x - q_{i-1})^2 p(x) dx + \int_{c_i}^{c_{i+1}} (x - q_i)^2 p(x) dx] = 0$$

$$\Rightarrow \quad c_{i,opt} = \frac{1}{2}(q_{k,opt} + q_{k-1,opt}) \quad i = 2, 3, \cdots, n \quad (7)$$

$$\frac{\partial}{\partial q_i}[\int_{c_k}^{c_{k+1}}(x-q_i)^2 p(x)dx]=0$$

$$\Rightarrow \quad q_{k,opt}=\frac{\int_{c_{k,opt}}^{c_{k+1,opt}} xp(x)dx}{\int_{c_{k,opt}}^{c_{k+1,opt}} p(x)dx}, \quad i=1,2,\cdots,n \tag{8}$$

Formula (7) and (8) show that the best boundary (i.e., cut-points) is at the midpoint of two adjacent output levels (quanta) and the best quantum is the centroid of zone lying between two adjacent boundaries, respectively.

5 Experimental Results

All experimental data sets are from the UCI repository (http://www.ics.uci.edu/~mlearn/MLRepository.html). Detailed descriptions of the data set are shown in Table 1. It is assumed that each continuous attribute of the data sets in Table 1 has normal distribution when applying GUDA-CCC algorithm. In real life, before the GUDA-CCC algorithm is applied, the probability density functions of continuous attributes can be estimated. Based on the given probability density functions, the quanta $q_{k,opt}$ and cut-points $c_{k,opt}$ can be accordingly obtained with the formula (7) and (8).

Table 1. Experimental data sets

Data set	Instances	Continuous attribute	classes
Bupa	345	6	2
Glass	214	9	6
Ionosphere	351	33	2
Iris	150	4	3
Musk	476	166	2
Musk2	6598	166	2
Pima Indians Diabetes	768	8	2
Vehicle	846	18	4
Waveform	5000	21	3
Wine	178	13	3

Here, GUDA-CCC algorithm is used as a preprocessing step for the classification method (C4.5 decision tree).

For comparison purposes, we apply four discretization methods, including two unsupervised approaches (equal-width and equal-frequency) and two supervised approaches (Fayyad & Irani's MDL criterion [12] and Kononenko's MDL criterion [13]) as preprocessing for the classification algorithms (C4.5). For the two unsupervised approaches, we use 10 as preset for the number of discretization intervals. Table 2 shows the predictive error rates using C4.5.

Table 2. Experimental results in error rates based on C4.5 classifier (the best results in bold)

Data set	Equal Width	Equal Frequency	Fayyad & Irani's MDL criterion	Kononenko's MDL criterion	Continuous (non-discretization)	GUDA-CCC
Bupa	40.35%	42.61%	42.09%	41.28%	37.39%	**35.23%**
Glass	41.03	44.67	32.15	31.68	32.06	**27.57**
Ionosphere	12.25	11.68	11.34	11.63	11.23	**9.02**
Iris	5.73	**5.47**	6.40	6.40	6.14	5.73
Musk	27.52	30.00	17.69	17.69	19.37	**17.27**
Musk2	4.30	5.75	4.26	4.24	3.49	**3.06**
Pima Indians Diabetes	25.86	26.17	25.34	25.86	25.55	**23.08**
Vehicle	31.35	33.12	32.13	31.49	**27.66**	28.82
Waveform	26.67	28.92	24.82	25.19	23.91	**22.10**
Wine	16.74	18.31	10.90	11.68	8.09	**8.08**

All data sets are divided into the training and test data by a 5-trial, 4-fold stratified sampling cross-validation test method. In each fold, the discretization is learned on the training data and the resulting bins are applied to the test data. All these experiments are implemented with Matlab and Weka [14].

From Table 2, we can see that C4.5 combined with the GUDA-CCC has the lowest error rates in Bupa, Glass, Ionosphere, Musk, Musk2, Pima Indians Diabetes, Waveform and Wine.

A one-sided t-statistic hypothesis test indicates that GUDA-CCC is significantly better than Equal width, Equal Frequency, Fayyand & Irani's MDL criterion, Kononenko's MDL criterion and Continuous (non-discretization) in most data sets: i.e. Bupa, Glass, Ionosphere, Musk2, Pima Indians Diabetes and Waveform. As for the data set Iris, the difference between GUDA-CCC and Equal Frequency is not significant though the error rate of Equal Frequency is lower than that of GUDA-CCC. When the data set Vehicle is processed, although the result of Continuous (non-discretization) is better than that of GUDA-CCC, the difference between them is not significant. Likewise, for the data sets: Musk and Wine, even though the error rates from GUDA-CCC are lower than those of Kononenko's MDL criterion and Continuous (non-discretization) respectively, the differences between them are not significant either.

The experimental results imply our algorithm could work much better on the data set with high dimensions, such as Musk2 with 166 continuous attributes and Ionosphere with 33 continuous attributes. The reason behind it could be that both the rank of attribute importance calibrated by CCC values and the discretization errors measured by the MSE are more suitable for data sets with more continuous attributes.

To sum up, Table 2 suggests that GUDA-CCC is superior to any other in these six algorithms in classification tasks. This strongly supports our hypothesis that when other available information is not reliable enough to be leveraged in the discretization, sticking closely to an original data set, just like honesty, might be the best policy. Moreover, preserving the rank of attribute importance together with minimizing discretization errors measured by the MSE seems to be an effective method to the end.

6 Conclusion

In this paper, we present a new data discretization algorithm which requires no class labels or few predefined parameters. The key idea behind the method is that to stick closely to an original data set by preserving the rank of attribute importance together with minimizing discretization errors measured by the MSE might be the best policy. As an important practical advantage, the algorithm does not require class labels, but can outperform some supervised methods. The algorithm can discretize all attributes simultaneously, rather than one attribute at a time, which improves the efficiency of unsupervised discretization. Experiments on benchmark data sets demonstrate that the GUDA-CCC could be a rewarding discretization approach to improve the quality of data mining results, especially when dealing with high dimension data sets.

Acknowledgments. This study was supported by NFS of Guangdong grant 06300252.

References

1. Zeng, A., Pan, D., Zheng, Q.L., Peng, H.: Knowledge Acquisition based on Rough Set Theory and Principal Component Analysis. IEEE Intelligent Systems 21, 78–85 (2006)
2. Lloyd, S.P.: Least Squares Quantization in PCM. IEEE Transactions on Information Theory 28(2), 129–137 (1982)
3. Liu, H., Hussain, F., Tan, C., Dash, M.: Discretization: An Enabling Technique. Data Mining and Knowledge Discovery 6(4), 393–423 (2002)
4. Kurgan, L.A., Cios, K.J.: CAIM Discretization Algorithm. IEEE Transactions on Knowledge and Data Engineering 16, 145–153 (2004)
5. Tsai, C.J., Lee, C.I., Yang, W.P.: A Discretization Algorithm based on Class-attribute Contingency Coefficient. Information Sciences 178, 714–731 (2008)
6. Yang, Y., Webb, G.I.: Discretization for Naïve-Bayes Learning: Managing Discretization Bias and Variance. Machine Learning 74, 39–74 (2009)
7. Au, W.H., Chan, K.C.C., Wong, A.K.C.: A Fuzzy Approach to Partitioning Continuous Attributes for Classification. IEEE Transactions on Knowledge and Data Engineering 18, 715–719 (2006)
8. Bondu, A., Boulle, M., Lemaire, V., Loiseru, S., Duval, B.: A Non-parametric Semi-supervised Discretization Method. In: Proceedings of 2008 Eighth International Conference on Data Mining, pp. 53–62 (2008)
9. Mehta, S., Parthasarathy, S., Yang, H.: Toward Unsupervised Correlation Preserving Discretization. IEEE Transactions on Knowledge and Data Engineering 17, 1174–1185 (2005)
10. Li, X.L., Shao, Z.J.: An Optimizing Method base on Autonomous Animals: Fish-Swarm Algorithm. Systems Engineering-Theory & Practice 11, 32–38 (2002) (in Chinese)
11. Reynolds, C.W.: Flocks, Herds, and Schools: a Distributed Behavioral Model. Computer Graphics 21, 25–34 (1987)
12. Fayyad, U.M., Irani, K.B.: On the Handling of Continuous-Valued Attributes in Decision Tree Generation. Machine Learning 8, 87–102 (1992)
13. Kononenko, I.: On Biases in Estimating Multi-Valued Attributes. In: Proceedings of 14th International Joint Conference on Artificial Intelligence, pp. 1034–1040 (1995)
14. Weka, http://www.cs.waikato.ac.nz/ml/weka/

A Heuristic Data-Sanitization Approach Based on TF-IDF

Tzung-Pei Hong[1,4], Chun-Wei Lin[1,3], Kuo-Tung Yang[1], and Shyue-Liang Wang[2]

[1] Department of Computer Science and Information Engineering
[2] Department of Information Management
[3] General Education Center
National University of Kaohsiung, Kaohsiung, Taiwan
[4] Department of Computer Science and Engineering
National Sun Yat-sen University, Kaohsiung, Taiwan
{tphong,cwlin,kdyang,slwang}@nuk.edu.tw

Abstract. Data mining technology can help extract useful knowledge from large data sets. The process of data collection and data dissemination may, however, result in an inherent risk of privacy threats. In this paper, the SIF-IDF algorithm is proposed to modify original databases in order to hide sensitive itemsets. It is a greedy approach based on the concept of the Term Frequency and Inverse Document Frequency (TF-IDF) borrowed from text mining. Experimental results also show the performance of the proposed approach.

Keywords: Privacy preserving, data mining, data sanitization, TF-IDF.

1 Introduction

In recent years, the privacy-preserving data mining (PPDM) has become an important issue due to the quick proliferation of electronic data in governments, corporations and non-profit organizations. Verykios *et al.* [10] thus proposed a data sanitization process to hide sensitive knowledge by item addition or deletion. Zhu *et al.* [11] then discussed what kind of public information type was suitable for not revealing sensitive data and insinuated that the k-anonymity technique might still have security problems.

In text mining, the technique of term frequency–inverse document frequency (TF-IDF) [9] is usually used to evaluate how relevant a word in a corpus is to a document. It may be thought of as a statistical measure. In this paper, a novel greedy-based approach called sensitive items frequency - inverse database frequency (SIF-IDF) algorithm is thus designed to modify the TF-IDF [9] approach. It evaluates the degrees of transactions associated with given sensitive itemsets, reducing the frequencies of sensitive itemsets for data sanitization. Based on the SIF-IDF algorithm, the user-specific sensitive itmesets can be completely hidden with reduced side effects. Experimental results are also used to evaluate the performance of the proposed approach.

2 Review of Related Works

In this section, we respectively review data mining process and general concept of data sanitization.

2.1 Data Mining Process

Data mining is most commonly used in attempts to induce association rules from transaction data, such that the presence of certain items in a transaction will imply the presence of some other items. To achieve this purpose, Agrawal et al. proposed several mining algorithms based on the concept of large itemsets to find association rules in transaction data [3-4]. In the first phase, candidate itemsets were generated and counted by scanning the transaction data. If the count of an itemset appearing in the transactions was larger than a pre-defined threshold value (called the minimum support), the itemset was considered a large itemset. Large itemsets containing only single items were then combined to form candidate itemsets containing two items. This process was repeated until all large itemset had been found. In the second phase, association rules were induced from the large itemsets found in the first phase. All possible association combinations for each large itemset were formed, and those with calculated confidence values larger than a predefined threshold (called the minimum confidence) were output as association rules.

2.2 Data Sanitization

Years of effort in data mining have produced a variety of efficient techniques, which have also caused the problems of security and privacy threats [7]. In the past, Atallah et al. first proposed the protection algorithm for data sanitization to avoid the inference of association rules [2]. Dasseni et al. then proposed a hiding algorithm based on the hamming-distance approach to reduce the confidence or support values of association rules [5]. Amiri then proposed three heuristic approaches to hide multiple sensitive rules [1]. Pontikakis et al. [8] then proposed two heuristics approaches based on data distortion.

The optimal sanitization of databases is, in general, regard as an NP-hard problem. Atallah et al. [2] proved that selecting which data to modify or sanitize was also NP-hard. Their proof was based on the reduction from the NP-hard problem of hitting-sets [6].

3 The Proposed Heuristic Approach

In this paper, a greedy-based approach called *sensitive items frequency - inverse database frequency* (*SIF-IDF*) is proposed to hide given sensitive itemsets. It uses and modifies the concept of TF-IDF [9] in text mining to evaluate the degrees of transactions associated with given sensitive itemsets. The measure for the *SIF-IDF* value of a transaction T_i is defined as follows:

$$SIF-IDF(T_i) = \sum_{j=1}^{n}\left(\frac{|si_{ij}|}{|T_i|} \times \sum_{k=1}^{p} \log\frac{|n|}{|f_k - MRC_k|}\right),$$

where $|si_{ij}|$ is the number of sensitive items contained in the j-th sensitive itemset in T_i, and $|T_i|$ is the number of items in transaction T_i, $|n|$ is the number of records in a database, $|f_k|$ is the frequency count of each item, and $|MRC_k|$ is the maximum reduced count of each item.

The *SIF-IDF* value of each transaction is calculated and is used to measure whether a transaction has a large number of sensitive items but with less influence to other transactions. The transactions with high *SIF-IDF* values are considered to be processed with high probabilities for sanitization. The transactions are sorted in a descending order of their *SIF-IDF* values. The order is used as the processing order of the transactions for the proposed algorithm. In data sanitization, an item with a higher occurrence frequency in the sensitive itemsets may be considered to have a larger influence than the ones with a lower occurrence frequency. The sensitive items in the processed transactions are then deleted according to the ordering of their occurrence frequencies. This procedure is repeated until the set of sensitive itemsets becomes *null*, which indicates all the supports of the sensitive itemsets are under the user-specific minimum support threshold. The proposed algorithm is the described as follows.

The proposed SIF-IDF algorithm:
INPUT: A transaction dataset $D = \{T_1, T_2, \ldots, T_i, \ldots, T_n\}$ with a set of p items $I = \{i_1, i_2, \ldots, i_k, \ldots, i_p\}$, a user-specific minimum support threshold s, and a set of m user-specific sensitive itemsets $S = \{si_1, si_2, \ldots, si_j, \ldots, si_m\}$.
OUTPUT: A sanitized database with no sensitive rules mined out.
STEP 1: Find the transactions with sensitive itemsets in the database D.
STEP 2: Calculate the *sensitive items frequency* (SIF_{ij}) value of each sensitive itemset si_j in each transaction T_i as:

$$SIF_{ij} = \frac{|si_{ij}|}{|T_i|},$$

where $|si_{ij}|$ is the number of sensitive items in T_i which appears in si_j, and $|T_i|$ is the number of items in T_i.
STEP 3: Calculate the value of the *inverse database frequency* (*IDF*) of each sensitive itemset in each transaction by the following substeps.

Substep 3-1: Calculate the reduced count value (RC_{kj}) of each item i_k for each sensitive itemset si_j as $f_j - s*n + 1$ if si_j includes i_k and as 0 otherwise, where f_j is the occurrence frequency of the sensitive itemset si_j in the database, s is the minimum support threshold, and n is the number of transactions in the database, $1 \leq j \leq m, 1 \leq k \leq p$.

Substep 3-2: Calculate the maximum reduced count value (MRC_k) of each item i_k as:

$$MRC_k = \max_{j=1}^{m} RC_{kj}.$$

Substep 3-3: Calculate the inverse database frequency (IDF_k) value of each items i_k as follows:

$$IDF_k = \log \frac{|n|}{|f_k - MRC_k|},$$

where f_k is the occurrence count of item i_k in the database.

Substep 3-4: Sum the *IDF* values of all sensitive items within sensitive itemsets and calculate the *SIF-IDF* value for each transaction as follows:

$$SIF - IDF(T_i) = \sum_{i=1}^{n} \left(\frac{|si_{ij}|}{|T_i|} \times \sum_{k=1}^{p} \log \frac{|n|}{|f_k - MRC_k|} \right).$$

STEP 4: Find the transaction (T_b) which has the best *SIF-IDF* value.
STEP 5: Process the transaction T_b to prune appropriate items by the following substeps.
  Substep 5-1: Sort the items in a descending order of their occurrence frequencies within the sensitive itemsets.
  Substrp 5-2: Find the first sensitive item ($item_o$) in T according to the sorted order obtained in Substep 5-1.
  Substep 5-3: Delete the item ($item_o$) from the transaction.
STEP 6: Update the occurrence frequencies of the sensitive itemsets.
STEP 7: Repeat STEPS 2 to 6 until the set of sensitive itemsets is ***null***, which indicates that the supports of all the sensitive itemsets are below the user-specific minimum support threshold *s*.

4 An Example

In this section, an example is given to demonstrate the proposed *sensitive items frequency - inverse database frequency* (*SIF-IDF*) algorithm for privacy preserving data mining (PPDM). Assume a database shown in Table 1 is used as the example. It consists of 10 transactions and 9 items, denoted *a* to *i*.

Table 1. A database example with 10 transactions

TID	Item
1	a, b, c, d, f, g, h
2	a, b, d, e
3	b, c, d, f, g, h
4	a, b, c, f, h
5	c, d, e, g, i
6	a, c, f, i
7	b, c, d, e, f, g
8	c, d, f, h, i
9	a, d, e, f, i
10	a, c, e, f, h

Assume the set of user-specific sensitive itemsets S is $\{cfh, af, c\}$. Also assume the user-specific minimum support threshold is set at 40%, which indicates that the minimum count is 0.4*10, which is 4. The proposed approach proceeds as follows to hide the sensitive Itemsets for avoiding being mined from the database.

STEPs 1 & 2: The transactions with sensitive itemsets in the database are found and kept. The *sensitive items frequency* (*SIF*) value of each sensitive itemset in each transaction is calculated. The results are shown in Table 2.

Table 2. The *SIF* values of each sensitive itemset in each transaction

TID	Item	SIF_{cfh}	SIF_{af}	SIF_c
1	a, b, c, d, f, g, h	3/7	2/7	1/7
2	a, b, d, e	0/4	1/4	0/4
3	b, c, d, f, g, h	3/6	1/6	1/6
4	a, b, c, f, h	3/5	2/5	1/5
5	c, d, e, g, i	1/5	0/5	1/5
6	a, c, f, i	2/4	2/4	1/4
7	b, c, d, e, f, g	2/6	1/6	1/6
8	c, d, f, h, i	3/5	1/5	1/5
9	a, d, e, f, i	1/5	1/5	0/5
10	a, c, e, f, h	3/5	2/5	1/5

STEP 3: The *inverse database frequency* (*IDF*) value of each sensitive itemset in each transaction is calculated. In this step, the *reduced count* (*RC*) of each item for each sensitive itemset is first calculated and the maximum of the *RC* values of each item is found as the *MRC* value. The *IDF* value of each item is then calculated. The *MRC* and *IDF* values of all items are shown in Table 3.

Table 3. The *IDF* value of each item

Item	Count	MRC	IDF
a	6	2	0.398
b	5	0	0.301
c	8	5	0.523
d	7	0	0.155
e	5	0	0.301
f	8	2	0.222
h	5	2	0.523
i	4	0	0.398

After that, the *SIF-IDF* value of a sensitive itemset in each transaction is then calculated as the *SIF* value of the sensitive itemset multiplied by its *IDF* value in the transaction. After that, the results are shown in Table 4.

Table 4. The *SIF-IDF* values for all the transactions

TID	SIF_1	IDF_1	SIF_2	IDF_2	SIF_3	IDF_3	*SIF-IDF*
1	3/7	1.268	2/7	0.62	1/7	0.523	0.795
2	0/4	0	1/4	0.398	0/4	0	0.099
3	3/6	1.268	1/6	0.222	1/6	0.523	0.758
4	3/5	1.268	2/5	0.62	1/5	0.523	1.113
5	1/5	0.523	0/5	0	1/5	0.523	0.209
6	2/4	0.745	2/4	0.62	1/4	0.523	0.813
7	2/6	0.523	1/6	0.222	1/6	0.523	0.299
8	3/5	1.268	1/5	0.222	1/5	0.523	0.901
9	1/5	0.222	1/5	0.62	0/5	0	0.168
10	3/5	1.268	2/5	0.62	1/5	0.523	1.113

STEPs 4 & 5: The transactions in Table 4 are sorted in the descending order of their *SIF-IDF* values. The transactions are processed in the above descending order to prune appropriate items. In this example, the item *c* in transaction 4 is then first selected to be deleted.

STEP 6: After item *c* is deleted from the fourth transaction, the new occurrence frequencies of the sensitive itemsets in the transactions are updated. The sensitive itemsets with their occurrence frequencies are then updated from {*cfh*:5, *af*:5, *c*:8} to {*cfh*:4, *af*:5, *c*:7}.

STEP 7: STEPs 2 to 6 are then repeated until the supports of all the sensitive itemsets are below the minimum count. The results of the final sanitized database in the example are shown in Table 5.

Table 5. The result of the final sanitized database in the example

TID	Item
1	a, b, d, f, g, h
2	a, b, d, e
3	b, c, d, g, h
4	a, b, h
5	c, d, e, g, i
6	a, f, i
7	b, c, d, e, f, g
8	d, f, h, i
9	a, d, e, f, i
10	a, e, h

5 Experimental Results

Two datasets called BMSPOS and Webview-1 are respectively used to evaluate the performance of the proposed algorithm. The numbers of sensitive itemsets for

BMSPOS and WebView_1 were set at 4, 4, and 8, respectively. For the proposed SIF-IDF algorithm, the relationships between the numbers of iterations and the *EC* values are compared to indicate the proposed algorithm could completely hide all user-specific sensitive itemsets. The results for BMSPOS and WebView_1 are shown in Figures 1 and 2, respectively.

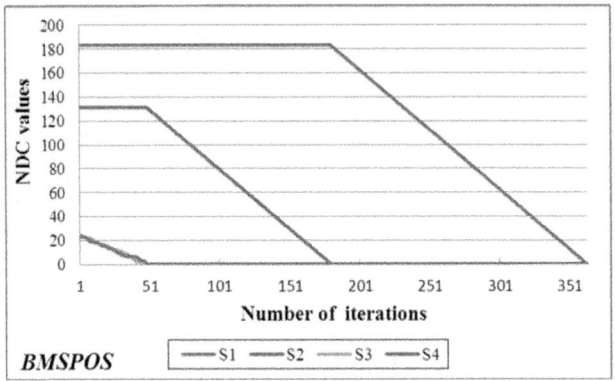

Fig. 1. The relationships between the *EC* value and the number of iterations in the BMSPOS database

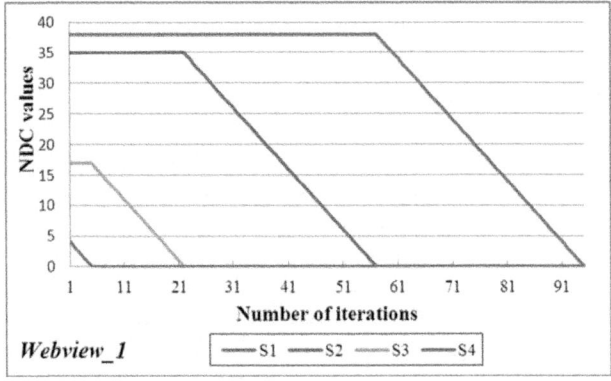

Fig. 2. The relationships between the *EC* values and he number of iterations in the Webview_1 database

From Figure 1, it can be seen that the sensitive itmesets *S3* and *S4* were alternatively processed to be hidden. From Figure 2, the sensitive itemsets *S4*, *S3*, *S2* and *S1* are sequentially processed to be hidden. The execution time to hide the four sensitive itemsets for two databases was also compared and shown in Figure 3. Note that the WebView_1 database required more execution time to hide the sensitive itemsets than the BMSPOS dataset.

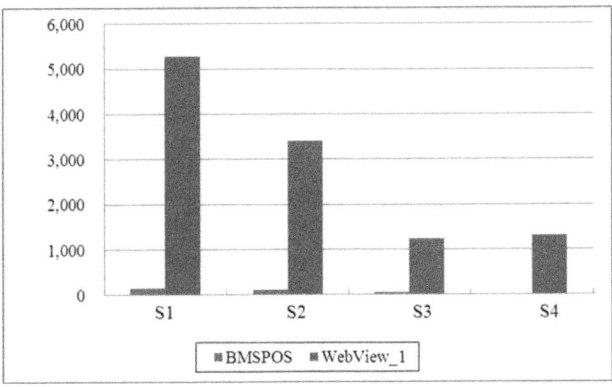

Fig. 3. The execution time to hide the four sensitive itemsets in the two databases

6 Conclusion

In this paper, the SIF-IDF algorithm is proposed to evaluate the similarity between sensitive itemsets and transactions for minimizing the side effects, which inherits the properties from TF-IDF algorithm in information retrieval. Based on the user-specific sensitive itemsets in the experiments, the proposed SIF-IDF algorithm can process all defined sensitive itemsets without any side effects in three databases. In the experimental results, the proposed algorithm has a good performance without any side effect, which is efficient and effective to hide the sensitive itemsets.

References

1. Amiri A.: Dare to share: Protecting sensitive knowledge with data sanitization. Decision Support Systems, 181–191 (2007)
2. Atallah, M., Bertino, E., Elmagarmid, A., Ibrahim, M., Verykios, V.S.: Disclosure limitation of sensitive rules. In: Knowledge and Data Engineering Exchange Workshop, pp. 45–52 (1999)
3. Agrawal, R., Imielinski, T., Sawmi, A.: Mining association rules between sets of items in large databases. In: The ACM SIGMOD Conference on Management of Data, pp. 207–216 (1993)
4. Agrawal, R., Srikant, R.: Fast algorithm for mining association rules. In: The International Conference on Very Large Data Bases, pp. 487–499 (1994)
5. Dasseni, E., Verykios, V.S., Elmagarmid, A.K., Bertino, E.: Hiding Association Rules by Using Confidence and Support. In: The 4th International Workshop on Information Hiding, pp. 369–383 (2001)
6. Garey, M.R., Johnson, D.S.: Computers and Intractability: A Guide to the Theory of NP-Conpleteness. W. H. Freeman, New York (1979)
7. Leary D. E. O.: Knowledge Discovery as a Threat to Database Security. Knowledge Discovery in Databases, 507–516 (1991)

8. Pontikakis, E.D., Tsitsonis, A.A., Verykios, V.S.: An experimental study of distortion-based techniques for association rule hiding. In: 18th Conference on Database Security, pp. 325–339 (2004)
9. Salton, G., Fox, E.A., Wu, H.: Extended Boolean information retrieval. Communications of the ACM 26(2), 1022–1036 (1983)
10. Verykios, V.S., Elmagarmid, A., Bertino, E., Saygin, Y., Dasseni, E.: Association Rule Hiding. IEEE Transactions on knowledge and Data Engineering 16(4), 434–447 (2004)
11. Zhu, Z., Wang, G., Du, W.: Deriving Private Information from Association Rule Mining Results. In: IEEE International Conference on Data Engineering, pp. 18–29 (2009)

Discovering Patterns for Prognostics: A Case Study in Prognostics of Train Wheels

Chunsheng Yang and Sylvain Létourneau

Institute for Information Technology, National Research Council Canada
1200 Montreal Road, Ottawa, Ontario K1A 0R6, Canada
{Chunsheng.Yang,Sylvain.Letourneau}@nrc.gc.ca

Abstract. Data-driven prognostic for system health management represents an emerging and challenging application of data mining. The objective is to develop data-driven prognostic models to predict the likelihood of a component failure and estimate the remaining useful lifetime. Many models developed using techniques from data mining and machine learning can detect the precursors of a failure but sometimes fail to precisely predict time to failure. This paper attempts to address this problem by proposing a novel approach to find reliable patterns for prognostics. A reliable pattern can predict state transitions from current situation to upcoming failures and therefore help better estimate the time to failure. Using techniques from data mining and time-series analysis, we developed a KDD methodology for discovering reliable patterns from multi-stream time-series databases. The techniques have been applied to a real-world application: train prognostics. This paper reports the developed methodology along with preliminary results obtained on prognostics of wheel failures on train.

Keywords: Data Mining, Time-Series, Reliable Patterns, Utility, Prognostics.

1 Introduction

Prognostic(s) has recently attracted much attention from researchers in different areas such as sensor, reliability engineering, machine learning/data mining, and machinery maintenance. It is defined as the detection of precursors of a failure and accurate prediction of remaining useful life (RUL) or time-to-failure (TTF) before a failure [1], and it is expected to provide a paradigm shift from traditional reactive-based maintenance to proactive-based maintenance. The final aim is to perform the necessary maintenance actions "just-in-time" in order to minimize costs and increase availability.

Many researchers have focused on developing knowledge-based prognostic [2-6]. With this approach, the prognostic models are built using physics and materials knowledge [7]. Data-driven prognostic is an alternative approach in which the models are developed using huge amounts of historical data and techniques from data mining [8-9]. The data-driven models can predict the likelihood of a component failure and are applicable to various applications such as aircraft maintenance [8] and train wheel failure predictions [9]. However, they may sometimes fail to precisely predict the remaining useful life or time-to-failure due to limitations of the techniques used with respect to important variations in the data. To address this issue, we propose to focus on the identification of reliable patterns for prognostics. A reliable pattern identifies the evolution of states such as degradation of the physical component and provides an onset for TTF estimation. In this paper, we propose a novel KDD methodology that uses historic operation and sensor data to automatically identify such patterns. After describing the methodology, we present results from the application on a real-world problem: train wheel failure prognostics.

The rest of the paper is organized as follows: Section 2 introduces the KDD methodology; Section 3 presents the application domain: train wheel prognostics; Section 4 presents the experiments and some preliminary results; Section 5 discusses the results and limitations; and the final section concludes the paper.

2 Methodology

In modern operation of complex electro-mechanical systems, a given component or subsystem generates operational or sensor data periodically. When we combine all sensor data for a given component, we obtain multivariate time-series that characterizes the evolution of that component from its installation to its removal. Generalizing to a fleet of vehicles, we obtain a series of multivariate time-series, each characterizing an individual component. For example, the monitoring of train wheels can generate thousands of multivariate time-series (Section 3). Each of these time-series, noted as s_i, is associated with a unique component *ID*, noted *SequenceID*. We denote an instance within a time-series as x_{ij}, referring the jth instance of the ith time-series in the multivariate time-series. An instance (x_{ij}) is represented by a set of attributes ($\vec{a} \in \{a_1, a_2, ... a_n\}$) and is associated with a timestamp (t_j). Therefore, an instance can be noted as $x_{ij} = \{a_1, a_2, ... a_m, ..t_j\}$ and a time-series as $s_i = \{x_{i1}, x_{i2}, ... x_{ij} \cdots\}$.

The operational database for a complex system, noted as *DS*, consists of a set of multivariate time-series, i.e., $DS \in \{s_1, s_2, s_3 ... s_i ... s_n\}$. A pattern, p_k, is defined as a combination of multiple observations. Its length, m, is the number of observations in a given time period starting from the kth observation. As a result, p_k is denoted as $p_k = \{x_{ik}, x_{ik+1}, ... x_{ik+m}\}$, i.e., $p_k = \{(\vec{a}_{ik}, t_k), (\vec{a}_{ik+1}, t_{k+1})...(\vec{a}_{ik+m}, t_{k+m})\}$. Here, t_k is the TTF estimate associated with pattern p_k.

The challenge is to find the reliable patterns (p_k) from a given dataset ($DS \in \{s_1, s_2, s_3 ... s_i ... s_n\}$). A reliable pattern can be expressed using numerical values or symbols representing various states. In this paper, we focus on symbolic patterns as they allow us to represent the evolution of states in an eay to understand manner. For example, a symbolic pattern "*DEF*" contains three statuses along with information on transitions that happened. It shows that a monitored component changes from status "*D*" to "*E*" to "*F*". To find such symbolic patterns, we developed the KDD methodology shown in Table 1. The methodology consists of three main processes: (1) Symbolizing time-series data to generate a symbolic dataset

Table 1. The KDD methodology for pattern discovery

Input: A given time-series dataset $DS = \{s_1, s_2, s_3...s_i...s_n\}$
and a given attribute vector $\vec{a} = \{a_1, a_2...a_i...\}$;
Output: Ranked patterns ($p_1, p_2,...p_k$)
Process: {
   CS = creatingSymbolicSequence(DS); // Subsection 2.1, $CS = \{c_1, c_2, c_3...c_i...c_n\}$
   Split CS into train (S) and test (T) datasets;
   P = findingPatterns(S);   // Subsection 2.2, P is a set of patterns
   P' = evaluatingPatterns(P, T); // Subsection 2.3, P' is a set of ranked patterns
}

(*creatingSymbolicSequence()*); (2) Discovering a set of patterns from symbolic sequences (*findingPatterns()*); (3) Evaluating the patterns to select the reliable patterns (*evaluatingPatterns()*).

2.1 Symbolizing Time-Series Data

Symbolizing data means re-representing the initial numerical time-series into sequences of symbols. Two steps are involved in this process: reducing the dimensionality and converting the dimensionality-reduced data into symbolic sequences.

A. Dimensionality reduction

The dimensionality reduction process compresses all instances within a given window size into a new observation. Several techniques such as Fourier transformations, wavelets, and Piecewise Aggregate Approximation (PAA) [10-11] could be used for this task. In this work, we opted to use PAA as it could be directly applied without further processing of the real world equipment data available for this project. Most of the other techniques would have required additional pre-processing to account for data issues such as missing values and irregular data sampling. Following PAA, we convert each time-series of length (m_i) into a new time-series of length n_i ($n_i < m_i$). Precisely, the initial sequence $s_i = \{x_{i1}, x_{i2},...,x_{ij},...(j=1,2,...,m_i)\}$ is converted into a new dimensionality-reduced sequence $\hat{s}_i = \{\hat{x}_{i1}, \hat{x}_{i2},...\hat{x}_{ij},...(j=1,2,...,n_i)\}$ by using the following equation:

$$\hat{x}_{ij} = \frac{n_i}{m_i} \sum_{l=k\bullet j-k}^{k\bullet j} x_{il} \tag{1}$$

where,

    $\hat{x}_{ij}$:   the jth observation in $\hat{s}_i$;

    x_{il} :   the lth original observation in s_i ;

  k :   the window size for dimensionality reduction;

m_i: the length of s_i; and

n_i: the length of $\hat{s}_i$.

B. Symbolic sequence creation

This second step transforms the dimensionality-reduced sequences into symbolic sequences such that each symbol in the sequence represents a unique system or component state. The full pattern for a sequence clearly shows the system state transitions. As part of the dimensionality reduction process, the original time-series data is usually normalized to a standardized distribution. Since normalized data follows a Gaussian distribution, we can easily set a group of breakpoints and assign a unique symbol to each area defined by these breakpoints. In a Gaussian distribution, the area under the distribution curve is divided into N areas, each represented with a unique symbol or letter. In this work, we divide the area into N unequal areas because we have to pay more attention to the observations which are close to failure events. In other words, we increase granularity as we get closer to the failure events in order to increase precision in that critical zone of the time-series. In practice, the breakpoints are determined based on the requirements of the application at-hand. Figure 1 illustrates the symbolization process for a given distribution. In this example, the dimensionality-reduced data, which is transformed from a time-series of 42 observations using a window size of 3 (i.e., by combining 3 original observations into a new one), contains 14 new observations. The application requires unequal areas and 8 symbols were deemed adequate. In this case, the seven breakpoints required were set as: (1) -1.24, (2) -1.19, (3) -1.03, (4) -0.27, (5) 0.66, (6) 1.25, (7) 2.40. Using these breakpoints, we map the data to corresponding symbols. The 8 symbols used are A, B, C, D, E, F, G and H. The data below -1.24 is mapped to the symbol "A", and all data which are greater than -1.24 and less than -1.19 is mapped to the symbol "B", etc. Finally, all 14 observations are symbolized as a sequence with 14 symbols: "$AABCDDDDEEEFGH$".

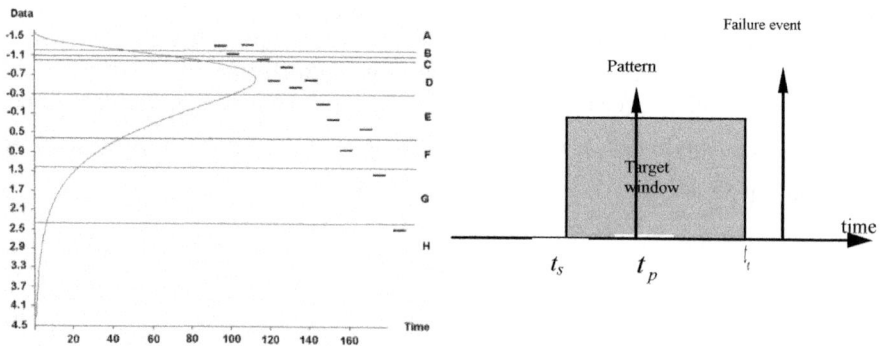

Fig. 1. The breakpoints for symbolization   **Fig. 2.** The target window for patterns

2.2 Searching Patterns from Symbolic Sequences

Having obtained a symbolic sequence dataset (*CS*), we can initiate the search for patterns. For this purpose, we divide *CS* into a training dataset (*S*) and a testing dataset (*T*). To search patterns from *S* for a given pattern size, a Full Space Search (FSS) algorithm is developed. FSS consists of the following 5 steps:

- *Step 1:* parse a symbolic sequence into a list of strings (potential patterns) according to the given pattern length. For example, when the pattern length is 3, the symbolic sequence, "*AABCDDDDEEEFGH*", is parsed into a list of 3-symbol strings, i.e., *AAB, ABC, BCD, CDD, DDD, DDD, DDE, DEE, EEE, EEF, EFG,* and *FGH*.
- *Step 2:* remove the duplicated strings from the list. In the above list, the string "*DDD*" is duplicated, only one of them would be kept.
- *Step 3:* calculate the frequencies for all potential patterns over all sequences in the training dataset. For example, if there are 100 symbolic sequences in the training dataset and a string appears in 75 sequences, then its frequency is 0.75 or 75%.
- *Step 4:* select a number of potential patterns based on the frequency. We select patterns having highest frequency in the list of strings. For example, if an application needs 3 patterns, we select top 3 strings as the potential patterns.
- *Step 5:* decide the onset of the time for the selected patterns. We take the onset of the original data in a pattern as TTF. This is a reverse engineering process from dimensionality reduction. This onset is used as TTF estimation for prognostics.

2.3 Evaluating Patterns

Pattern evaluation is similar to model evaluation in the development of predictive models for prognostic applications. From the processes above, it is possible to find a large number of patterns by varying the related parameters such as the breakpoints, the window size of the dimensionality reduction, and the pattern length. The discovered patterns must be evaluated before applying them to prognostics. In this work, pattern evaluation is to verify if the patterns discovered from the training dataset are reliable for unseen dataset. For prognostics, it is desirable that a pattern should cover the failure events (or appear in those time-series) as often as possible and appear in the right onset. Existing pattern evaluation methods such as pattern ordering [12] and pattern extracting [13] are not adequate for prognostic applications. The main reason is that those evaluation methods do not take TTF estimations and problem coverage into account. Therefore, we proposed a utility-based method for pattern evaluation. Ideally, a reliable pattern should cover all failure events and appear in the target time window such as shown in Figure 2. This window, which is pre-determined based on application requirements, corresponds to the critical time interval during which the prognostic models are required to perform as accurately as possible.

We define the utility for a pattern as a function of the time window and TTF. As shown in Figure 2, the target window is determined by a start time (t_s) and an end time (t_e). Using the window size (w), the utility (u) for a pattern is computed by Equation 2.

$$u = \begin{cases} w + t_p & \text{if } t_p \in [t_s, t_e] \\ t_p & \text{if } t_p \notin [t_s, t_e] \end{cases} \quad (2)$$

where,

w: window size ($w = t_e - t_s$);

t_p: TTF estimation of a pattern.

and the total utility for a set of patterns is computed as follows.

$$U = \sum_{i=1}^{k} \sum_{j=1}^{N_i} (\alpha_i \bullet u_{ij}) \quad (3)$$

where,

k: the number of reliable patterns;
N_i: the number of failure events where pattern i appears;
u_{ij}: the utility of pattern i in the jth time-series;
α_i: the confidence of pattern i based-on pattern frequency.

Using Equations 2 and 3, we evaluate the reliability of patterns by computing a total utility on the testing dataset (T). The larger the total utility is, the higher the patterns reliability is and the better the pattern performance is.

3 The Application Domain: Train Wheel Prognostics

Train wheel failures are the cause of half of all train derailments and cost the global rail industry billions of dollars each year. Wheel failures also contribute significantly to rail deterioration, ultimately resulting in broken rails. These breaks are dangerous and very expensive to repair. The risk of wheel failures is increasing as global competitiveness pushes railways to use larger and heavier cars. To prevent failures, thereby avoiding catastrophic events such as derailments, railways now closely monitor the wheels by using "Wheel Impact Load Detector" (WILD) systems. These monitoring systems, installed at strategic locations on the rail network, measure the dynamic impact of each wheel. When the measured impact exceeds a fixed threshold (e.g., 140 kips) the train must immediately reduce its speed and then stop at the nearest siding. The offending wheel must be replaced before the car that it belongs to can be used again. Although this method reduces some of the costs associated with a wheel failure, such as those due to derailment and rail damage, it increased other costs. Replacing a wheel that has exceeded the threshold can be a costly event, particularly if the siding is remotely situated.

To reduce these extra costs, railways are looking for a new approach for train prognostics by developing data mining models for prognostics. The objective is to develop models from WILD data and apply them to a real-time monitoring system for prognostics of train wheels [9]. We now propose to evaluate the techniques presented above to WILD data in order to see if they could be used for prognostics of train wheel failures.

For this study, we used data collected over a period of 17 months from a fleet of 804 large cars with 12 axles each. After data pre-processing, we ended up with a dataset containing 2,409,696 instances grouped in 9906 time-series (one time-series for each axle in operation during the study). Out of the 9906 time-series, only 218 are associated with an occurrence of a wheel failure. Therefore, we selected these 218 time-series for pattern discovery and pattern evaluation. We divided these 218 time series into a training dataset (106 time-series) and a testing dataset (112 time-series).

4 Experiments and Results

This section details the application of the proposed method as well as preliminary results for the prognostics of wheel failures on trains. We first describe the parameter settings; we then present the experimentation process and discuss the results.

4.1 Parameter Settings

Conducting experiments to find the patterns following the methodology involves many parameters or factors, which have great impact on the results. There is yet no guideline on how to predetermine these parameters so we need to proceed empirically. In this section, we describe the settings of these parameters as well as the process leading to the chosen values.

- *The dimensionality reduction size (D#)*

This parameter determines how many original observations are to be combined into a new observation during the PAA process. It depends fully on application requirements and characteristics of the available time-series, including the sampling rate, noise, outlier, and volume of data. For instance, if the sampling rate is high, a larger size should be used for dimensionality reduction. After a pilot experiment, we found 1, 3 and 4 are good numbers for dimensionality reduction for the given application ("1" means there is no dimensionality reduction in experiments), i.e., $D\#$ = 1, 3, or 4.

- *The pattern length (P#)*

The pattern length parameter must be chosen so that it represents an adequate period of time in terms of state transitions relevant to the component of interest. The longer the pattern is, the wider the time span. In cases where state transitions or component deterioration takes a long time, a rather long pattern length would be required. On the other hand, a long pattern is likely to degrade the precision of TTF predictions; it is also difficult to find reliable long patterns. For this work, we experimented with pattern length of 3 and 4 symbols, i.e., $P\# = 3$ or 4.

- *The number of symbols (S#)*

The number of symbols will directly influence the complexity of the pattern discovery process and the quality of patterns. Generally speaking, a larger number of symbols is required when a more subtle modeling of the state transitions is required. However, using too many symbols could make it very difficult to find reliable patterns. Therefore, we need to set a reasonable number of symbols. In our experiments, we used 8, 10, 12 and 14 when converting the numeric value to symbols (nominal value), i.e., $S\#$ = 8, 10, 12, or 14.

- **The breakpoints (BKs)**

The breakpoints determine how we map the original numerical values to the selected symbols. This directly affects the quality of the resulting patterns. We need to take into account the distribution and characteristics of original time-series as well as the particularities of the application. As discussed above, we adopted an unequal area for the symbol definitions. For each selected number of symbols, several breakpoints were tried to improve the quality of patterns. We use BK-x to note the setting for BK. Here K is a value of symbol number ($S\#$) and x is values for K. Corresponding to the symbol number, 8, 10, 12, and 14, the breakpoints are chosen as follows (the unit is percentage):

For 8 symbols: $B8$-1 ={ 2.5, 5, 7.5, 44.38, 81.26, 91.26, 97.51}
$B8$-2 ={ 6.25, 12.5, 25, 43.75, 62.5, 81.25, 93.75}
$B8$-3 ={ 2.5, 6.25, 11.25, 17.5, 51.25, 85, 97.5}

For 10 symbols: $B10$-1 ={ 2, 4, 6, 25.75, 45.5, 65.25, 85, 93, 98}
$B10$-2 ={ 5, 10, 20, 30, 40, 55, 70, 85, 95}
$B10$-3 ={ 2, 5, 9, 14, 32.5, 51, 69.5, 88, 98}

For 12 symbols: $B12$-1 ={ 1.67, 3.34, 5.01, 18.76, 32.51, 46.01, 59.76, 73.51, 87.26, 93.93, 98.1}
$B12$-2 ={ 4.17, 8.34, 16.67, 25, 33.33, 41.66, 49.99, 62.49, 74.99, 87.49, 95.82}
$B12$-3 ={ 1.67, 4.17, 7.5, 11.67, 24.68, 37.69, 50.7, 63.71, 76.72, 89.73, 98.06}

For 14 symbols:
$B14$-1 ={ 1.43, 3.57, 6.43, 10, 20.18, 30.36, 40.54, 50.72, 60.9, 71.08, 81.26, 91.44, 98.58}
$B14$-2 ={ 3.57, 7.14, 14.28, 21.42, 28.56, 35.7, 42.84, 49.98, 57.12, 67.83, 78.54, 89.25, 96.39}
$B14$-3 ={1.43, 2.86, 4.29, 14.92, 25.55, 36.18, 46.81, 57.44, 68.07, 78.7, 89.33, 95.04, 98.61}

- **The number of reliable patterns**

In practice, there might be several patterns representing relevant state transition trends. In order for the discovered patterns to cover more failure events, a reasonable number is needed for selecting reliable patterns. In our experiment, we choose the top 3 patterns as the reliable patterns from each experiment for evaluation.

- **The target window size of time to failure**

The target window size impacts directly the pattern evaluation criteria. For this application, we used 20 (days) as the target window size of time to failure. Therefore, t_s equals -22 and t_e equals -2 (the unit of time is day in this application).

4.2 Experiments and Results

Using those parameters, we conducted a fairly comprehensive experiment. To streamline the process, we implemented all steps of the proposed KDD methodology (symbolic sequence generation, pattern discovery, and result analysis) into an easily-configurable SAS application. We used training dataset for pattern discovery and testing dataset for pattern evaluation. First, we reduced the dimensionality and created the symbolic sequences. Then, we generated a list of patterns from the training dataset using the FSS search algorithm. Finally, we evaluated the patterns on the testing dataset by computing the total utility for each experiment, which is configured with the different combination of parameter setting. For example, we can run an experiment using a parameter combination: $D\# = 4$, $P\# = 3$, $S\# = 8$, $BK = B8$-1.

Table 2 presents results for a number of runs. The results are obtained by applying the discovered patterns on the test dataset which contains 112 wheel failures (time-series). The total utility for each experiment is computed using the key parameters and Equations 2 and 3. Table 2 also shows the mean and deviation of the TTF estimates as well as the problem coverage information. The problem coverage is computed as a problem detection rate (*PDR*) which is the number of detected problems over the total number of problems in the test dataset. Whenever one of the reliable patterns appears in a time-series, the failure is considered as detected. The mean and standard deviation information for the TTF estimates have been obtained using the differences between TTF estimates of a pattern and the actual failure time.

5 Discussions and Limitations

The results in Table 2 demonstrate the applicability of the proposed approach to find reliable patterns from the multivariate time-series data. The patterns discovered can be used to predict the likelihood of the problems and estimate TTF. For example, the patterns ("*GGF*", "*GHH*", and "*HHH*") obtained in Exp 2 cover 92% (103 cases over 112) of wheel failure cases in the test set and produce TTF estimates with an error of 19.7 ±12.24 days while achieving a relatively high total utility.

Table 2. Some results selected from the experiments (# of problems (time-series) = 112)

Exp(s)	Parameter Setting	Reliable Patterns	Detected Problems	PDR	Total Utility	TTF Estimate
Exp1	D# =3,P# =4, S# =8,BK =B8-1	DDDD, DEEE, EEEE	112	100%	-904.16	-31.47±18.31
Exp2	D# =4,P# =3, S# =8,BK =B8-1	GGF, GHH, HHH	103	92%	1,527.21	-19.7±12.24
Exp3	D# =4,P# =3, S# =10,BK =B10-1	GGF, GHH, HHH	103	92%	1,469.42	-20.18±13.32
Exp4	D# =3,P# =3, S# =12,BK =B12-2	JJI, JKK, KKK	98	88%	2,236.56	-20.07±15.77
Exp5	D# =3,P# =3, S# =10,BK =B10-2	EEE, EEF, EFF	104	93%	-4,031.81	-58.68±26.26
Exp6	D# =4,P# =3, S# =10,BK =B10-1	EEE, EEF, EFF	99	88%	-5,346.32	-68.12±26.11

These results strongly support the applicability and feasibility of the proposed approach. However, the proposed method is empirical. Many factors such as the parameter setting, data quality, and requirements of applications, affect directly the reliability of patterns. To find reliable patterns, we needed to conduct a large number of experiments using a trial an error process. It is also worth noting that the patterns found using the FSS algorithm from the symbolic sequence data do not have transition probabilities from one state to another. It would be desirable that a pattern provides practitioners with additional information on state transitions. For example, pattern "*BDF*" should provide the transition probability from "*B*" to "*D*" and from "*D*" to "*F*", such that the prognostic decision could be effectively made for applications.

To this end, we could use the Gibbs algorithm [14] during the search for patterns and directly obtain state transition probabilities.

It is worth noting that the pattern length has a significant impact on the prognostic performance. Generally speaking, the longer the pattern is, the higher the problem detection rate is. However, the precision of TTF estimation will deteriorate with increasing of the pattern length, which was demonstrated in the results from Exp 1. With a relatively long pattern, the total utility becomes lower.

We also noted that the propose pattern evaluation method appear adequate for selecting reliable patterns. The higher the total utility is, the more precise the TTF estimation is. For example, Exp 4 and Exp 5 with higher utility have more accurate TTF estimates compared to the other experiments.

6 Conclusions

In this paper, we presented a KDD methodology to discover reliable patterns from multivariate time-series data. The methodology consists of three main processes: creating symbolic sequences from time-series, searching for patterns from the symbolic sequences, and evaluating the patterns with a novel utility-based method. The developed techniques have been applied to a real-world application: prognostics of train wheel failures. The preliminary results demonstrated the applicability of the developed techniques. The paper also discussed the results and the limitations of the methodology.

Acknowledgment

Special thanks go to Kendra Seu, and Olena Frolova for their hard work when they were at NRC. We also extend our thanks to Chris Drummond and Marvin Zaluski for their valuable discussions.

Reference

[1] Schwabacher, W., Goebel, K.: A survey of artificial intelligence for prognostics. In: The AAAI fall Symposium on Artificial Intelligence for Prognostics, pp. 107–114. AAAI Press, Arlington (2007)
[2] Brown, E.R., McCollom, N.N., Moore, E.E., Hess, A.: Prognostics and health management: a data-driven approach to supporting the F-35 lightning II. In: IEEE Aerospace Conference (2007)
[3] Camci, F., Valentine, G.S., Navarra, K.: Methodologies for integration of PHM systems with maintenance data. In: IEEE Aerospace Conference (2007)
[4] Daw, C.S., Finney, E.A., Tracy, E.R.: A review of symbolic analysis of experimental data. Review of Scientific Instruments 74(2), 915–930 (2003)
[5] Przytula, E.W., Chol, A.: Reasoning framework for diagnosis and prognosis. In: IEEE Aerospace Conference (2007)
[6] Usynin, A., Hines, J.W., Urmanov, A.: Formulation of prognostics requirements. In: IEEE Aerospace Conference (2007)

[7] Luo, M., Wang, D., Pham, M., Low, C.B., Zhang, J.B., Zhang, D.H., Zhao, Y.Z.: Model-based fault diagnosis/prognosis for wheeled mobile robots: a review. In: The 31st Annual Conference of IEEE Industry Electronics Society 2005, New York (2005)

[8] Létourneau, S., Yang, C., Drummond, C., Scarlett, E., Valdés, J., Zaluski, M.: A domain independent data mining methodology for prognostics. In: The 59th Meeting of the Machinery Failure Prevention Technology Society (2005)

[9] Yang, C., Létourneau, S.: Learning to predict train wheel failures. In: The 11th ACM SIGKDD International Conference on Knowledge Discovery and Data Mining (KDD 2005), pp. 516–525 (2005)

[10] Keogh, E., Lin, J., Fu, A., Herle, H.V.: Finding the unusual medical time series: algorithms and applications. IEEE Transactions on Information Technology in Biomedicine (2005)

[11] Keogh, E., Chu, S., Hart, D., Pazzani, M.: Segmenting time series: A survey and novel approach. In: Kandel, Bunke (eds.) Data Mining in Time Series Databases, pp. 1–44. World Scientific Publishing, Singapore (2004)

[12] Mielikäinen, T., Mannila, H.: The pattern ordering problem. In: Lavrač, N., Gamberger, D., Todorovski, L., Blockeel, H. (eds.) PKDD 2003. LNCS (LNAI), vol. 2838, pp. 327–338. Springer, Heidelberg (2003)

[13] Xin, D., Cheng, H., Yan, X., Han, J.: Extracting redundancy-aware top-k patterns. In: ACM KDD 2006, Philadelphia, Pennsylvania, USA, pp. 444–453 (2006)

[14] Lawrence, C.E., Altschul, S.F., Boguski, M.S., Liu, J.S., Neuwald, A.F., Wootton, J.C.: Detecting subtle sequence signals: A Gibbs sampling strategy for multiple alignments. Science 262(8), 208–214 (1993)

Automating the Selection of Stories for *AI in the News*

Liang Dong[1], Reid G. Smith[2,*], and Bruce G. Buchanan[3]

[1] School of Computing, Clemson University, SC, USA
  ldong@clemson.edu
[2] Marathon Oil Corporation
  rgsmith@marathonoil.com
[3] Computer Science Department, Univ. of Pittsburgh, PA, USA
  buchanan@cs.pitt.edu

Abstract. It is relatively easy, albeit time-consuming, for a person to find and select news stories that meet subjective judgments of relevance and interest to a community.NewsFinder is an AI program that automates the steps involved in this task, from crawling the web to publishing the results.NewsFinder incorporates a learning program whose judgment of interestingness of stories can be trained by feedback from readers.Preliminary testing confirms the feasibility of automating the service to write *AI in the News* for the AAAI.

Keywords: News crawler, machine learning, supervised classification, SVM, artificial intelligence, AAAI, AITopics.

1 Introduction

Selecting interesting news stories about AI, or any other topic, requires more than searching for individual terms. The AAAI started collecting current news stories about AI and making them available to interested readers several years ago, with manual selection and publishing by an intelligent webmaster.

Current news stories from credible sources that are considered relevant to AI and interesting to readers are presented every week in five different formats: (i) posting summarized news stories on the *AI in the News* page of the AITopics web site [2],(ii) sending periodic email messages to subscribers through the "AI Alerts" service, (iii) posting RSS feeds for stories associated with major AITopics,(iv) archiving each month's collection of stories for later reference, and(v) posting each news story into a separate page on the AITopics web site.[1]

Manually finding and posting stories that are likely to be interesting is time-consuming.Therefore, we have developed an AI program, NewsFinder, that collects news stories from selected sources, rates them with respect to a learned measure of goodness, and publishes them in the five formats mentioned. Off-the-shelf implementations of several existing techniques were integrated into a working system for the AAAI.

---

[1] Anyone may view current and archived stories and subscribe to any of the RSS feeds; alerts are available only to AAAI members.

Traditional recommender systems [9]require recording a user's preference and using techniques such as non-negative matrix factorization [12] to find users with similar tastes. Then, recommendations are based on the preferencesof similar users.In our approach, we learn the characteristics of the items preferred by users and classify new items with respect to those.

The NewsFinder Program

The work of NewsFinder is implemented in four loosely-coupled program modules as in Fig. 1: (A) Crawling; (B) Training; (C) Ranking; (D) Publishing. The first three are independent from each other and the last two usually run together.

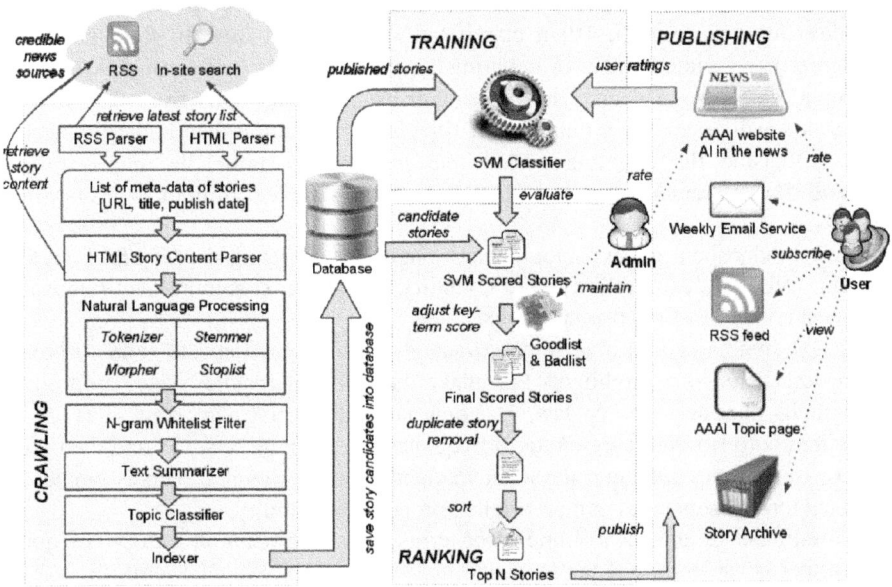

Fig. 1. NewsFinder Procedure Diagram

1.1 Crawling

In the crawling phrase, the program collects a large number of recent news stories about AI.Since crawling the entire web for stories mentioning a specific term like 'artificial intelligence' brings in far too many stories, we restrict the crawling to about two dozen major news publications. Thismakes a story more credibleand more likely to interest an international audience.The system administrators (AI subject matter experts) maintain a list of news sources, chosen for their international scope, credibility, and stability. These include The BBC, The New York Times, Forbes, The Wall Street Journal, MIT Technology Review, CNET, Discovery, Popular Science, Wired, The Washington Post, and The Guardian. Others can be added to the list.

NewsFinder collects the latest news stories via either in-site search or RSS feeds from those sources and filters out blogs, press releases, and advertisements.If a source has a search function, then the program uses it to find stories that contain 'artificial intelligence' or 'robots' or 'machine learning'. If a source has RSS feeds, then NewsFinder selects those feed labeled as 'technology' or 'science'.

In order to parse the text to retrieve the content of candidate pages, we write a specific HTML parser for each news source to identify and extract the news content from its news web pages. The advantage of this method is precision in that it can accurately extract news text stories and eliminate advertisements, user comments, navigation bars, menus and irrelevant in-site hyperlinks. The disadvantage of writing separate parsers for each news source is somewhat offset by starting with a generic template. We have written a dozen specific source parsersas modifications of code inherited from a base parser. Each parser is specifically designed for one news source web site since different sites use different HTML/CSS tags. We are also investigating an alternative method [5, 10] a classification method is used to train parsers to recognize news content either by counting hyperlinked words or by visual layout.

For a typical news source the parser will extract three items from the metadata associated with each news item: URL, title, and publication date. If the publication date is outside the crawling period (currently seven days), the news story is skipped.[2] For the remaining stories, the parser extracts the text from of each story from its URL.

NewsFinder then processes the natural language text, using the Natural Language Toolkit (NLTK)[7] to perform word counting, morphing, stemming, and removal of the most common words from a stoplist.[3]

A text summarization algorithm extracts 4-5 sentences from the story to build a short description—the highlights that make the story interesting—since an arbitrary paragraph, like the first or last, is often not informative. The main idea of the algorithm is to first measure the term frequency over the entire story, and then select the 4-5 sentences containing the most frequent terms. In the end, it re-assembles the selected top 4-5 sentences in their original order for readability.

NewsFinder references a Whitelist of terms whose inclusion in a story is required for further consideration. If the extracted text contains no Whitelist term, the story is skipped. In addition to the term 'artificial intelligence', Whitelist includes several dozen other words, bigrams and trigrams that indicate a story has additional relevance and interest beyond the search term used to find it in the first place. For example, stories are retrieved from RSS feeds for the topic 'robots' but an additional mention of 'autonomous robots,' or 'unmanned vehicles' suggests that AI is discussed in sufficient detail to interest AAAI readers.

The program then determines the main topic of each story. It uses the traditional Salton tf-idf cosine vector algorithm [8, 11] to measure the similarity of a story to the

---

[2] When using Google News, we also skip stories originating from a URL on our list of inappropriate domains. We set up the list initially to block formerly legitimate domains that have been purchased by inappropriate providers, but it can be used to block any that are known to be unreliable or offensive.

[3] The program also includes a Name Entity recognition algorithm, but it is not used routinely becauseit runs slowly. Instead, we check for names of particular interest, like "Turing", by adding them to the Goodlist described in the Ranking section.

introductory pages of each of the major topics on the AITopics web site.[4,5] Each document is treated as a vector with one component corresponding to each term and its tf-idf weight.Thus, we can measure the similarity of two documents by measuring the dot product of their normalized vectors, which produces the cosine of the vectors' angle in a multi-dimensional space[8].

The story is then linked to the AITopics page with the highest similarity so that readers wanting to follow up on a story with background information on that topic.The story is also added to a list for the RSS feed for the selected topic.At publication time the topic is shown with the story and the RSS feed that contains it.

Finally, NewsFinder saves the candidate news stories and their metadata into a database for subsequent processing.

1.2 Training

In order to train NewsFinder's classifier to recognize stories that the readers of AITopics want to see, we collect ratings from them.The classifier is retrained periodically (currently every week), when an old set of stories is archived and a new set is about to be collected.

Readers are asked to rate the relevance and interest of a story for the AITopics readership as "not relevant to AI" (0), or 1-5 for a degree of relevance and interest.

The rating system is modeled after the five-star rating system used by Netflix [6], although our purpose is to classify unseen items with respect to their likely interest to other readers, and not just their interest to the specific individual doing the rating.While individualized suggestions may be added in the future, for nowwe assume that the aggregate of many ratings reflects the opinion of the community at large.

After each story we show the rating as in Fig. 2, including the average rating of other readers during the week, both as a number and as a row of stars, for readers who may wish to focus first on stories that others have rated highly.

Fig. 2. Rating Interface

The PmWiki Cookbook StarRater[1] is used to collect Users' ratings.We record each user's rating for every news story together with IP address and username.The IP

---

[4] The current major topics are: AI Overview, Agents, Applications / Expert Systems, Cognitive Science, Education, Ethical & Social, Games & Puzzles, History, Interfaces, Machine Learning, Natural Language, Philosophy, Reasoning, Representation, Robots, Science Fiction, Speech, Systems & Languages, Vision.

[5] The topic assignment algorithm was originally written in PHP by Tom Charytoniuk and rewritten in Python by Liang Dong.

address is a proxy for a user ID and allows us to record just one vote per news item per IP address.[6]

During training, all the readers' ratings are collected and averaged.If a news story has fewer ratings than a specified number, the average rating is ignored (unless it is from one of the administrators).If the standard deviation of a news story's ratings is greater than a cutoff (default 2.0), the ratings are discarded as well.This way, a news story is only added to the training set if there is general consensus among several raters about it (or if one of the administrators ranks it).

The Support Vector Machine (SVM) [3] is a widely used supervised learning method which can be used for classification, regression or other tasks by constructing a hyperplane or set of hyperplanes in a high dimensional space.An SVM from a python library LibSVM[4] has been trained with manually scored stories from the web to classify the goodness of each story into one of three categories:(a) high – interesting enough to *AI in the News* readers to publish, (b) medium – relevant but not as interesting to readers as the first group, and (c) low – not likely to interest readers.Currently, we build three 'one against the rest' classifiers to identify these three sets.

1.3 Ranking

After crawling and training, the next step is ranking the candidate stories during the current news period by computing and comparing the scores of all news stories crawled during the period.The score for each news story is computed in two steps: (i) assign an SVM score and (ii) adjust it using a key term score.

The SVM score is assigned based on which of the three SVM categories has the highest probability: high interest = 5, medium = 3, low or no interest = 0.If none of the classifiers assigns a 50% or greater probability of the story being in its category, the default score for the story is 1.The probability is based on the tf-idf measure of interestof all non-stop words in the document, typically about 200 words.

NewsFinder performs an adjustment to the SVM scoreby first retrieving every recent news story containing a term from a list called Goodlist.Terms on Goodlist are those whose inclusion in a story signals higher interest, as determined by subject-matter experts.

NewsFinder then measures the tf-idf score for each Goodlist term.All the term scores are accumulated and normalized across the recent stories.

When a new topic of interest first appears in AI, as "semantic web" did several years ago, the SVM canautomatically recognize its importance as readers give high ratings to stories on this topic.Normal practice is for authors of stories on a new topic to tie the topic to the existing literature.However, an administrator (who is a subject matter expert) mayalso add new terms to Goodlist to jump-start this practice.Although one can imagine many dozen key terms on Goodlist, the initial two tests reported here used only 12 terms.

---

[6] As with Netflix, if there are multiple ratings for the same story from the same reader (moreprecisely, from the same IP address), only the last vote is used.

The same process is executed for terms on a list called Badlist. Terms on Badlist are those whose inclusion in a story signals lower interest. Initial testing was done using 12 Badlist terms. Both Goodlist and Badlist are easily edited in the setup file.

The key term score from Goodlist lies in [0, +1], which boosts the final score. The key term score from Badlist, which reduces the final score, is unbounded. Unlike the terms on Whitelist, whose omission forces exclusion of a story from further consideration, the terms on Goodlist and Badlist merely add or subtract from the initial SVM score based on the number of terms appearing and their frequency of occurrence. Multi-word terms on Goodlist, such as 'unmanned vehicle', have been manually selected as signals of increased interest. Badlist terms such as 'ceo', 'actor', and 'movie' can reduce the score for including unrelated news such as gossip about actors who appeared in Spielberg's movie "Artificial Intelligence." The terms 'tele-operated' and 'manually operated' similarly reduce the score on many stories about robots that are less likely to involve AI.

The computation of the key term score is as follows: given a key term such as 'automated robot', the program first finds all the recent stories containing both 'automated' and 'robot'. Then it computes the tf-idf score for each term, and adds all the tf-idf scores for this story.

After NewsFinder obtains the trained SVM score and key term score, each news story's final score is a weighted sum of its SVM score and its key term score, where the weight of the weight term, w, was selected heuristically to be 3.0:

$$Score = SVMScore_w \cdot KeyTermScore$$

Currently, both Goodlist and Badlist are manually maintained by the webmaster, in order to control quality during startup. When the size of the training set reaches about 500 stories, we plan to remove both lists.

It is worth-noting that the length of the story doesn't affect the SVM score since each story's tf-idf is normalized before being classified. But it affects the key term scores since each term's tf-idf depends on the number of terms in the document. However, longer stories are *prima facie* more likely to include more key terms. In addition, when selecting from among similar stories, the program prefers longer ones.

After all the potential candidates have been scored, NewsFinder measures the text similarity to eliminate duplicate stories. The program clusters all the news candidates to identify news about the same event. These may be exact duplicates (e.g., the same story from one wire service used in different publications), or they may be two reports of the same event (e.g., separately written announcements of the winner of a competition). Again, NewsFinder measures the cosine of the angle between the two documents' tf-idf to determine their similarity in the vector space. If the computed similarity value is greater than a cutoff (0.33 by default), these stories are clustered together. If there is more than one story in a group, the story with the highest final score is kept for publishing.

The N-highest-scoring stories are selected for publishing each week. At the current time, these are the N (or fewer) "most interesting" stories with final scores above a threshold of 3.0; i.e., ranked "medium" to "very high." For the initial testing, N=20; in the last test and current version, N=12.

1.4 Publishing

The stories selected for publishing are those with the highest final scores from the ranking phase, but these still need to be formatted for publishing in different ways: (i) posting summarized news stories on the Latest *AI in theNews* page of the AITopics web site, (ii) sending periodic email messages to subscribers through the "AI Alerts" service, (iii) posting RSS feeds for stories associated with major AITopics,(iv) archiving each month's collection of stories for later reference, and (v) posting each news story into a separate page on the AITopics web site.

2 Validation

2.1 SVM Alone

After training on the first 100 cases scored manually, we determined the extent to which the selections of the SVM part of NewsFinder matched our own.For a new set of 49 stories retrieved from Google News by searching for 'artificial intelligence', we marked each story as "include" or "exclude" from the stories we would want published, and we matched these against the list of stories NewsFinder would publish, without use of the additional knowledge of terms on Goodlist and Badlist.On the unseen new set of 49 recent stories crawled from Google News, the SVM put 46 of 49 stories (94%) into the same two categories – include as "relevant and interesting" or exclude – as we did.Five stories would have been included for the 10-day period, which we take to be about right (but on the low side) for weekly email alerts.

This was not a formal study with careful controls since the person rating the stories could see the program's ratings, and the SVM was retrained using some of the same stories it then scored again.Nevertheless, it did suggest that the SVM was worth keeping.It also suggested that merely using an RSS feed or broad web search with a term like 'artificial intelligence' would return many more irrelevant and low-interest stories than we wanted.In a one week period Google News returned 400 candidate stories mentioning the term 'artificial intelligence', 88 mentioning 'machine learning', 8,195 mentioning 'robot', and 2,264 mentioning 'robotics.'We concluded that not all would be good to publish in *AI in the News*, nor would readers want this many.

2.2 Adjusted Scores

In a subsequent test, we used a specified set of credible news sources, a training set of 265 stories (including the 149 from the initial test), and a test set of 69 new stories.The full NewsFinder programwas used, with scores from the SVM adjusted by additional knowledge of good and bad terms to look for.We compared the program's decision to include or exclude from the published set against the judgment of one administrator (BGB) that was made before looking at the program's score.

We accumulated scores and ratings by the administrator for 3-4 stories per day that were not in each previous night's training set, a total of 69 stories in the first three weeks of September, 2010.Although the SVM is improving (or at least changing) each night, these stories are truly "unseen" in the sense that they did not yet appear in the training set used to train the classifier that scored them. Among 42 stories that the program scored above the publication threshold (≥ 3.0), the administrator rated 33 (78.6%) above threshold.

Out of 27 candidate stories that the program rated below the publication threshold (< 3.0), the administrator rated 11 (40.1%) below threshold.Thus the program is publishing mostly stories that the administrator agrees should be published but is omitting about half the likely candidates that the administrator rates above threshold.The 27 candidates in this study that were not published were mostly "near misses." Many were rated 3 by the administrator, indicating that they were OK, but not great.Also, a few of the stories the administrator would have published may be selected on a later day, after retraining or when their normalized scores rise above threshold because the best story in the new set is not as good as in the previous set.Given a limit of twelve stories, the tradeoff between false positives and false negatives weighs in favor of omitting some good stories over including uninteresting or marginal ones.

We conducted a 5-fold cross validation for 218 stories with administrator ratings to validate the performance of the SVM classifier (before adjustment).As above, each of the tests was on "unseen" stories.For these 218 valid ratings, we counted the times that the administrator and the SVM classified a story in the same way.The accuracy of the "high" predictions was 66.5%, of the "medium" ratings 72.9%, and of the "low" ratings 74.3%.

2.3 Final Test

After the completion of these tests, some adjustments were made to correct occasional problems noted during testing.

- A story categorized as low or no interest by the SVM (category 0) is not published, regardless of its adjusted score.
- The threshold for similarity of two news stories was lowered from 0.4 to 0.33 to reduce the number of duplicates.
- Whitelist and Goodlist were made to contain the same terms, though their uses remain different.Thus a story must contain at least one of several dozen terms to be considered at all (Whitelist), and the more occurrences of these terms that are found in a story, the more its score will be boosted (Goodlist).Three new terms were added to Whitelist and Goodlist.
- Upward adjustments to the score from the key terms on Goodlist are now normalized to the highest adjustment in any period because adding a larger number of Goodlist terms created uncontrollably large adjustments.(Unbounded downward adjustments do not concern us because stories containing Badlist terms are unwanted anyway.)
- Terms having to do with tele-operated robots and Hollywood movies were added to the Badlist, thus downgrading stories that are about manually controlled robots or movie personalities.
- The frequency with which the program searches for stories and publishes a new *AI in the News* page was changed from daily to weekly.

- The number of stories published in any period has been changed from 20 to 12, since that will reduce the false positives and also reduce the size of weekly email messages to busy people.
- Stories can be added manually to be included in the current set of stories to be ranked. Thus when an interesting story is published in a source other than the ones we crawl automatically, it can be considered for publication. It will also be included in subsequent training, which may help offset the inertia of training over the accumulation of all past stories and the lag time in recognizing new topics.

A follow-up test was performed on 118 unseen stories to confirm that the changes we had made were not detrimental to performance. We also gathered additional statistics to help us understand the program's behavior better. Two-thirds of the stories were at or above the program's publication threshold of 3.0 (80/118), based on their initial SVM and adjustment scores).

Among 118 stories that passed the relevance screening and duplicate elimination, and thus were scored with respect to interest, the overall rate of agreement between the program and an administrator is 74.6% on decisions to publish or not (threshold $\geq$ 3.0), with Precision = 0.813, Recall = 0.813, and F1 = 4.92. Both the program and the administrator recommend publishing about two-thirds of the stories passing the relevance filters, just not the same two-thirds.

Table 1. Decisions on 118 Stories Rated by Both Admin and NewsFinder

	Admin:	Admin:
NewsFinder: Publish	65 (55%)	15 (13%)
NewsFinder: Don't Publish	15 (13%)	23 (20%)

3 Conclusions

Replacing a time-consuming manual operation with an AI program is an obvious thing for the AAAI to do. Although intelligent selection of news stories from the web is not as simple to implement as it is to imagine, we have shown it is possible to integrate many existing techniques into one system for this task, at low cost. There are many different operations, each requiring several parameters to implement the heuristics of deciding which stories are good enough to present to readers. The two-step scoring system appears to be a conceptually simple way of combining a trainable SVM classifier based on term frequencies with prior knowledge of semantically significant term relationships.

NewsFinder has not been in operation for long, but it appears to be capable of providing a valuable service. We speculate that it could be generalized to alert other groups of people to news stories that are relevant to the focus of the group and highly interesting to many or most of the group. The program itself is not specific to AI, but the terms on Goodlist and Badlist, the terms used for searching news sites and RSS feeds, and to some extent the list of sources to be scanned, are specific to AI.

Learning how to select stories that the group rates highly adds generality as well as flexibility to change its criteria as the interests of the group change over time.

Acknowledgments. We are grateful to Jim Bennett for many useful comments especially on the Netflix rating system after which the NewsFinder rating is modeled, and to Tom Charytoniuk for implementing the initial prototype of this system.

References

1. 5 star rating system for Pmwiki, http://www.pmwiki.org/wiki/Cookbook/StarRater
2. Buchanan, B.G., Glick, J., Smith, R.G.: The AAAI Video Archive. AI Magazine 29(1), 91–94 (2008)
3. Burges, J.C.C.: A tutorial on support vector machines for pattern recognition. Data Mining and Knowledge Discovery 2(2), 121–167
4. Chang, C.-C., Lin, C.-J.: LIBSVM: a library for support vector machines, Software (2001), http://www.csie.ntu.edu.tw/~cjlin/libsvm
5. Gupta, S., Kaiser, G., Neistadt, D., Grimm, P.: DOM-based Content Extraction of HTML Documents. In: Proceedings of the 12th International World Wide Web Conference(WWW2003), Budapest, Hungary (2003)
6. Herlocker, J., Konstan, J., Terveen, L., Riedl, J.: Evaluating Collaborative Filtering Recommender Systems. ACM Transactions on Information Systems 22 (2004)
7. Loper, E., Bird, S.: NLTK: The Natural Language Toolkit. In: Proceedings of the ACL Workshop on Effective Tools and Methodologies for Teaching Natural Language Processing and Computational Linguistics. Association for Computational Linguistics, Philadelphia (2002), http://www.nltk.org/
8. Manning, C., et al.: Intro. to Information Retrieval. Cambridge University Press, Cambridge (2008)
9. Melville, P., Sindhwani, V.: Recommender System. In: Sammut, C., Webb, G. (eds.) Encyclopedia of Machine Learning, Springer, Heidelberg (2010)
10. Song, R., Liu, H., Wen, J.-R., Ma, W.-Y.: Learning Block Importance Models for Web Pages. In: Proceedings of the 13th International World Wide Web Conference (WWW 2004), New York (2004)
11. Salton, G., Gerard, Buckley, C.: Term-weighting approaches in automatic text retrieval. Information Processing & Management 24(5), 513–523 (1988), doi:10.1016/0306-4573(88)90021-0
12. Zhang, S., Wang, W., Ford, J., Makedon, F.: Learning from incomplete ratings using non-negative matrix factorization. In: Proc. of the 6th SIAM Conference on Data Mining (2006)

Diagnosability Study of Technological Systems

Michel Batteux[1], Philippe Dague[2], Nicolas Rapin[3], and Philippe Fiani[1]

[1] Sherpa Engineering, La Garenne Colombe, France
{m.batteux,p.fiani}@sherpa-eng.com
[2] LRI, Univ. Paris-Sud & CNRS, INRIA Saclay – Île-de-France, Orsay, France
philippe.dague@lri.fr
[3] CEA, LIST, Laboratory of Model driven engineering for embedded systems,
Point Courier 94, Gif-sur-Yvette, 91191, France
nicolas.rapin@cea.fr

Abstract. This paper describes an approach to study the diagnosability of technological systems, by characterizing their observable behaviors. Due to the interaction between many components, faults can occur in a technological system and cause hard damages not only to its integrity but also to its environment. Though a diagnosis system is a suitable solution to detect and identify faults, it is first important to ensure the diagnosability of the system: will the diagnosis system always be able to detect and identify any fault, without any ambiguity, when it occurs? In this paper, we present an approach to identify and integrate faults in a model of a technological system. Then we use these models for the diagnosability study of faults by characterizing their observable behaviors.

Keywords: faults modeling, diagnosability, model-based diagnosis.

1 Introduction

Technological systems are complex systems constituted with many components interacting with each other and combining multiple physical phenomena: thermodynamic, hydraulic, electric, etc. Faults, which are un-observable damages affecting components of a system, can occur due to many causes: wear, dirtying, breakage, etc. Some are serious and must require to stop the system, or to put it in a safety mode; while others have minor impact and should only be reported for being repaired off-board. Thus, it is necessary to achieve on-board the detection of these faults and to identify them the most precisely; this in order to take the appropriate decision. An embedded diagnosis system, completing the controller, is a suitable solution to do this ([1]). However, the problem is then to ensure that this diagnosis system will always be able not only to detect any fault when it occurs (does the fault induce an observable behavior distinct from the normality?), but also to assign a unique listed fault to a divergent observable behavior (do some faults induce the same observable behavior?). This problem is known as diagnosability ([2]).

A way to handle this diagnosability, with respect to a system, is to augment the model of this system (the normal model) with faults (producing faulty models); and to exploit them to characterize observable behaviors of the system, under or not a fault, by a specific property which will be verified by the diagnosis system. This approach, called diagnosability study of faults, inheres in the analysis process of the diagnosis system. It requires, by definition, all faulty models to produce, for each one, a specific fault characterization according to its observable behaviors. All these faults characterizations will then be used by the embedded diagnosis system to detect and identify faults.

In this paper, we present an approach to study the diagnosability of faults of a technological system by exploiting its observable behaviors. These observable behaviors are obtained by using the normal and all faulty models of the system. In the second part, we present the framework to model a technological system and show how to integrate faults in it. In the third part, we exploit these produced models in order to define observable behaviors of the system in the normal and all faulty cases. In the fourth part, we study diagnosability of faults by producing their characterizations. In the fifth part, we apply this theoretical framework to a practical application: a fuel cell system. Finally, the last part concludes by summarizing results and outlining interesting directions for future works.

2 System and Faults Modeling

In order to study faults diagnosability of a system, it is necessary to get the normal model and all faulty models of the system. In this part, we present the framework to obtain a model of a system and show how to integrate faults into it.

2.1 Normal Model of the System

Classical works found in literature for diagnosis ([1], [3] and [4]) are based on a representation of the system in open-loop. But for the majority of industrial applications, the system is inserted in a closed-loop and its controller computes system inputs by taking into account its outputs; this to increase system performances and to maintain them in spite of unknown perturbations affecting it. In this context fault detection and isolation are more difficult because of the contradiction between control objectives and diagnosis objectives. In fact, control objectives are to minimize disturbances or faults effects; whereas diagnosis objectives are precisely to bring out these faults. The considered solution, to take into account this problem, is to model the system with its controller in closed-loop. Fig. 1 below represents the complete structure of the system: the system and its controller in closed-loop.

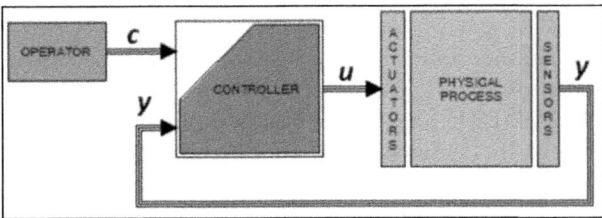

Fig. 1. Complete structure of the system

We model a controlled system with two parts: the controller and the system itself composed of the physical process, actuators and sensors. We consider a continuous model in discrete time T with a state space representation, described by the set (1) of equations:

$$\begin{aligned} x(t+1) &= f(x(t),\theta,u(t),d(t)),\ x(0) = x_{init} \\ y(t) &= g(x(t),\theta,u(t),d(t)) \\ a(t+1) &= h(c(t),a(t),y(t)),\ a(0) = a_{init} \\ u(t) &= k(c(t),a(t),y(t)) \end{aligned} \quad (1)$$

where the two first equations model the system and the two other ones model the controller (more precisely its control laws). Variables u, x, θ, d and y are respectively input, state, parameter, disturbance and output vectors of the system; c and a are respectively order (from the operator) and state vectors of the controller. We denote by $V = \{c;a;u;x;y;d\}$ the set of all variables of the model, with respective domains of values C, A, U, X, Y and D, and by $V_{Obs} = \{c;u;y\}$ the set of observable variables.

2.2 Faults Modeling

Faults in the system can cause failures or malfunctions, resulting in serious damages not only to the system integrity but also to its environment. It is therefore important to identify and classify all potential faults of the system in order to ensure their integration in the model.

By using methodologies of safety engineering ([5] and [4]), an identification of all important faults of the system can be made. Thanks to faults analysis techniques, such as Failure Mode and Effects Analysis or Fault Tree Analysis, we can identify most of potential faults which could occur in the system; and analyze their causes and impacts. Therefore, the set $\Gamma = \{F_0, F_1, \ldots, F_k\}$ of potential faults, that must be taken into account by a diagnosis system, is identified and defined during this safety analysis, where to simplify the presentation, the fault F_0 represents the normal case.

Potential faults can be classified in order to ensure their integration in the model. Various classifications of faults can be found in literature ([1], [4] and [6]); but all of them differentiate the behavior of the fault and its effects on the system ([7]). Fault behavior is characterized by its occurrence time (randomly, at a specific time or from a specific event), its appearance (abruptly or progressively) and its form (permanent, transient or intermittent). Fault effects consist in its location inside the system and its disturbance induced. For our purpose, we do not consider faults occurring in the controller. Thus, there are sensor faults (perturbing the output vector y), actuator faults (perturbing the input vector u) and faults in the process (perturbing the state vector x or the parameter vector θ). The disturbance can be additive, multiplicative, sinusoidal or limitative.

Thus, for a fault $F \in \Gamma \setminus \{F_0\}$ and a time instant $t_n \in T$, the faulty model of the system under F is obtained by considering the set of equations (1) where the considered disturbed variable v is replaced by its disturbance $v_F = dist(t,v(t),flt(t,t_n))$, with flt the fault behavior and $t_n \in T$ its occurrence time. For example with a sensor fault F, represented by $y_F(t) = dist(t,y(t),flt(t,t_n))$, the faulty model of the system is described by the set (2) of equations:

$$\begin{aligned} x(t+1) &= f(x(t),\theta,u(t),d(t)), \; x(0) = x_{init} \\ y(t) &= \mathit{dist}(t,g(x(t),\theta,u(t),d(t)),\mathit{flt}(t,t_n)) \\ a(t+1) &= h(c(t),a(t),y(t)), \; a(0) = a_{init} \\ u(t) &= k(c(t),a(t),y(t)) \end{aligned} \quad (2)$$

3 Observable Behaviors

By adding faults in the model of the system, we have produced all faulty models requested for the diagnosability study. We can now exploit them to characterize observable behaviors in the normal and all faulty cases. A behavior of the system represents its way of operation, according to an instruction (a sequence of orders) from the operator. An observable behavior is therefore obtained from the behavior by restricting it to the only observable variables. Observable means visible from an external observer, such as a diagnosis system for example.

3.1 System Behaviors

A behavior of the system is represented by the set of values of variables during its operation and according to an instruction from the operator. This operation can be under the presence, or not, of a fault. A behavior is therefore specified according to an instruction and a fault.

An instruction is the evolution of orders from the operator during the time. Formally, it is a sequence cs from a temporal window $I_{cs} \subseteq T$, assumed beginning from the time instant 0, to the domain C. In the following, we consider a set *Cons* of instructions the most representative, i.e.: providing all operation ranges of the system. Thus, though each instruction cs is defined from its own temporal window I_{cs}, we assume that all instructions are defined from a same temporal window $I = \max\{I_{cs}\}$, by extending any instruction cs, where $I_{cs} \subset I$, with its last value $cs(\max(I_{cs}))$.

For a vector $v = (v_1,\ldots,v_n)$ and for an index $i \in \{1,\ldots,n\}$, we denote by $p_i(v) = v_i$ the i-th element of v. For a set E, constituted by a direct product $E = E_1 \times \ldots \times E_n$, for a subset $G \subseteq E$ and for indexes $i_1,\ldots,i_k \in \{1,\ldots,n\}$ with $k \leq n$; the projection of G onto the $E_{i_1} \times \cdots \times E_{i_k}$ is the set $\Pr_{E_{i_1} \times \cdots \times E_{i_k}}(G) = \{(p_{i_1}(v),\cdots,p_{i_k}(v)) \in E_{i_1} \times \cdots \times E_{i_k} \setminus v \in G\}$.

Normal Behaviors. For an instruction $cs \in Cons$, the normal behavior of the system, according to cs, is the set $B(cs,F_0)$ of vectors of data $(t,v(t)) \in I \times C \times A \times U \times X \times Y \times D$, ordered by time t, with $v(t) = (c(t),a(t),u(t),x(t),y(t),d(t))$. This set of data vectors satisfies the following:

a. Existence and uniqueness in time: for any time instant $t \in I$, it exists a unique vector $v(t) \in C \times A \times U \times X \times Y \times D$ such as $(t,v(t)) \in B(cs,F_0)$.
b. Construction according to the instruction cs: for any time instant $t \in I$, $p_1(v(t)) = cs(t)$.
c. Satisfaction of system equations in normal case: for any time instant $t \in I$,
   1. $p_4(v(t+1)) = f(p_4(v(t)),\theta,p_3(v(t)),p_6(v(t)))$ and $p_4(v(0)) = x_{init}$
   2. $p_5(v(t)) = g(p_4(v(t)),\theta,p_3(v(t)),p_6(v(t)))$

3. $p_2(v(t+1)) = h(p_1(v(t)),p_2(v(t)),p_5(v(t)))$ and $p_2(v(0)) = a_{init}$
4. $p_3(v(t)) = k(p_1(v(t)),p_2(v(t)),p_5(v(t)))$

Faulty Behaviors. A characteristic of faults behaviors, defined in the above part, is the occurrence time (randomly, at a specific time, or from a specific event). In the following, we only consider faults occurring at a specific time; in fact our focus is only to ensure diagnosability of a fault when it occurs, not to predict its occurrence. Thus, for each fault $F \in \Gamma \setminus \{F_0\}$ and each instruction $cs \in Cons$, we consider a set $\Omega_{(F,cs)}$ of time occurrence $t_n \in I$ of the fault F according to the instruction cs.

Therefore, for an instruction $cs \in Cons$ and a fault $F \in \Gamma \setminus \{F_0\}$ occurring at a time instance $t_n \in \Omega_{(F,cs)}$, the faulty behavior of the system $B(cs,F,t_n)$ is defined as the normal one above where points (c) is replaced with the satisfaction of system equations in the considered faulty case.

3.2 Observable Behaviors of the System

An observable behavior of the system represents its visible, from an external observer, way of operation according to an instruction from the operator. It is obtained by projecting the behavior, according to the considered instruction, onto the set of observable variables.

In addition, detection and isolation of a fault require, for industrial applications, to be made in bounded time b after the fault occurrence. This bound can be assumed more than the response time δ of the system. Thus, as a behavior could be defined for a fault occurring at the time instant $t_n = \max(I)$, it could not consider onto the time interval $[t_n, t_n+b]$ as it is not defined. Therefore, for any fault $F \in \Gamma \setminus \{F_0\}$ and any instruction $cs \in Cons$, we only consider time occurrences $t_n \in \Theta_{(F,cs)} = \Omega_{(F,cs)} \cap [0; \max(I) - b]$. F_0 can be considered as a fault always occurring at the time instant $t_n = 0$; thus, $\Theta_{(F_0,cs)} = \{0\}$ for any instruction $cs \in Cons$.

For a faulty behavior $B(cs,F,t_n)$, according to an instruction $cs \in Cons$ and under a fault $F \in \Gamma \setminus \{F_0\}$ occurring at a time instance $t_n \in \Theta_{(F,cs)}$, the underlying faulty observable behavior $ObsB(cs,F,t_n)$ is the projection of $B(cs,F,t_n)$ onto the direct product $I \times C \times U \times Y$ of observable variables: $ObsB(cs,F,t_n) = \Pr_{I \times C \times U \times Y}(B(cs,F,t_n))$.

For a normal behavior $B(cs,F_0)$, according to an instruction $cs \in Cons$, and the time instant $t_n \in \Theta_{(F_0,cs)}$ (={0}), the underlying normal observable behavior $ObsB(cs,F_0,t_n)$ is the projection of $B(cs,F_0)$ onto the direct product $I \times C \times U \times Y$ of observable variables: $ObsB(cs,F_0,t_n) = \Pr_{I \times C \times U \times Y}(B(cs,F_0))$. The parameter t_n, which is always equal to 0, is added to harmonize with the notation of faulty observable behaviors $ObsB(cs,F,t_n)$.

An observable behavior is defined according to an instruction $cs \in Cons$, a fault $F \in \Gamma$ and an occurrence time $t_n \in \Theta_{(F,cs)}$ ($t_n = 0$ for F_0). Therefore the set of observable behaviors, according to the set $Cons$ of instructions and under a fault $F \in \Gamma$, is the union of all observable behaviors for all instructions cs of $Cons$ and all occurrence time $t_n \in \Theta_{(F,cs)}$ of F: $ObsBeh_{Cons}(F) = \bigcup_{cs \in Cons} \bigcup_{t_n \in \Theta_{(F,cs)}} \{ObsB(cs,F,t_n)\}$. Finally, the set of observable behaviors, according to the set $Cons$ of instructions, is the union of all sets of observable behaviors for all faults: $ObsBeh_{Cons} = \bigcup_{F \in \Gamma} ObsBeh_{Cons}(F)$. We also define the time domain $Tdom(ob)$ of an observable behavior $ob \in ObsBeh_{Cons}$ as I and a set $J = [b;max(I)] \subset I$.

4 Diagnosability Study of the System

Intuitively a fault $F \in \Gamma \setminus \{F_0\}$ is said diagnosable if observable behaviors of the system under this fault are not the same that observable behaviors of the system under another fault $F' \in \Gamma \setminus \{F\}$. This other fault can be F_0, or another one $F' \in \Gamma \setminus \{F_0;F\}$, which expresses the two ideas: the fault detectability and the fault isolability.

The diagnosability study inheres in the analysis process of the diagnosis system. In fact, during its operation, the diagnosis system will check if the observed behavior of the system, provided by data of observable variables, satisfies a specific property characterizing the normal operation of the system. If this property is not satisfied, it will search which property, characterizing an abnormal operation, is satisfied. Consequently, the diagnosability study will be made from a set $\Lambda = (P_F)_{F \in \Gamma}$ of properties characterizing the most precisely observable behaviors of the system under the considered fault. Λ is said a faults characterization.

4.1 Faults Characterization

For a fault $F \in \Gamma$, its characterization is a property P_F which must be satisfied at each time instant by at least observable behaviors under this considered fault ($P_F : ObsBeh_{Cons} \times J \to \{true;false\}$). We propose two kinds of faults characterization.

The Perfect Characterization (PC). The most natural solution to characterize observable behaviors is to consider them restricted to the temporal windows $[t - b;t]$, with b the bound presented above.

- For the fault $F_0 \in \Gamma$, the set of bounded normal observable behaviors is
  $ObsBeh^{bd}_{Cons}(F_0) = \bigcup_{cs \in Cons} \bigcup_{t \in J} \{Pr_{[t-b;t] \times C \times U \times Y}(ObsB(cs,F_0,0))\}$.
- For a fault $F \in \Gamma \setminus \{F_0\}$ the set of bounded faulty observable behaviors is
  $ObsBeh^{bd}_{Cons}(F) = \bigcup_{cs \in Cons} \bigcup_{t_n \in \Theta_{(F,cs)}} \bigcup_{t \in [t_n;t_n+b]} \{Pr_{[t-b;t] \times C \times U \times Y}(ObsB(cs,F,t_n))\}$.

For a fault $F \in \Gamma$, its perfect characterization is the property P_F defined as follow: for an observable behavior $ob \in ObsBeh_{Cons}$ and a time instant $t \in J$, $P_F(ob,t)$ is true iff $Pr_{[t-b;t] \times C \times U \times Y}(ob) \in ObsBeh^{bd}_{Cons}(F)$.

Intuitively, an observable behavior $ob \in ObsBeh_{Cons}$, restricted to a temporal window $[t_i - b;t_i]$, is an element of the set $ObsBeh^{bd}_{Cons}(F)$ iff it exists an instruction $cs \in Cons$, an occurrence time $t_n \in \Theta_{(F,cs)}$ of F and a time instant $t_d \in [t_n;t_n+b]$, such as data vectors of ob (which are restricted to the temporal window $[t_i - b;t_i]$) are equal to data vectors of $ObsB(cs,F,t_n)$ restricted to the temporal window $[t_d - b;t_d]$. I.e., if for any time instant $k \in [0;b] \subseteq T$, the data vector $v(t_i+k)$ of ob at the time instant t_i+k is equal to the data vector $v'(t_d+k)$ of $ObsB(cs,F,t_n)$ at the time instant t_d+k.

It is a perfect characterization because it is not possible to specify, in a best way, observable behaviors. In fact, the sets $ObsBeh^{bd}_{Cons}(F)$, constructed by restricting elements of $ObsBeh_{Cons}(F)$, are the nominal definitions of observable behaviors of the system under a fault; it is therefore not possible to do it better. Nevertheless, not only the set *Cons* of instructions must be the most representative: the set C^I of all functions from the temporal window I to the domain C; but also all sets $\Theta_{(F,cs)}$ of occurrence time of faults must be equal to J.

The Temporal Formulas Characterization (TFC). This faults characterization describes how the system operates under a fault thanks to a temporal formula. We will use an adaptation of the metric interval temporal logic ([8]), well adapted to specify bounded real-time properties.

The syntax of temporal formulas is classically defined by induction. The set of terms, representing arithmetic formulas, is built from the set $V_{Obs}=\{c;u;y\}$ of observable variables of the system, a set K of constants, arithmetic operators (+, −, · and ÷) and a temporal operator $V_{[\alpha]}$, where $\alpha \in T$ is a positive or negative time instant. Atomic formulas, expressing a comparison (equality or inequality) between two arithmetic formulas, are built from the set of terms and by using comparison operators (=, ≠, <, >, ≥ and ≤). Temporal formulas are therefore built, recursively, with operators ¬ (not), ∧ (and) $G_{[\alpha;\beta]}$ (globally during a temporal window bounded by α and β) and $E_{[\alpha;\beta]}$ (eventually during a temporal window bounded by α and β); where $\alpha,\beta \in T$ are two positive or negative time instants. Classical operators ∨ (or) and ⇒ (implication) and ⇔ (equivalent) are built from the above operators ¬ and ∧.

Temporal formulas are interpreted by observable behaviors $ob \in ObsBeh_{Cons}$ at each time instant $t \in J$: the satisfaction of φ by *ob* at the time instant t, denoted by $(ob,t) \models \varphi$, is classically defined by induction onto the set of temporal formulas:

- $(ob,t) \models atom$, iff $t \in Dom(atom) = Tdom(ob)$ and *atom* is true when all observable variables are replaced by their values from the vector of *ob* at the time instant t. If *atom* contains a term of the form $V_{[\alpha]}v$, its satisfaction is obtained if $t+\alpha \in Dom(atom)$ and by considering the value of v at the time instant $t+\alpha$.
- $(ob,t) \models \neg\varphi$ iff $t \in Dom(\varphi)$ and (ob,t) does not satisfy φ.
- $(ob,t) \models \varphi \wedge \psi$ iff $t \in Dom(\varphi) \cap Dom(\psi)$ and $(ob,t) \models \varphi$ and $(ob,t) \models \psi$.
- $(ob,t) \models G_{[\alpha;\beta]}\varphi$ iff $[t+\beta;t+\alpha] \subseteq Dom(\varphi)$ and for all t' in $[t+\beta;t+\alpha]$ we have $(ob,t') \models \varphi$ (if $[t-\beta;t-\alpha] \not\subseteq Dom(\varphi)$ the value of $G_{[\alpha;\beta]}\varphi$ is not defined).
- $(ob,t) \models E_{[\alpha;\beta]}\varphi$ iff $[t+\beta;t+\alpha] \subseteq Dom(\varphi)$ and there exists t' in $[t+\beta;t+\alpha]$ such that $(ob,t') \models \varphi$ (if $[t-\beta;t-\alpha] \not\subseteq Dom(\varphi)$ the value of $E_{[\alpha;\beta]}\varphi$ is not defined).

For example, the formula $\varphi_{ex}: G_{[-3;0]}((c \in [0;1[) \wedge (c - V_{[-0.1]}c = 0))$, where $c \in [0;1[$ means $(0 \leq c) \wedge (c < 1)$, asserts that since 3 time units, the variable c is in the interval $[0;1[$ and has not changed.

Fault Characterization Formulas. All data sets $ObsBeh_{Cons}(F)$ of observable behaviors under a fault are obtained by simulation. By analyzing them, for each fault $F \in \Gamma$, a specific temporal formula φ_F is generated.

The normal formula φ_{F_0} consists in checking thresholds of the gap $|c - y|$ according to changes of orders during the time:

- Firstly, φ_{Fo} is divided into sub-formulas $\varphi^i{}_{Fo}$ according to a partition $C = U_{i\in E}C_i$ of the domain C. This partition represents all operating points of the system. Thus, when the order c is in a part C_i, the sub-formula $\varphi^i{}_{Fo}$ is checked.
- Secondly, each sub-formula $\varphi^i{}_{Fo}$ is divided into two sub-formulas: a sub-formula $\varphi^{i,h}{}_{Fo}$ for high changes of order; and another $\varphi^{i,l}{}_{Fo}$ for low changes. For a high change of order, more than a threshold, the sub-formula $\varphi^{i,h}{}_{Fo}$ is checked from the time of the change and during the response time δ of the system. Otherwise, for low changes of order less than the threshold, the sub-formula $\varphi^{i,l}{}_{Fo}$ is checked.

All faulty formulas are elaborated by taking into account behaviors and effects of faults. Thus, these characteristics are transformed, if it is possible, into temporal formulas describing how the fault disturbs the gap $|c - y|$ and all observable variables u and y.

For a fault $F \in \Gamma$, its temporal formula characterization is the property P_F defined as follow: for an observable behavior $ob \in ObsBeh_{Cons}$ and a time instant $t \in J$, $P_F(ob,t)$ is true iff $(ob,t) \models \varphi_F$. Temporal formulas are checked, thanks to the ARTiMon© tool, from the CEA, LIST ([9]), interfaced to MATLAB/Simulink©.

4.2 Diagnosability Study

Whatever is the considered faults characterization $\Lambda = (P_F)_{F \in \Gamma}$, we can give a general definition of diagnosability. Although we have presented two kinds of such characterizations (PC and TFC); another kind could be used.

Formal definitions. Given a faults characterization $\Lambda = (P_F)_{F \in \Gamma}$, a fault $F \in \Gamma$ is said *diagnosable* if it is eligible, detectable and isolable.

- A fault $F \in \Gamma$ is *eligible* iff for any instruction $cs \in Cons$ and for any occurrence time $t_n \in \Theta_{(F,cs)}$ of F ($t_n = 0$ for F_0), it exists a time instant $t_e \in [t_n;t_n+b]$ such that for all time instant $t \in J$ with $t \geq t_e$: $P_F(ObsB(cs,F,t_n),t)$ is true.
- A fault $F \in \Gamma \setminus \{F_0\}$ is *detectable* iff for any instruction $cs \in Cons$ and for any occurrence time $t_n \in \Theta_{(F,cs)}$, it exists a time instant $t_d \in [t_n;t_n+b]$ such that for all time instant $t \in J$ with $t \geq t_d$: $P_{F_0}(ObsB(cs,F,t_n),t)$ is false.
- A fault $F \in \Gamma \setminus \{F_0\}$ is *isolable* iff for any instruction $cs \in Cons$ and for any occurrence time $t_n \in \Theta_{(F,cs)}$, it exists a time instant $t_i \in [t_n;t_n+b]$ such that for all time instant $t \in J$ with $t \geq t_i$: $P_{F'}(ObsB(cs,F,t_n),t)$ is false for any other fault $F' \in \Gamma \setminus \{F_0;F\}$.

Analysis According to the Kind of Characterization. We have defined the diagnosability for any faults characterization $\Lambda = (P_F)_{F \in \Gamma}$; we can therefore point up some remarks.

The diagnosability is defined for all faults $F \in \Gamma$; thus for F_0, there is just to check its eligibility. Furthermore, this eligibility notion is added because as it is defined for any faults characterization Λ, it could be possible to consider one where some faults are not eligible. Obviously, it is not the case with PC: by definition of all sets $ObsBeh^{bd}{}_{Cons}(F)$, all faults are eligible.

Since the diagnosability is defined by the conjunction of three notions, we could study them independently. But in practical applications for example the isolability study of a fault $F \in \Gamma\backslash\{F_0\}$ will be considered only if F is detectable and with respect to other detectable faults $F' \in \Gamma\backslash\{F_0;F\}$. In fact, if a fault is not detectable for the diagnosability study, it means that it will not be detected by the diagnosis system when it will occur. This is due to the fact that the diagnosability study inheres in the analysis process of the diagnosis system.

We can observe a useful result: if a fault is not diagnosable with PC, it will not be diagnosable with TFC. In fact, TFC is obtained by exploiting the data sets $ObsBeh_{Cons}(F)$, thus it can be considered as a reduction of these sets. TFC is therefore a reduction of PC and its diagnosability requirements are thus stronger.

A brief complexity analysis shows the advantage of using TFC for an embedded diagnosis system.

- Space complexity: first, at each t_k sample, the diagnosis system has to store the received observable data vector. In addition, it has to keep only previous required data vectors during a temporal window bounded according to the considered faults characterization: the bound b for PC and β for TFC (the size of the longest temporal window appearing in all formulas). Thus, it is a number $n_{Perf} = \#v \cdot (b/u)$ for PC and $n_{Form} = \#v \cdot (\beta/u)$ for TFC, where u is the time unit (e.g.: $u = 0.01$) and $\#v$ is the number of observable variables. Then, the diagnosis system must store all formulas for TFC; whereas all data sets $ObsBeh^{bd}_{Cons}(F)$ for PC. It means a number of data $m_{Form} = n_{Form} + \Pi_{F \in \Gamma}(2^{h(F)})$ for TFC and $m = \Sigma_{F \in \Gamma}(\Sigma_{cs \in Cons}(m_{F,cs}))$ for PC (where $h(F)$ is the level of the formula F, $m_{Fo,cs} = (\#J \cdot n_{Perf})$ and $m_{F,cs} = (\#\Theta_{(F,cs)}) \cdot (b/u) \cdot n_{Perf})$, $\#J$ is the number of time instants between b and max(I) and $\#\Theta_{(F,cs)}$ is the number of time occurrences F according to cs).
- Time complexity: for each t_k sample the embedded diagnosis system must check the real observed behavior of the system during a temporal window of operation, according to the considered faults characterization, against all data of observable behavior previously stored, also according to the kind of faults characterization. This time complexity is therefore proportional to the space complexity.

Therefore, PC could not be exploited by an embedded diagnosis system. First, the diagnosis system should have enough memory to store all these data sets $ObsBeh^{bd}_{Cons}(F)$ for each fault $F \in \Gamma$, and enough computational power to make all comparisons. Second, as we have seen, the set $Cons$ of instructions must be the most representative: the set C^I of all functions from $I \subseteq T$ to C. Therefore this solution would be unusable for complex system with large ranges of operation. Nevertheless, it is useful during design and development phases of the system to intrinsically ensure the diagnosability of faults identified during the safety analysis.

Finally, TFC is perfectly adapted to be embedded inside a diagnosis system. In fact, conception and development phases are actually more and more accomplished with simulation tools which are able to generate the source code of control laws directly for a controller. We could add the embeddable ARTiMon© technology (involved in the diagnosability study), with all temporal formulas, during the source code generation; it will be the embedded diagnosis system completing the controller.

5 Practical Applications

With the above theoretical framework, we have defined the diagnosability study of faults by analyzing their observable behavior obtained by models. By using simulation tools, such as MATLAB/Simulink$^©$, we can apply this theoretical framework in practical applications.

First of all, we require a simulation model of the system. Simulation models used in control design could be a first solution; nevertheless, as showed in [10], we have to keep in mind that models needed for diagnosis are not the same that models needed for control. A model for control is generally less complex than a model for diagnosis.

We assume to have a simulation model of the system (the process and its controller) and furthermore we assume the model of the process represents perfectly the real system.

5.1 A Simulation Model Example

For example, we consider a part of a simulation model, developed in Matlab/Simulink$^©$, of a fuel cell system embedded in an electric vehicle ([7]). The concerned part is the air alimentation line of the fuel cell stack. To summarize, thanks to a compressor at the beginning and a valve at the end of the line, this air line has to provide air to the fuel cell stack at specific mass flow rates and pressures supplied by the global controller of the fuel cell system. This model can be described, in a simplified manner for the system part, by the set (3) of equations:

$$\begin{aligned}
P(t+1) &= k(Q_{in}(t) - Q_{out}(t)), \; P(0) = 1 \\
Q_{in}(t) &= f_{in}(W(t), P(t)) \\
Q_{out}(t) &= f_{out}(X(t), P(t)) \\
W(t) &= \alpha_W \cdot u_W(t) \\
X(t) &= \alpha_X \cdot u_X(t) \\
y_P(t) &= \lambda_P \cdot P(t) \\
y_Q(t) &= \lambda_Q \cdot Q_{in}(t)
\end{aligned} \quad (3)$$

P is the air pressure in the line. Q_{in} and Q_{out} are air mass flow rates respectively before and after the stack. W is the compressor speed rotation and X is the valve opening. u_W and u_X are respectively compressor and valve orders from the air line controller; y_P and y_Q are respectively air pressure and air mass flow rate measures. k, α_W, α_X, λ_P, λ_Q are constants.

These mass flow rate and pressure are controlled according to orders (c_Q for mass flow rate and c_P for pressure) supplied by the global controller of the fuel cell system. The supplied mass flow rate order c_Q is computed, for the most part, according to the electrical power needed by the vehicle controller (the operator). The pressure order c_P is then deduced from this mass flow rate according to pressure requirements in the stack and in the line. Thus, we can assume the mass flow rate order c_Q is the main order. Therefore, the set of observable variables of this air line is $V_{Obs} = \{c_Q; c_P; u_W; u_X; y_Q; y_P\}$.

All important faults of the air line have been identified and integrated in the simulation model by using an adapted faults library developed in MATLAB/Simulink© ([7]). For example, we consider a lock of the compressor F_{lock} which causes an abrupt decrease of the mass flow rate output of the compressor; and during the fault presence, this mass flow rate output stays equal to 0 gram per second. Within the model, variable Q_{in} is therefore disturbed by a multiplicative perturbation and is described by $Q_{in,Flock}(t) = (1 - flt(t,t_n)) \cdot Q_{in}(t)$, where flt is the behavior of F_{lock}, parameterized to occur abruptly at a time instant t_n and stay permanent (so $flt(t,t_n) = 0$ before t_n and 1 after).

5.2 Observable Behaviors Obtained by Simulations

We have considered a set *Cons* of instructions representing all operation ranges of the air line. By simulating the normal and all faulty models, according to the set *Cons* of instructions, we have obtained all data sets $ObsBeh_{Cons}(F)$ of observable behaviors for all identified faults $F \in \Gamma$ parameterized according to their behaviors and effects ([7]).

Fig. 2 illustrates two observable behaviors obtained for a random instruction cs during the temporal window [0;20]. In the two figures, first and second graphs show the mass flow rate and the pressure in the air line: dotted lines represent orders c_Q and c_P whereas plain lines represent measures y_Q and y_P from sensors. The third graph shows orders from the air line controller: compressor orders u_W in plain line and valve orders u_X in dotted line. Fig. 2 on left shows the normal observable behavior $ObsB(cs,F_0,0)$ whereas Fig. 2 on right shows the faulty observable behavior $ObsB(cs,F_{lock},13)$, for the lock of the compressor occurring at the time instant 13. During the time interval [0;13[, the air line operates correctly, as the normal behavior; but from the time instant 13, we can see disturbances in all graphs. Mass flow rate measures (first graph) decrease abruptly to 0 gram per second; compressor orders (third graph) are therefore maximal (equal to 1) in order to compensate the difference between orders and measures. Pressure measures (second graph) are thus equal to 1 bar.

5.3 Results on the Air Line Model

The diagnosability study of faults in the air line model was achieved with use of the temporal formulas characterization.

For example, the conjunction φ_{Fo}, of the two following formulas, characterizes the observable behavior of the air line in the normal case:

φ^1_{Fo}: $(c_Q \in [0;25[) \Rightarrow$
$((\,(\,G_{[-3;0]}(c - V_{[-0.01]}c = 0) \Rightarrow ((|c_Q - y_Q| \leq 0.5) \wedge (|c_P - y_P| \leq 0.1))\,)$
$\vee\,(\,E_{[-3;0]}(c - V_{[-0.01]}c \neq 0) \Rightarrow ((|c_Q - y_Q| > 0) \wedge (|c_P - y_P| > 0))\,)\,)$

φ^2_{Fo}: $(c_Q \in [25;30]) \Rightarrow$
$((\,(\,G_{[-3;0]}(c - V_{[-0.01]}c = 0) \Rightarrow ((|c_Q - y_Q| \leq 1) \wedge (|c_P - y_P| \leq 0.5))\,)$
$\vee\,(\,E_{[-3;0]}(c - V_{[-0.01]}c \neq 0) \Rightarrow ((|c_Q - y_Q| > 0) \wedge (|c_P - y_P| > 0))\,)\,)$

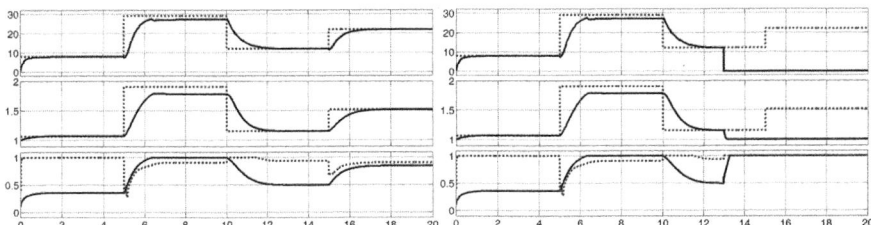

Fig. 2. Observable behaviors obtained by simulations (left: normal; right: lock of compressor)

The following formula φ_{lock} characterizes the observable behavior of the air line under the lock of the compressor:

$\varphi_{lock}: (\ G_{[-1;0]}(\ (y_Q = 0) \wedge (y_P = 1) \wedge (u_W = 1) \wedge (u_X = 1)\ ))$

We consider the set of faults $\Gamma = \{F_0; F_{lock}\}$, where F_0 is the normal case and F_{lock} is the lock of the compressor only occurring at the time instant $t_n = 13$. We suppose the set *Cons* of instructions is reduced to $\{cs\}$ and we consider a bound $b = 5$. Thus, $ObsBeh_{\{cs\}} = \{ObsB(cs, F_0, 0); ObsB(cs, F_{lock}, 13)\}$, with $\Theta_{(F_0, cs)} = \{0\}$ and $\Theta_{(F_{lock}, cs)} = \{13\}$.

Firstly, the normal observable behavior $ObsB(cs, F_0, 0)$ satisfies the formula φ_{F_0}. Therefore, the fault F_0 is eligible.

Secondly, the faulty observable behavior $ObsB(cs, F_{lock}, 13)$ satisfies the faulty formula φ_{lock} from the time instant $t_e = 14, 3 \in [13; 18]$; therefore, the fault F_{lock} is eligible. In addition, from the time instant 13, this faulty observable behavior $ObsB(cs, F_{lock}, 13)$ does not satisfy the normal formula φ_0; the fault F_{lock} is therefore detectable.

Of course, it is just an example. Firstly, the set of instructions is not reduced to only one but contains several ones representing all operation ranges of the system. Moreover, all identified faults have been taken into account with several occurrence times: before or after a change of orders and according the response time δ of the system. Furthermore all real temporal formulas obtained are more elaborated.

6 Conclusions and Perspectives

In this paper, our goal was to exploit faulty models of a technological system to study faults diagnosability. We have first presented the theoretical framework to define observable behaviors of a system under, or not, a fault. It considers a model of the system with its controller and integrates faults, preliminary identified and classified, in this model. Then, according to a set of instructions representing all operation ranges of the system, we have defined observable behaviors in the normal and all faulty cases. They consist in sets of vectors of observable data ordered by time according to a given instruction of the set of instructions.

The notion of faults diagnosability was defined regardless of a considered faults characterization, describing, the most precisely, how the system operates under or not a fault. Two faults characterizations were proposed. A perfect one, well adapted for the study during design and development phases of the system, simply considers all sets of data vectors, restricted to temporal windows. The other one uses temporal logic formalism to express the temporal evolution of observable data of the system and is adapted for an embedded diagnosis system.

Finally, for diagnosable faults, their characterization will then be embedded inside the diagnosis system in order to detect and identify faults on line. It could be combined with an embedded model of the system simulated by the controller and temporal formulas could take into account the temporal evolution of the difference between real and model outputs data. These future works will be presented in forthcoming papers.

References

1. Venkatasubramanian, V., Rengaswamy, R., Yin, K., Kavuri, S.N.: A review of process fault detection and diagnosis. 'part I to III'. Computers and Chemical Engineering 27, 293–346 (2003)
2. Travé-Massuyès, L., Cordier, M.O., Pucel, X.: Comparing diagnosability in CS and DES. In: International Workshop on Principles of Diagnosis, Aranda de Duero, Spain, June 26-28 (2006)
3. Blanke, M., Kinnaert, M., Lunze, J., Staroswiecki, M.: Diagnosis and Fault-tolerant Control. Springer, Berlin (2003)
4. Isermann, R.: Fault Diagnosis Systems. Springer, Berlin (2006)
5. Stapelberg, R.F.: Handbook of Reliability, Availability, Maintainability and Safety in Engineering Design. Springer, London (2009)
6. Basseville, M., Nikiforov, I.V.: Detection of Abrupt Changes: Theory and Application. Prentice-Hall, Englewood Cliffs (1993)
7. Batteux, M., Dague, P., Rapin, N., Fiani, P.: Fuel cell system improvement for model-based diagnosis analysis. In: IEEE Vehicle Power and Propulsion Conference, Lille, France, September 1-3 (2010)
8. Alur, R., Feder, T., Henzinger, T.A.: The Benefits of Relaxing Punctuality. Journal of the ACM 43, 116–146 (1996)
9. Rapin, N.: Procédé et système permettant de générer un dispositif de contrôle à partir de comportements redoutés spécifiés, French patent n°0804812 pending, September 2 (2008)
10. Frank, P.M., Alcorta García, E., Köppen-Seliger, B.: Modelling for fault detection and isolation versus modelling for control. Mathematics and Computers in Simulation 53(4-6), 259–271 (2000)

Using Ensembles of Regression Trees to Monitor Lubricating Oil Quality[*]

Andres Bustillo[1], Alberto Villar[2], Eneko Gorritxategi[2],
Susana Ferreiro[2], and Juan J. Rodríguez[1]

[1] University of Burgos, Spain
{abustillo,jjrodriguez}@ubu.es
[2] Fundación TEKNIKER, Eibar, Guipúzcoa, Spain
{avillar,egorritxategi,sferreiro}@tekniker.es

Abstract. This work describes a new on-line sensor that includes a novel calibration process for the real-time condition monitoring of lubricating oil. The parameter studied with this sensor has been the variation of the Total Acid Number (TAN) since the beginning of oil's operation, which is one of the most important laboratory parameters used to determine the degradation status of lubricating oil. The calibration of the sensor has been done using machine learning methods with the aim to obtain a robust predictive model. The methods used are ensembles of regression trees. Ensembles are combinations of models that often are able to improve the results of individual models. In this work the individual models were regression trees. Several ensemble methods were studied, the best results were obtained with Rotation Forests.

1 Introduction

One of the main root causes of failures within industrial lubricated machinery is the degradation of the lubricating oil. This degradation comes mainly by the oxidation of the oil caused by overheating, water and air contamination [1]. The machinery stops and failures coming from oil degradation reduces its life and generates important maintenance costs. Traditionally, the lubricating oil quality was monitored by traditional off-line analysis carried out in the laboratory. This solution is not valid to detect the early stages of degradation of lubricating oil because it is expensive and complex to extract oil from the machine and deliver it to a laboratory where this test can be performed. Therefore, the machine owner usually changes the lubricating oil without any real analysis of its quality, just following the specifications of the supplier which are always highly conservative.

The advances in micro and nano-technologies in all fields of engineering have opened new possibilities for the development of new equipment and devices and therefore have permitted the miniaturization of such systems with an important reduction of manufacturing costs.

---

[*] This work was supported by the vehicle interior manufacturer, Grupo Antolin Ingenieria S.A., within the framework of the project MAGNO2008 - 1028.- CENIT Project funded by the Spanish Ministry of Science and Innovation.

Taking all this into account the use of on-line sensor for lubricating oil condition monitoring will permit in a short period of time the optimization of its life and result in significant savings in operational costs by solving problems in industrial machinery, increasing reliability and availability [2]. The main advantage of using this type of sensor is that real time information could be extracted in order to establish a proper predictive and proactive maintenance. Therefore early stages of lubricating oil degradation could also be detected [3]. To evaluate the degradation status of a lubricating oil, the most often used parameter is the Total Acid Number (TAN). This parameter accounts for acid compounds generated within the oil and, therefore, about oil degradation status.

Although there are some examples of artificial intelligence techniques applied to oil condition monitoring, specially Artificial Neural Networks [4], no previous application of ensembles techniques has been reported for this industrial task. There are a few examples of the use of ensembles to predict critical parameters of different manufacturing processes. Mainly, ensembles have been used for fault diagnosis in different milling operations [5,6]. This lack of previous work is mainly due to the traditional gap between data mining researchers and manufacturing engineers and is slowly disappearing thanks to the development of comprehensible software tools that implement the main ensembles techniques like WEKA [7] that can be used directly by manufacturing engineers without a deep knowledge of ensembles techniques.

The purpose of this work is to describe a new on-line sensor that includes a novel calibration process for oil condition monitoring. The new sensor should not require the same accuracy and precision as laboratory equipment, because its aim is to avoid the change of lubricating oil simply because it already has lasted the mean life specified by the oil supplier, although its condition could still be suitable for further operation. Therefore the new sensor should be able to detect medium degradation of lubricating oil working under industrial conditions and its cost should be considerable lower than laboratory equipment.

The paper is structured as follows: Section 2 explains the experimental procedure followed to obtain the dataset to validate the new calibration process proposed in this work; Section 3 reviews the ensemble learning techniques tested for the calibration process; Section 4 shows the results obtained with all the tested ensembles techniques. Finally, conclusions and future lines of work are summarized in Section 5.

2 Experimental Set Up and Data Acquisition

2.1 On-Line Sensor Description

Lubricating oil suffers a colour change while it is degrading which is possible to detect by means of optical techniques. Figure 1 shows this process with a collection of samples of a commercial lubricating oil in different stages of degradation. It can be seen that, in the initial stage, the colour of the lubricating oil is yellow and while the lubricating oil continues to degrade, the colour turns brown. Finally the lubricating oil is black when the oil is fully degraded [8].

Considering the fact of colour change in lubricating oils during its operation life, Tekniker has developed a micro spectrometer sensor system to monitor oil condition. The spectrometer is designed to detect visible light absorption in the range from 320

Fig. 1. Fuchs Renolyn Unysin CLP 320 samples with different degradation stages

to 780 nm. The sensor has been developed taking into account end user requirements. Various knowledge fields, optic, electronic and fluidic, were necessary in order to obtain a robust equipment, Figure 2. Finally, a software tool manages the calibration and the data treatment [9].

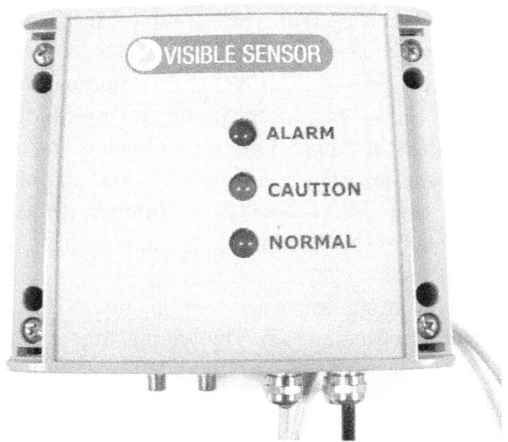

Fig. 2. On-line visible sensor used to obtain the spectrum data

2.2 Data Acquisition

179 samples of 14 different lubricating oils have been considered for the definition and validation of the new sensor's calibration process. The lubricating oils are used in hydraulic systems like presses, cutters, windmills and lubricated systems like gears. The samples suffer two different analysis.

First of all, they were analysed with a laboratory standard equipment at Wearcheck laboratory in Tekniker. TAN was analysed according to the American Society for Testing and Materials (ASTM) D974-04. This method covers the determination of acid compounds in petroleum products by titration against colour indicator [10]. TAN's variation was calculated considering the TAN value of the same commercial oil but new.

Then, the samples were analysed with the new on-line sensor and the generated spectra was used as input data for the calibration process. The output variable was the variation of the TAN value. Three measurements per sample have been carried out in order to obtain an average measurement in order to minimize the error. The oil samples have been introduced in the sensor through a 5 ml syringe and between every measurement the fluidic cell was cleaned twice with petroleum ether. The fluidic cell is the compartment where the light source is directed on the oil sample. Some time is necessary in order to obtain proper stabilization of the oil sample in the fluidic cell. All samples were measured in a temperature range of 21-22°C, due to the effect of variation of signal with temperature in the visible spectra.

3 Ensembles of Regression Trees

3.1 Regression Trees

Regression trees [11] are simple models for predicting. Figure 3 shows a regression tree obtained for the considered problem. In order to predict with these trees, the process starts in the root node. There are two types of nodes. In internal nodes a condition is evaluated: an input variable is compared with a threshold. Depending of the result of the evaluation, the process selects one of the node children. The leaves have assigned a numeric value. When a leaf is reached, the prediction assigned is the value of the leaf.

The displayed variables in Figure 3 are the following: intensities of the emission's spectrum at a certain wavelength (e.g., 431nm, 571nm...), intensity of the maximum of the emission's spectrum (**Max**), Total Acid Number of the analysed oil before its use (**Tan_ref**) and the wavelength (expressed in nanometres) of the maximum of the emission's spectrum (**Pos**).

In order to avoid over-fitting, it is usual to prune trees. The pruning method used in this work is Reduced Error Pruning [12]. The original training data is partitioned in two subsets, one is used to construct the tree, the other to decide which nodes should be pruned.

Trees are specially suitable for ensembles. First, they are fast, both in training and testing time. This is important when using ensembles because several models are constructed and used for prediction. Second, they are unstable [13]. This means that small changes in the data set can cause big differences in the obtained models. This is desirable in ensembles because combining very similar models is not useful.

3.2 Ensemble Methods

Several methods for constructing ensembles are considered. If the method used to construct the models to be combined is not deterministic, an ensemble can be obtained based only on Randomization: different models are obtained from the same training data.

In Bagging [13], each model in the ensemble is constructed using a random sample, with replacement, from the training data. Usually the size of the sample is the same as the size of the original training data, but some examples are selected several times in

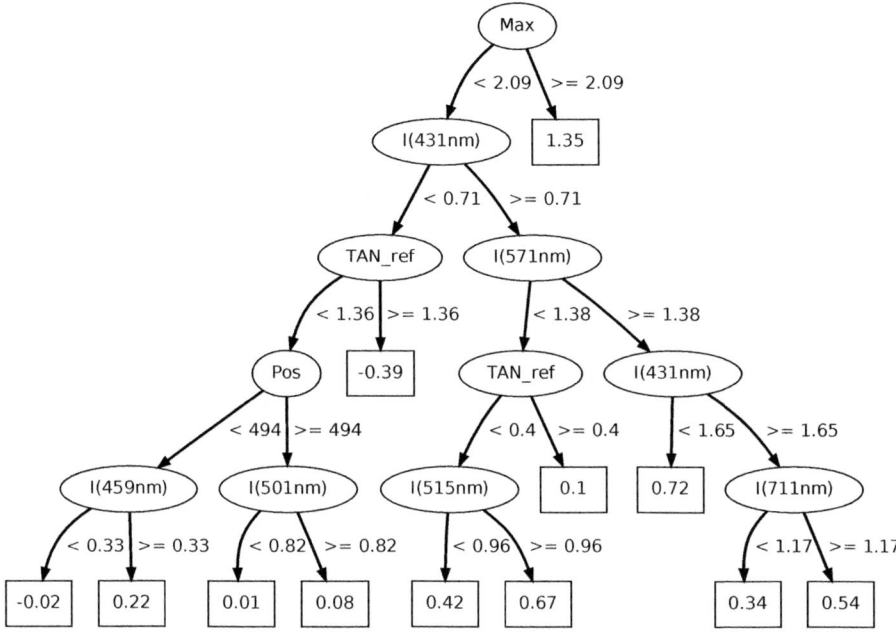

Fig. 3. Example of regression tree generated by WEKA

the sample, so other examples will not be included in this sample. The prediction of the ensemble is the average of the models predictions.

Random Subspaces [14] is based on training each models using all the examples, but only a subset of the attributes. The prediction of the ensemble is also the average of the models predictions.

Iterated Bagging [15] can be considered an ensemble of ensembles. First, a model based on Bagging is constructed using the original training data. The following Bagging models are trained for predicting the residuals of the previous iterations. The residuals are the difference between the predicted values and the actual values. The predicted value is the sum of the predictions of the Bagging models.

AdaBoost.R2 [16] is a boosting method for regression. Each training example has assigned a weight. The models have to be constructed taking these weights into account. The weights of the examples are adjusted depending on the predictions of the previous model. If the example has a small error, its weight is reduced, otherwise its weight is augmented. The models also have weights and better models have greater weights. The prediction is obtained as the weighted median of the models' predictions.

Rotation Forest [17,18] is an ensemble method. Each model is constructed using a different rotation of the data set. The attributes are divided into several groups and for each group Principal Component Analysis is applied to a random sample from the data. All the obtained components from all the groups are used as new attributes. All the training examples are used to construct all the models and the random samples are only used for obtaining different principal components. The ensemble prediction is obtained as the mean value from the predictions of the models.

4 Results and Discussion

4.1 Settings

The models were constructed and evaluated using WEKA [7]. As performance measures, the Mean Absolute Error (MAE) and Root Square Mean Error (RMSE) were used.

The results were obtained using 10-fold cross validation, repeated 10 times. In 10-fold cross validation, the data set is randomly divided in 10 disjoint folds of approximately the same size. For each fold, a model is constructed using the 9 other folds and this model is evaluated on this fold. The performance of the method is obtained as the average performance for 10 folds. In order to reduce the variability of this estimator, cross validation is repeated 10 times. Hence, the reported results are the average of 10 cross validations. This is the same as the average of the performance of 100 models.

Several methods have options or parameters. For regression trees a pruning process can be used with the objective of reducing over-fitting. To denote if the trees are pruned or unpruned the suffixes *(P)* and *(U)* are used.

The number of trees in the ensembles was set to 100. All the considered ensemble methods are used with pruned and unpruned trees, with only one exception: the method named as Randomization. In this method the only difference among the different trees in the ensemble is that they are constructed using a different random seed. In pruned trees, the original data is randomly divided in two parts, one for constructing the tree and the other for pruning. Nevertheless, the construction of pruned trees is deterministic. Hence, with this method only pruned trees are considered.

For the Random Subspaces method, two subspace sizes were considered: 50% and 75% of the original space size. For Iterated Bagging two configurations were considered 5×20 (Bagging of 20 trees is iterated 5 times) and 10×10 (Bagging of 10 trees is iterated 10 times). In both cases the number of trees in the ensemble is 100. For AdaBoost.R2, three loss functions were considered: linear, square and exponential. They are denoted, respectively, with the suffixes "-Li", "-Sq" and "-Ex".

Other methods were also included in the study as base line methods: predicting the average value, Linear Regression, Nearest Neighbour, Multilayer Perceptrons and Support Vector Machines.

4.2 Results

Table 1 shows the obtained results for the different methods. The results are sorted according to the average MAE and RMSE. The best results are from Rotation Forests, combining Regression Trees without pruning. The next methods are AdaBoost.R2 with unpruned trees and Rotation Forests with pruned trees.

In general, ensembles of unpruned trees have better results than ensembles of pruned trees. In fact, a single unpruned regression tree has better MAE than all the ensembles of pruned trees, with the only exception of Rotation Forest.

The best results among the methods that are not ensembles of regression trees are for the Nearest Neighbour method.

As expected, the worst results are obtained with the simplest model, predicting the average value.

Table 1. Experimental results

(a) Mean Absolute Error			(b) Root Mean Squared Error		
Method	Average	Deviation	Method	Average	Deviation
Rotation Forest (U)	0.1103	0.0298	Rotation Forest (U)	0.1708	0.0621
AdaBoost.R2-Sq (U)	0.1190	0.0323	AdaBoost.R2-Sq (U)	0.1793	0.0649
AdaBoost.R2-Ex (U)	0.1221	0.0332	AdaBoost.R2-Ex (U)	0.1813	0.0663
AdaBoost.R2-Li (U)	0.1223	0.0366	Rotation Forest (P)	0.1828	0.0611
Rotation Forest (P)	0.1230	0.0313	AdaBoost.R2-Li (U)	0.1832	0.0730
Nearest Neighbour	0.1241	0.0453	AdaBoost.R2-Li (P)	0.1843	0.0562
Subspaces-50% (U)	0.1263	0.0401	AdaBoost.R2-Ex (P)	0.1880	0.0549
Bagging (U)	0.1273	0.0375	AdaBoost.R2-Sq (P)	0.1886	0.0540
Subspaces-75% (U)	0.1295	0.0416	Bagging (U)	0.1908	0.0649
Iterated Bag. 10×10 (U)	0.1308	0.0370	Subspaces-50% (U)	0.1941	0.0666
Regression tree (U)	0.1341	0.0445	Iterated Bag. 10×10 (U)	0.1966	0.0641
Iterated Bag. 5×20 (U)	0.1347	0.0390	Subspaces-75% (U)	0.1975	0.0698
AdaBoost.R2-Li (P)	0.1348	0.0278	Bagging (P)	0.1992	0.0643
Iterated Bag. 5×20 (P)	0.1356	0.0425	Subspaces-50% (P)	0.2003	0.0613
Subspaces-50% (P)	0.1359	0.0379	Iterated Bag. 5×20 (U)	0.2010	0.0652
Iterated Bag. 10×10 (P)	0.1361	0.0417	Subspaces-75% (P)	0.2010	0.0642
Subspaces-75% (P)	0.1362	0.0389	Randomization (P)	0.2021	0.0647
Bagging (P)	0.1363	0.0389	Iterated Bag. 5×20 (P)	0.2023	0.0810
Randomization (P)	0.1383	0.0388	Iterated Bag. 10×10 (P)	0.2032	0.0767
AdaBoost.R2-Sq (P)	0.1388	0.0280	Nearest Neighbour	0.2086	0.0773
AdaBoost.R2-Ex (P)	0.1410	0.0267	Regression tree (U)	0.2086	0.0790
Support Vector Machines	0.1498	0.0371	Support Vector Machines	0.2182	0.0630
Regression tree (P)	0.1571	0.0470	Multilayer Perceptron	0.2349	0.1010
Multilayer Perceptron	0.1661	0.0590	Regression tree (P)	0.2350	0.0802
Linear Regression	0.1750	0.0400	Linear Regression	0.2410	0.0660
Mean value	0.3193	0.0684	Mean value	0.4324	0.1103

5 Conclusions

This work describes a novel calibration process for an on-line sensor for oil condition monitoring. The calibration process is done using ensembles of regression trees. Several ensemble methods were considered, the best results were obtained with Rotation Forest.

The validation of the new calibration process with a dataset of 179 samples of 14 different commercial oils in different stages of degradation shows that the calibration can assure an accuracy of ±0.11 in TAN prediction. This accuracy is worse than accuracy of laboratory equipment, which is usually better than ±0.03, but it is enough for the new on-line sensor because it achieves an accuracy better than ±0.15 , This fact is due to the sensor's purpose: to avoid the change of lubricating oil simply because it already has lasted the mean life specified by the oil supplier, although its condition could still be suitable for further operation.

Future work will focus on the study and application of the calibration process to on-line sensors devoted to a certain application. These sensors will achieve a higher accuracy but with a smaller variation range in the measured spectra. The ensembles used in this work combine models obtained with regression trees and ensembles combining models obtained with other methods will be studied. Moreover, it is also possible to combine models obtained from different methods; this heterogeneous ensembles could improve the results.

References

1. Noria Corporation: What the tests tell us, http://www.machinerylubrication.com/Read/873/oil-tests
2. Gorritxategi, E., Arnaiz, A., Spiesen, J.: Marine oil monitoring by means of on-line sensors. In: Proceedings of MARTECH 2007- 2nd International Workshop on Marine Technology, Barcelona, Spain (2007)
3. Holmberg, H., Adgar, A., Arnaiz, A., Jantunen, E., Mascolo, J., Mekid, J.: E-maintenance. Springer, London (2010)
4. Yan, X., Zhao, C., Lu, Z.Y., Zhou, X., Xiao, H.: A study of information technology used in oil monitoring. Tribology International 38(10), 879–886 (2005)
5. Cho, S., Binsaeid, S., Asfour, S.: Design of multisensor fusion-based tool condition monitoring system in end milling. International Journal of Advanced Manufacturing Technology 46, 681–694 (2010)
6. Binsaeid, S., Asfour, S., Cho, S., Onar, A.: Machine ensemble approach for simultaneous detection of transient and gradual abnormalities in end milling using multisensor fusion. Journal of Materials Processing Technology 209(10), 4728–4738 (2009)
7. Hall, M., Frank, E., Holmes, G., Pfahringer, B., Reutemann, P., Witten, I.H.: The WEKA data mining software: An update. SIGKDD Explorations 11(1) (2009)
8. Terradillos, J., Aranzabe, A., Arnaiz, A., Gorritxategi, E., Aranzabe, E.: Novel method for lube quality status assessment base on-visible spectrometric analysis. In: Proceedings of the International Congress Lubrication Management and Technology LUBMAT 2008, San Sebastian, Spain (2008)
9. Gorritxategi, E., Arnaiz, A., Aranzabe, E., Aranzabe, A., Villar, A.: On line sensors for condition monitoring of lubricating machinery. In: Proceedings of 22nd International Congress on Condition Monitoring and Diagnostic Engineering Management COMADEM, San Sebastian, Spain (2009)
10. Mang, T., Dresel, W.: Lubricants and lubrication. WILEY-VCH Verlag GmbH, Weinheim, Germany (2007)
11. Witten, I.H., Frank, E.: Data Mining: Practical Machine Learning Tools and Techniques, 2nd edn. Morgan Kaufmann, San Francisco (2005)
12. Elomaa, T., Kääriäinen, M.: An analysis of reduced error pruning. Journal of Artificial Intelligence Research 15, 163–187 (2001)
13. Breiman, L.: Bagging predictors. Machine Learning 24(2), 123–140 (1996)
14. Ho, T.K.: The random subspace method for constructing decision forests. IEEE Transactions on Pattern Analysis and Machine Intelligence 20(8), 832–844 (1998)
15. Breiman, L.: Using iterated bagging to debias regressions. Machine Learning 45(3), 261–277 (2001)
16. Drucker, H.: Improving regressors using boosting techniques. In: ICML 1997: Proceedings of the Fourteenth International Conference on Machine Learning, pp. 107–115. Morgan Kaufmann Publishers Inc., San Francisco (1997)
17. Rodríguez, J.J., Kuncheva, L.I., Alonso, C.J.: Rotation forest: A new classifier ensemble method. IEEE Transactions on Pattern Analysis and Machine Intelligence 28(10), 1619–1630 (2006)
18. Zhang, C., Zhang, J., Wang, G.: An empirical study of using rotation forest to improve regressors. Applied Mathematics and Computation 195(2), 618–629 (2008)

Image Region Segmentation Based on Color Coherence Quantization

Guang-Nan He[1], Yu-Bin Yang[1], Yao Zhang[2], Yang Gao[1], and Lin Shang[1]

[1] State Key Laboratory for Novel Software Technology, Nanjing University,
Nanjing 210093, China
[2] Jinlin College, Nanjing University, Nanjing 210093, China
yangyubin@nju.edu.cn

Abstract. This paper presents a novel approach for image region segmentation based on color coherence quantization. Firstly, we conduct an unequal color quantization in the HSI color space to generate representative colors, each of which is used to identify coherent regions in an image. Next, all pixels are labeled with the values of their representative colors to transform the original image into a "Color Coherence Quantization" (CCQ) image. Labels with the same color value are then viewed as coherent regions in the CCQ image. A concept of "connectivity factor" is thus defined to describe the coherence of those regions. Afterwards, we propose an iterative image segmentation algorithm by evaluating the "connectivity factor" distribution in the resulted CCQ image, which results in a segmented image with only a few important color labels. Image segmentation experiments of the proposed algorithm are designed and implemented on the MSRC datasets [1] in order to evaluate its performance. Quantitative results and qualitative analysis are finally provided to demonstrate the efficiency and effectiveness of the proposed approach.

Keywords: Visual coherence; color coherence quantization; connectivity factor; color image segmentation.

1 Introduction

To segment an image into a group of salient, coherent regions is the premise for many computer vision tasks and image understanding applications, particularly for object recognition, image retrieval and image annotation. There are already many segmentation methods currently available. If an image contains only several homogeneous color regions, clustering methods such as mean-shift [2] is sufficient to handle the problem. Typical automatic segmentation methods include stochastic model based approaches [4,14], graph partitioning methods [5,6], etc. However, the problem of automatic image segmentation is still very difficult in many real applications. Consequently, incorporation of prior knowledge interactively into automated segmentation techniques is now a challenging and active research topic. Many interactive segmentation methods have recently been proposed [7, 8, 9, 10,16]. Unfortunately, those kinds of method are still not suitable for segmenting a huge amount of natural images. In real situations, images such as natural scenes are rich in colors and textures, for which segmentation results achieved by the current available segmentation algorithms are still far away to match human's perception.

A typical way to identify salient image regions is to enhance the contrast between those regions and their surrounding neighbors [3][15]. In this work we aim to segment an image into a few coherence regions effectively, which is very useful for image retrieval, object recognition and image understanding purposes. The coherence regions are all semantically meaningful objects that can be discovered from an image. To achieve this goal, we propose a novel segmentation approach in which only low-level color features are needed. An image is firstly quantized into a CCQ image based on color consistency quantization. After that, an iteration is designed to segment the CCQ image to achieve the final segmentation result. The main contributions of this paper are listed as follows.

(1) We introduced a new color coherence quantization method in the HSI color space to generate the CCQ image, which has been validated by our experiments.
(2) We proposed a practical salient image region segmentation algorithm based on the concept of "coherence region", which involves maximizing "connectivity factor" and suppressing small connected component of each coherent region.

The remainder of the paper is organized as follows. Section 2 describes the process of color consistency quantization. The segmentation approach is then presented in Section 3. Experimental results are illustrated and analyzed in Section 4. Finally, concluding remarks with future work directions are provided in Section 5.

2 Generate the CCQ Image

The CCQ image is an approximate map to the original image, which reflects specific patterns of color or structure distribution in the original image. To generate the CCQ image, we can use clustering methods such as mean-shift or k-means. However, parameter selection is an unavoidable problem for those methods. If we use different parameters in these algorithms to generate the CCQ image, the results will be totally different as well. In order to improve this situation, we introduce a color quantization process to avoid the parameter selection problem.

The HSI color space describes perceptual color relationships more accurately than the RGB color space, while remaining computationally simple [11]. HSI stands for *hue*, *saturation*, and *lightness*, where *saturation* and *intensity* are two independent parameters. The idea of color quantization is originated from a common observation that there are usually several major coherent regions in a nature scene image, which denotes natural objects such as "sky", "grass", and "animals", etc. Based on that, we may quantize the HSI color representation into 66 types of representative colors empirically.

Let H, S and I be hue, saturation and intensity of each pixel respectively, and i denotes the color label. The quantization algorithm based on unequal intervals in the HSI color space is described as follows.

(1) *if* $S \leq 0.1$, partition the HSI color space into three types of color according to the intensity value by using Eq. (1).

$$\begin{cases} i = 0, if\ I \leq 0.2 \\ i = 1, if\ I > 0.2\ \&\ I \leq 0.75 \\ i = 2, if\ I > 0.75 \end{cases} \quad (1)$$

(2) otherwise, the hue, saturation and intensity value of each pixel are quantized by using Eq. (2), (3), and (4), respectively.

$$h' = \begin{cases} 0, if\ H \in (330, 360] \cup [0, 20] \\ 1, if\ H \in (20, 45] \\ 2, if\ H \in (45, 75] \\ 3, if\ H \in (75, 165] \\ 4, if\ H \in (165, 200] \\ 5, if\ H \in (200, 270] \\ 6, if\ H \in (270, 330] \end{cases} \quad (2)$$

$$s' = \begin{cases} 0, if\ S \in (0.1, 0.3] \\ 1, if\ S \in (0.3, 0.5] \\ 2, if\ S \in (0.5, 1] \end{cases} \quad (3)$$

$$i' = \begin{cases} 0, if\ I \in [0, 0.2] \\ 1, if\ I \in (0.2, 0.75] \\ 2, if\ I \in (0.75, 1] \end{cases} \quad (4)$$

Then, the color label i can be generated according to Eq. (5).

$$i = 3 + 9 * h' + 3 * s' + i' \quad (5)$$

Fig. 1. Color Coherence Quantization Result. (*Left*: the original image, *Right*: the quantized CCQ image.)

In this manner, all remaining colors in the HSI color space are partitioned into the other 63 types of representative colors. By applying the above quantization algorithm, a color image is firstly transformed from the RGB color space into the HSI color space. Then, all pixels are labeled with the values of their representative colors to obtain a "Color Coherence Quantization" (CCQ) image. Each representative color is labeled as i, where i is in the range [0, 65]. Figure 1 shows an example of our color coherence quantization result. The image shown in Figure 1 contains three coherence regions: "cow", "grass" and "land". In the CCQ image, each of the three regions has been labeled with several different numbers. Our aim in the next step is to label the CCQ image by using only three numbers, in which each number denotes a semantic region.

3 Salient CCQ Image Region Segmentation

In this section we describe our segmentation algorithm based on the generated CCQ images. Take Figure 1 as an example. The CCQ image has 33 types of color labels. However, the regions with the same color label vary in their sizes. Here the size of each region is calculated as the pixel number in the region divided by the size of the original image. To address this problem, we design a segmentation approach based on the "connectivity" of those regions in a region-growing manner. Even if an object in the image is described using different representative colors, the proposed algorithm can still segmented it as a whole.

3.1 The Connectivity Factor

In order to describe the connectivity of all color labels in a CCQ image, we define eight types of connectivity for each color label. The connectivity type of a color label i ($i \in$ [0,65]) is decided by the number of other label j in its 8-connectivity neighborhood in a 3×3 window. Take the color label "0" in Figure 2 as an example. The label "0" is in the window center, and the eight types of its connectivity are all shown in Figure 2. The notation '*' denotes color labels different from label "0". Rotating each type of connectivity does not change its connectivity factor value.

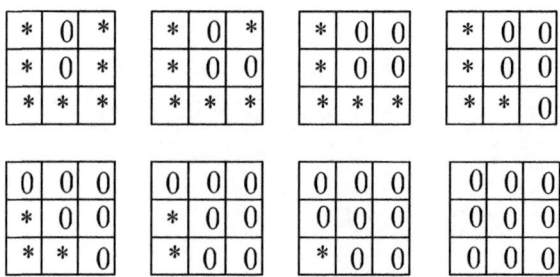

Fig. 2. The eight types of connectivity

Let a_i denote the size (number of pixels) of label i in a CCQ image; c_i^k denotes the total number of $k-$connectivity of label i, where $k \in [1,8]$, $i \in [0,65]$. Let σ_i denote the connectivity factor of label i, which is calculated as:

$$\sigma_i = \frac{\sum_{k=1}^{8} k \times c_i^k}{8 \times a_i}, \quad \sigma_i \in [0,1), \qquad (6)$$

According to the above definitions, the connectivity factors of label "0" in Figure 2 (image size: 3×3) can be easily calculated, as shown in Table 1.

Table 1. The connectivity factors of label "0"

k	1	2	3	4	5	6	7	8
σ_0	0.125	0.25	0.375	0.4	0.4167	0.4643	0.5313	0.5556

The connectivity factor of label i becomes larger as its size grows. Table 2 lists this changing trend of the connectivity factors of label i.

As we can see from Table 2, the connectivity factor value turns close to 1 as the size of label i grows. But it will rarely equal to 1 except in some limited situations. We find that this is very helpful for us to perform segmentation task in a region-growing manner. At the beginning, the generated CCQ image may have a lot of small regions labeled with different color values. If we try to maximize their connectivity factor by merging those small regions, they may finally merge into a few large and coherent regions, which can serve as a satisfied image segmentation result.

Table 2. Different connectivity factor σ_i of label i

Size of Label i	Connectivity factor σ_i
10×10	0.64
20×20	0.81
40×40	0.9025
80×80	0.9506
100×100	0.9604
200×200	0.9801
300×300	0.9867
400×400	0.99

An example is illustrated in Figure 3 to help understand the label's connectivity factor and its relationship between the label's size easily. For the circle in Figure 3, all pixels in it are represented as a symbol '#'. Now suppose the circle's radius is r, and in order to simplify the calculation, we can assume that each edge-point and its neighbors are all 4-connected, while the others are 8-connected. Therefore, the connectivity factor of the circle is $\sigma_{\#} = \frac{\pi r^2 \times 8 - 2\pi r \times 4 \times 2}{\pi r^2 \times 8} = 1 - \frac{2}{r}$. When $r = 20$, the circle's connectivity factor reaches 0.9, such that we can segment it out as an homogenous region.

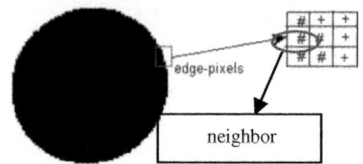

Fig. 3. An example of connectivity factor

In the segmentation process, two parameters of label i, σ_i and a_i, are calculated if and only if the label i appears in the CCQ image. As mentioned above, there are possibly many color labels with different sizes in a CCQ image, and a few types of color labels may occupy more than 90% of the whole CCQ image. Hence, their connectivity factors are possibly small due to their scattered distributions. Since small regions often have scattering distribution with very low connectivity factors, an extra step is added to maximize the region's connectivity factor by merging all small coherent regions before segmentation in order to solve this problem.

3.2 Merge Coherent Regions

In a CCQ image, each type of color label represents a coherent region. Let $R(n) = [i, \sigma_i, a_i]$ denote the nth coherent region, where i is the color label, σ_i is the corresponding connectivity factor, a_i is the region size, and $n = \{1, 2, ...N\}$. Here N denotes the total number of coherent regions in the CCQ image.

Most of the coherent regions will merge when the values of their connectivity factors are very small. The lower connectivity factors the regions have, the more scattering distributions they possess. A lower bound for coherent region's connectivity factor should be specified by user. For different categories of image, the low bound should be different. If a coherent region's connectivity factor is smaller than the lower bound, the coherent region will be merged. The regions with smaller connectivity factors are often incompact and usually have more than one connected components. Considering the spatial distribution of a coherent region, we need to merge the coherent region's connected components one by one. In order to find the connected components of a coherent region, we use the method outlined in Ref. [13] and assume each connected component in an 8-connectivity manner.

If region $R(n) = [i, \sigma_i, a_i]$ is merged, all of its connected components will be identified. First, let p be one of the connected components. Then, we need to decide which adjacent region of p should be merged with. One intuitive way is to merge

p into the nearest region. For each edge point of p, we construct a 3×3 window with the edge point located in the center of window. Those generated grid windows contain labels different from i. We then accumulate all the labels different from i in all those grid windows and find out the region with the largest number of those labeled points. It is then identified as the nearest region to p and merged with p.

To simplify the algorithm we only use the grid window centered at label i. When the total number of i is no less than 5 in the grid window, the point will be assumed as an edge point of region p, which means that label i at least has a connectivity of 4. The reason for choosing 5 as the threshold is to avoid some special situations, such as a region p with many holes (here "holes" indicates labels different from i). An example is illustrated in Figure 4. The region shown in it has a hole of label "0". The hole is usually detected as an edge of the region by using traditional techniques. However, when we calculate the edge point of label "0" by using our algorithm, the hole will not be identified as an edge. This rules out the unnecessary noises and helps to increase the robustness in the segmentation process by avoiding erroneous merges.

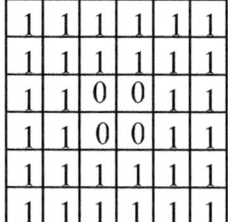

Fig. 4. Region with a hole

3.3 CCQ Image Segmentation

A practical iterative algorithm is then developed in this section to finally segment a CCQ image. Our algorithm receives a color image as an input, and transforms it into a CCQ image by performing the color coherence quantization. Next, we construct a description for the CCQ image by calculating every coherent region's connectivity factors and sizes.

The algorithm needs to set a threshold of connectivity factor, $\sigma_{threshold}$, so that the coherent region $R(n)$ will be merged if and only if $\sigma_i < \sigma_{threshold}$. The order to merge coherent regions is crucial to the final segmentation result, because if region i merges into region j, the connectivity factor of region j will increase significantly. Therefore, we firstly sort all coherent regions in ascending order according to their sizes. Then, coherent regions are strictly merged according to this order, starting from the smallest coherent region. In most cases, connectivity factors of the smallest region is smaller than $\sigma_{threshold}$. The process repeats until all of the remaining connectivity factors are no less than $\sigma_{threshold}$. In our algorithm, the size of a CCQ image is a constant. Therefore, when a coherent region is merged, size and connectivity factor of the new region will increase, which is also shown in Table 2. The algorithm will stop when no connectivity factor of coherent regions in the CCQ image less than $\sigma_{threshold}$. Figure 5 illustrates the flow of our algorithm.

By applying the algorithm to a color image, it achieves a segmented result which has only a few color labels, among which some regions may have more than one connected components, as shown in the middle image of Figure 6. Hence, we need to merge the small connected components in those regions to make the segmentation result more coherent and consistent. We refer to this step as *minor-component suppression*. The minor-component suppression is similar to the merging of coherent regions, and the only difference is that it merges the connect components of coherent regions, rather than regions themselves. When a connect component of a region is very small, we should merge it. For example, if the size of a region's connect component is less than 5% of the whole region's size, it will merge into the neighborhood region. The right image in Figure 6 illustrates the results after performing the suppression.

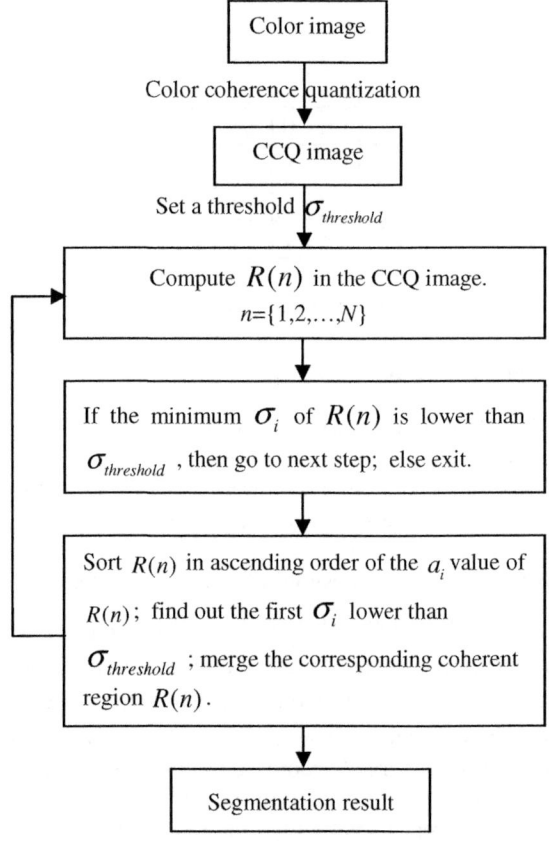

Fig. 5. Segmentation algorithm

It should be noted that the suppression step is optional. Whether use it or not makes no difference when an appropriate $\sigma_{threshold}$ is adopted in the algorithm. However, it is usually difficult to find an optimum $\sigma_{threshold}$ suitable for all kinds of images. Therefore, it is necessary to perform the suppression step in some applications.

Fig. 6. Segmentation results. *Left*: Original image; *Middle*: Results without performing minor-component suppression; *Right*: Results after suppression

4 Experimental Results

We have tested our algorithm on the MSRC challenging datasets, which contains many categories of high-resolution images with the size of 640 × 480 [1]. Different predetermined parameter $\sigma_{threshold}$ leads to different segmented results.

Firstly, we applied our algorithm on the "cow" images by using different $\sigma_{threshold}$. Examples of the segmentation results are shown in Figure 7.

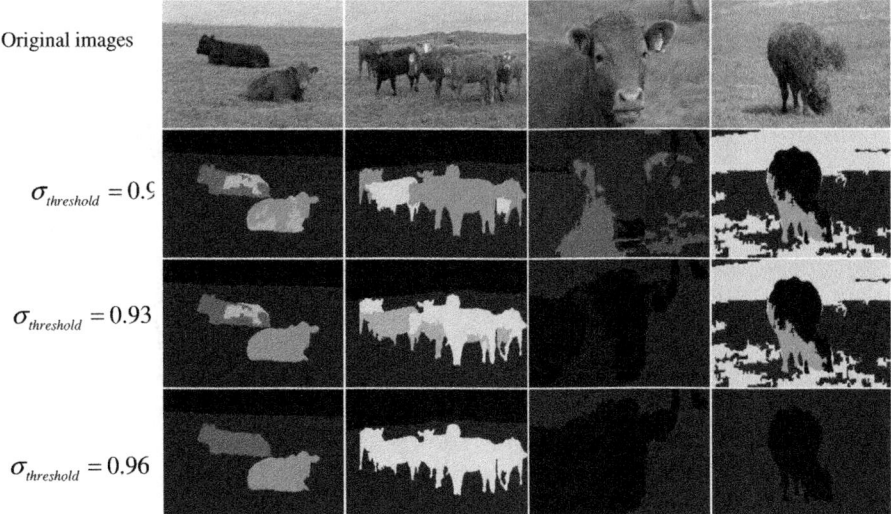

Fig. 7. Segmentation result by using different thresholds

As can be seen from Figure 7, the best result is obtained when $\sigma_{threshold} = 0.96$, and all the major objects have been segmented from the background. In order to evaluate the generalization ability of our algorithm, we also perform segmentation experiments on all other categories of images. Figure 8 illustrates the segmentation performance of our algorithm by setting $\sigma_{threshold} = 0.95$.

Fig. 8. Segmentation results on other categories of image

Fig. 9. Segmentation results on structured images

For images shown in Figure 7 and Figure 8, they are high-resolution and their contents are less structural, therefore we can set the threshold higher to get better results. On the contrary, for some low-resolution and highly structured images, we need to set the threshold lower in order to prevent over-segmentation. Figure 9 provides segmentation examples by applying our algorithm on low-resolution images with the size of 320×213 by setting $\sigma_{threshold} = 0.93$.

Since finding an appropriate threshold is important to achieve the best segmentation result, our method cannot segment some highly structured images as good as other categories of images. Take a "bicycle" image as an example. Figure 10 gives the segmentation results for an bicycle image. Our algorithm cannot segment the whole structure of the bicycle into a coherent region. However, the main parts of the bicycle, "tyre" and "seat", are both segmented. The segmentation result is still very useful for real applications.

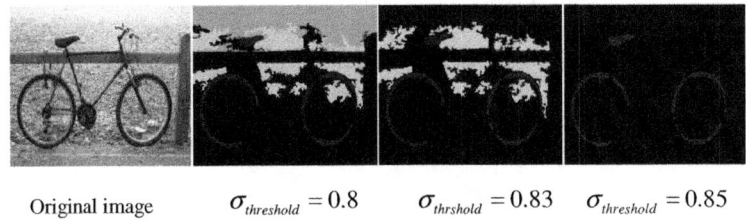

Original image $\sigma_{threshold} = 0.8$ $\sigma_{thrshold} = 0.83$ $\sigma_{threshold} = 0.85$

Fig. 10. Segmentation results on highly structured images

Furthermore, we also tested the widely-used mean-shift method [2] as a comparison with the proposed segmentation method in this paper. The results are shown in Figure 11.

As can be seen from Figure 11, the segmentation result is far from coherent and consistent according to our segmentation results. To obtain a more coherent and consistent result based on the mean-shift segmentation result, further processing, such as merge of small regions and small components, are necessary needed. Furthermore, the mean-shift algorithm usually takes more time than our algorithm under the same conditions.

Finally, in order to evaluate the general salient image region segmentation accuracy of our algorithm, we have tested our algorithm on all 21 object classes of MSRC database[12], composed of 591 photographs. The database has a large number of object categories and with dense labeling. Therefore, it is a good dataset for evaluating segmentation algorithms. We used six different thresholds and the corresponding accuracy rates (according to the dataset dense labeling) are shown in Figure 12. As can be seen from Figure 12, the algorithm works very well on image categories such as "flower", "building", "grass", "sky", "cow", and "sheep". It shows that our method can accurately segment the images into a few of salient regions which contain the objects, particularly for the natural images. For some structured object classes, such as "bike", "aeroplane", and "boat", the segmentation results is not that good. Therefore, it might be necessary to sufficiently address the structural characteristics of images in the future work.

Fig. 11. Mean-shift segmentation result

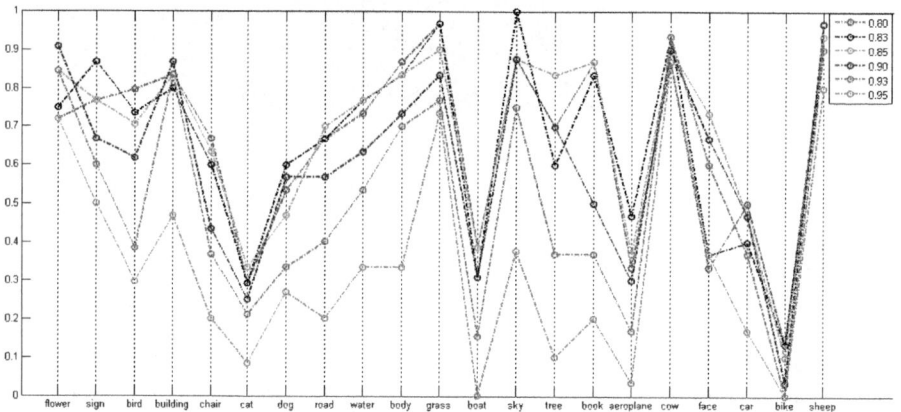

Fig. 12. Accuracy by using different thresholds on multiple-category images

5 Conclusions

In this paper, we present a novel approach for image region segmentation based on color coherence quantization. A concept of "connectivity factor" is defined, and an iterative image segmentation algorithm is proposed. Image segmentation experiments are designed and implemented on the MSRC datasets in order to evaluate the performance of the proposed algorithm. Experimental results show that the algorithm is able to segment a variety of images accurately by using a fixed threshold. Usually, the threshold suitable for high-resolution and unstructured object class images is larger than the threshold suitable for low-resolution and structured object class images. We also compare our algorithm with the mean-shift algorithm under the same conditions, and the results shows that our algorithm is much more preferable. Finally, quantitative experimental tests on multiple-category images of MSRC datasets have also been done to validate the algorithm.

The experiments also reveal the limitations of our algorithm in segmenting structured images. In the future we will improve our algorithm toward addressing the structural characteristics of images.

Acknowledgements

This work is supported by the "973" Program of China (Grant No. 2010CB327903), the National Natural Science Foundation of China (Grant Nos. 60875011, 60723003, 61021062, 60975043), and the Key Program of Natural Science Foundation of Jiangsu Province, China (Grant BK2010054).

References

1. Criminisi, A.: Microsoft research cambridge object recognition image database, http://research.microsoft.com/vision/cambridge/recognition/
2. Comaniciu, D., Meer, P.: Mean shift: A robust approach toward feature space analysis. IEEE Trans. Pattern Anal. Machine Intell. 24(5), 603–619 (2002)
3. Achanta, R., Estrada, F., Wils, P., Susstrunk, S.: Salient region detection and segmentation. In: International Conference on Computer Vision Systems (2008)
4. Wang, J.-P.: Stochastic relaxation on partitions with connected components and its application to image segmentation. PAMI 20(6), 619–636 (1998)
5. Shi, J., Malik, J.: Normalized cuts and image segmentation. IEEE Transactions on Pattern Analysis and Machine Intelligence 22(8), 888–905 (2000)
6. Tolliver, D.A., Miller, G.L.: Graph partitioning by spectral rounding: Applications in image segmentation and clustering. In: CVPR 2006: Proceedings of the 2006 IEEE Computer Society Conference on Computer Vision and Pattern Recognition, pp. 1053–1060. IEEE Computer Society Press, Washington, DC, USA (2006)
7. Blake, A., Rother, C., Brown, M., Perez, P., Torr, P.: Interactive image segmentation using an adaptive GMMRF model. In: Leonardis, A., Bischof, H., Pinz, A. (eds.) ECCV 2006. LNCS, vol. 3952, pp. 428–441. Springer, Heidelberg (2006)
8. Figueiredo, M., Cheng, D.S., Murino, V.: Clustering under prior knowledge with application to image segmentation. In: Schölkopf, B., Platt, J., Hoffman, T. (eds.) Advances in Neural Information Processing Systems, vol. 19. MIT Press, Cambridge (2007)
9. Guan, J., Qiu, G.: Interactive image segmentation using optimization with statistical priors. In: International Workshop on The Representation and Use of Prior Knowledge in Vision, In conjunction with ECCV 2006, Graz, Austria (2006)
10. Hduchenne, O.H., Haudibert, J.-Y.H., Hkeriven, R.H., Hponce, J.H., Hsegonne, F.H.: Segmentation by transduction. In: CVPR 2008, Alaska, US, June 24-26 (2008)
11. http://en.wikipedia.org/wiki/HSL_color_space
12. Malisiewicz, T., Efros, A.A.: Improving Spatial Support for Objects via Multiple Segmentations. In: BMVC (September 2007)
13. Haralick, R.M., Linda, G.: Shapiro, Computer and Robot Vision, vol. I, pp. 28–48. Addison-Wesley, Reading (1992)
14. Belongie, S., et al.: Color and texture based image segmentation using EM and its application to content-based image retrieval. In: Proc. of ICCV, pp. 675–682 (1998)
15. Matas, J., Chum, O., Urban, M., Pajdla, T.: Robust wide baseline stereo from maximally stable extremal regions. In: Proc. BMVC, pp. I384–I393 (2002)
16. Boykov, Y., Jolly, M.-P.: Interactive graph cuts for optimal boundary & region segmentation of objects in N-D images. In: Proc. ICCV, pp. 105–112 (2001)

Image Retrieval Algorithm Based on Enhanced Relational Graph

Guang-Nan He[1], Yu-Bin Yang[1], Ning Li[1], and Yao Zhang[2]

State Key Laboratory for Novel Software Technology, Nanjing University,
Nanjing 210093, China
Jinlin College, Nanjing University, Nanjing 210093, China
yangyubin@nju.edu.cn

Abstract. The "semantic gap" problem is one of the main difficulties in image retrieval task. Semi-supervised learning is an effective methodology proposed to narrow down the gap, which is also often integrated with relevance feedback techniques. However, in semi-supervised learning, the amount of unlabeled data is usually much greater than that of labeled data. Therefore, the performance of a semi-supervised learning algorithm relies heavily on how effective it uses the relationship between the labeled and unlabeled data. A novel algorithm is proposed in this paper to enhance the relational graph built on the entire data set, expected to increase the intra-class weights of data while decreasing the inter-class weights and linking the potential intra-class data. The enhanced relational matrix can be directly used in any semi-supervised learning algorithm. The experimental results in feedback-based image retrieval tasks show that the proposed algorithm performs much better compared with other algorithms in the same semi-supervised learning framework.

Keywords: Graph embedding; image retrieval; manifold learning; relevance feedback.

1 Introduction

In image retrieval, the most difficult problem is "semantic gap", which means that the features of different images are unable to discriminate from different semantic concepts. The relevance feedback technique provides a feasible solution to this problem [1][2]. Early relevance feedback technologies are mainly based on modifying the feedback information, that is, image features, such as re-weighting the query vectors [3], adjusting the query vector's positions, and etc. [4][5] In recent years, a large amount of image retrieval algorithms was proposed with the development of semi-supervised learning [6][7][8]. These algorithms generally use the feedback information to learn potential semantic distributions, so as to improve the retrieval performance. However, compared with the high dimensionality of image features, the information available for feedback is usually much less and inadequate. Manifold learning is a powerful tool for this kind of problem, with the goal of finding the regularity distribution in high dimensional data. It assumes that high-dimensional data lie on or distribute in a low dimensional manifold, and uses graph embedding technique to discover their low dimensions. The representative algorithms include ISOMAP [9], Local Linear Embedding [10], and Laplacian Eigenmap [11]. S. Yan et al. proposed a unified framework of graph embedding, under which most of data dimension reduction algorithms can be well integrated [12].

Manifold learning is proposed based on spectral graph theory. It firstly assumes that high-dimension data lie on or distribute in a low-dimension manifold, and then uses linear embedding methods to reduce their dimensions. Meantime, it can increase the distances of inter-class data points and reduce the distances of intra-class data points. In the above process, different data relationship matrices will influence the results of the corresponding algorithms greatly. Moreover, the performance of manifold learning also mainly depends on the relational graph built on data. In recent years, quite a few algorithms have been proposed to learn the image semantic manifold by using feedback information as the labeled data and the unlabeled data. Based on that, the relational graph can then be constructed simply by the K-nearest neighbor method, in which the connectivity value of the points belonging to the same K-nearest neighbor equals one. The representative algorithms are Augmented Relation Embedding (ARE)[6], Maximize Margin Projection (MMP)[7], and Semi-Supervised Discriminate Analysis (SDA)[8]. The MMP and ARE have made some improvements on the construction of the data's relational graph by modifying the weights in the local scope of the labeled data. But this improvement is still too limited. Take the MMP for an example, it uses the feedback information acquired from labeled data to increase the weights of intra-class data points and reduce the weights of inter-class data points. But that only changes the weights of the local scoped data points in the labeled data, which makes the improvement very trivial. To address the above issues, we proposed a novel algorithm to enhance the relational graph. It is proved to be capable of largely increasing the weights of intra-class data points while decreasing the weights of inter-class data points efficiently. Besides, the weights of the potential intra-class data points can be also increased as well. The algorithm outputs an enhanced relational graph, which possesses more information and is more instructive for feedback. Furthermore, we applied the enhanced relational graph in the framework of semi-supervised learning and improved the algorithm's performance effectively.

The rest of this paper is organized as follows. Section 2 briefly introduces the typical semi-supervised algorithms based on the framework of graph embedding. Section 3 presents the construction algorithm for enhanced relational graph. The experiment results are then illustrated in Section 4. Finally, Section 5 makes concluding remarks.

2 Data Dimension Reduction with Graph Embedding

In recent years, many semi-supervised algorithms have been proposed, such as ARE[6], MMP[7], SDA[8], and etc. A process of constructing the data relational graph is essential for those algorithms. After that, data relational matrix is then generated according to certain rules, in which the value of each element is calculated as a weight indicating the similarity of two points. The data relational matrix is very important to obtain better dimension reduction performance. In this section, we firstly discuss the problem of data dimension reduction based on graph embedding technique, and then make a brief introduction of ARE, MMP and SDA respectively.

2.1 Problem Definition

Data dimension reduction is naturally a manifold learning algorithm. It assumes that high dimensional data lie on a low dimensional manifold. After dimension reduction, data belonging to different classes will distribute in different manifolds. Let $X = [x_1, x_2, ... x_N]$, $x_i \in R^m$, denote the data sample, $G = \{X, W\}$ is a un-directional weighted graph, in which the vertices are the data sample X, and the similarity matrix of data points is denoted by $W \in R^{N \times N}$. Here W is a real symmetric matrix. The diagonal matrix D and Laplacian matrix L [11] can then be given as:

$$D_{ii} = \sum_j W_{ij}, \forall i \quad L = D - W \quad (1)$$

Then, a low dimension embedding can be achieved through a projection. Let A denote the projection matrix, it can be calculated by minimizing Eq. (2):

$$\sum_{ij} (A^T x_i - A^T x_j)^2 W_{ij} \quad (2)$$

Let a_j denote each column of A, Eq. (2) can be rewritten as $\arg\min_a \sum_{ij} (a^T x_i - a^T x_j)^2 W_{ij}$, where a denotes a projection vector.

Let $y_i = a^T x_i$, we have:

$$\sum_{ij} (y_i - y_j)^2 W_{ij} = \sum_{ij} y_i^2 W_{ij} - 2\sum_{ij} y_i y_j W_{ij} + \sum_{ij} y_j^2 W_{ij}$$
$$= 2\sum_i y_i^2 D_{ii} - 2\sum_{ij} y_i y_j W_{ij} = 2y^T (D - W) y \quad (3)$$
$$= 2y^T L y$$

Here y denotes the projection of all data on a, that is, $y = a^T X$; D_{ii} represents the number of points connected to the ith point, which to some extent reveals the importance of that point. By adding the constraints $y^T D y = 1$ and making coordinate transformation, the more important a point is, the closer it will be to the origin. Finally, the objective function is represented as:

$$a^* = \arg\min_a a^T XLX^T a, \quad a^T XDX^T a = 1 \quad (4)$$

From the above derivation, we can see that the similarity matrix W plays an important role in the procedure. The projective data point y is closely related to W as well. The larger W_{ij} is, the more similar x_i and x_j are. After dimension reduction, the distance between y_i and y_j should be smaller. Here the similarity relationship can be seen as the class label, and the intra-class point is more similar to each other. For those unlabeled data, the similarity can be measured by using the relationship of its neighboring points. Its neighboring points should be more similar than other points. For those data points neither belonging to the same class nor in the neighborhood, their similarity are generally set as $W_{ij} = 0$.

We will then introduce three representative semi-supervised learning algorithms proposed in recent years. They generally use the above dimension reduction method in relevance feedback process to solve image retrieval problem.

2.2 ARE (Augmented Relation Embedding)

In order to obtain the semantic manifold information preferably in feedback-based image retrieval, ARE method uses both image information and users' feedback information. It utilizes two relational graphs to record the positive and negative feedback sets, and take the local scoped relationships of the whole data as constraints. By maximizing the distance between the positive set and the negative set, the two relational graphs are calculated according to Eq. (5) and Eq. (6) respectively.

$$W_{ij}^P = \begin{cases} 1 & if \ x_i, x_j \in Pos \\ 0 & otherwise \end{cases} \quad (5)$$

$$W_{ij}^N = \begin{cases} 0 & otherwise \\ 1 & if \ x_i \in Pos \ and \ x_j \in Neg, \ x_j \in Pos \ and \ x_i \in Neg \end{cases} \quad (6)$$

where Pos and Neg denote the positive set and negative set respectively.

The relational matrix of the whole data set can then be denoted as:

$$W_{ij} = \begin{cases} 1, & if \ x_i \in N^k(x_j) \ or \ x_j \in N^k(x_i) \\ 0, & otherwise \end{cases} \quad (7)$$

where $N^k(x_i)$ denotes x_i's k-nearest neighbor set. Afterwards, we can use W to calculate the Laplacian matrix L. Similarly, W_{ij}^P and W_{ij}^N are used respectively to calculate L^N and L^P.

The objective function of ARE is then given as:

$$X[L^N - \gamma L^P]X^T a = \lambda XLX^T a \quad (8)$$

where γ is the ratio of the number of positive data points to the number of negative data points.

Eq. (8) finally provides the embedding vector of each image. Theoretically, those vectors represent the resulted images which match the user's search query better. Therefore, the feedback information can be used to modify W_{ij}^P and W_{ij}^N repetitively to finally improve the image retrieval accuracy.

2.3 MMP (Maximize Margin Projection)

The Maximize Margin Projection algorithm constructs two objective functions by using both the global data and the labeled data. One is used to maximize the distances of inter-class data points, and the other is used to preserve the global data structure information. It makes full use of each data point's neighboring information by redistricting the neighboring point set. The point x_i's neighbor set $N^k(x_i)$ is divided into two subsets: $N_b^k(x_i)$ and $N_w^k(x_i)$. The k-nearest neighbors of x_i belong to subset $N_b^k(x_i)$, and the other neighboring points with x_i are in subset $N_w^k(x_i)$. The global geometric information of the entire data set is described by using Eq. (7), and the relational matrix of MMP is given as:

$$W_{ij}^b = \begin{cases} 1 & \text{if } x_i \in N_b^k(x_j) \text{ or } x_j \in N_b^k(x_i). \\ 0 & \text{otherwise} \end{cases} \quad (9)$$

$$W_{ij}^w = \begin{cases} \gamma & \text{if } x_i \text{ and } x_j \text{ share the same label.} \\ 1 & \begin{array}{l} \text{if } x_i \text{ or } x_j \text{ unlabeled, and} \\ x_i \in N_w^k(x_j) \text{ or } x_j \in N_w^k(x_i) \end{array} \\ 0 & \text{otherwise} \end{cases} \quad (10)$$

The parameter γ is used to increase the weights between each positive point and each negative point. Suppose that the vector after a projection is $y=(y_1,y_2,...,y_m)^T$, the two objection functions are then calculated as follows:

$$\min \sum_{ij}(y_i - y_j)^2 W_{ij}^w \quad (11)$$

$$\max \sum_{ij}(y_i - y_j)^2 W_{ij}^b \quad (12)$$

Eq. (11) makes the data points belonging to the same class closer to each other after projection, while Eq. (12) makes the data points in different classes far away from each other. Through algebraic operations, the two objective functions can be derived as:

$$X(\alpha L_b + (1-\alpha)W^w)X^T a = \lambda X D^w X^T a \quad (13)$$

where L_b is the Laplacian matrix corresponding to W^b, D^w is the diagonal matrix corresponding to W^w. In feedback-based image retrieval, the feedback information in each iteration is used to modify W_{ij}^b and W_{ij}^w. α is the balance factor, generally set as 0.5. Similarly, in feedback-based image retrieval, repetitively modifying W_{ij}^b and W_{ij}^w will finally improve the image retrieval accuracy.

2.4 SDA (Semi-supervised Discriminate Analysis)

The SDA provides a learning framework for the feedback-based image retrieval. Generally, there is a great difference between the number of labeled data and unlabeled data. How to make full use of the unlabeled data is the key point of semi-supervised learning. The idea of SDA is similar to LDA, except that the LDA is a supervised learning algorithm and the SDA is a semi-supervised one. It utilizes a small number of labeled data to improve the discrimination of different class data. The key of this algorithm is to find the eigenvector by using the following equation:

$$XW_{SDA}X^T a = \lambda X(\tilde{I} + \alpha L)X^T a \qquad (14)$$

where L is the Laplacian matrix corresponding to the relational matrix defined in Eq. (7), W_{SDA} is the relational matrix for labeled images, and $W_{SDA} = \begin{bmatrix} W^l & 0 \\ 0 & 0 \end{bmatrix}$. The feedback information is preserved in W^l, where the number of positive images is l, and $W^l = \begin{bmatrix} W^{(1)} & 0 \\ 0 & W^{(2)} \end{bmatrix}$. The size of the matrix $W^{(1)}$ is $l \times l$, with all elements equal to $1/l$. Similarly, we can calculate the matrix $W^{(2)}$. $\tilde{I} = \begin{bmatrix} I & 0 \\ 0 & 0 \end{bmatrix}$, where I is a $l \times l$ unit matrix. Therefore, the rank of W_{sda} is 2, and the high dimensional image feature data is projected onto the two-dimensional data space. Repetitively modifying W_{SDA} and $\tilde{I}$ can improve the image retrieval accuracy.

Some researchers have summarized the above semantic learning methods as spectrum regression [13]. The objective functions of this kind of algorithms are a quadratic optimization problem in low dimensional space. Finally the optimization problem is transformed into the problem of solving eigenvalue. Generally it is similar to the spectrum clustering algorithm [14].

3 The Construction of Enhanced Relational Graph

The algorithms introduced above have different motivations, but their objective functions are similar: they are all general eigenvalue-solving problem. The only difference of them is that they use different algorithm to construct the relational graph, which is crucial for their performance. To address this issue, a novel algorithm for constructing enhanced relational graph is proposed in this paper. It utilizes the feedback information more comprehensively, and propagates this information both globally and locally to finally greatly improve the retrieval performance.

The enhanced relational graph is an improvement on $G = \{X, W\}$, which makes the projected data space more robust and accurate. The vertices of G are data points, and the connectivity between each pair of vertices is obtained by combing labeled information and neighboring information, as shown in Eq. (5), Eq. (6), Eq. (9), and Eq. (10). However, the neighboring points are not necessarily with the same class, and the points belonging to the same class may not be the neighbors. Therefore, our algorithm tries to build a more robust relational graph to simulate the semantic similarity between images more accurately.

3.1 Build Hierarchical Relational Matrix

In order to better distinguish the labeled data from the unlabeled data, we firstly build a hierarchical relational matrix given as:

$$W_{ij} = \begin{cases} \gamma & \text{if } x_i \text{ and } x_j \text{ share the same label} \\ 1 & \text{if } x_i \text{ or } x_j \text{ is unlabeled, and } x_i \in N^k(x_j) \text{ or } x_j \in N^k(x_i) \\ 0 & \text{if } x_i \text{ and } x_j \text{ have different label} \\ 0 & \text{otherwise} \end{cases} \quad (15)$$

where γ is an integer greater than 1, generally set as $\gamma = 30$.

The matrix contains more information including the information of labels, classes and neighbors. We then use the matrix W to construct the enhanced relational matrix. The basic idea here is how to propagate the neighbor information. The relationship representing "the neighbor's neighbor" is worth being enhanced, which means that the neighbor's neighbor has more chance to be similar and need to be enhanced. Since those points may have different label information, it will strengthen the relations between intra-class data points. Moreover, it will also enhance the relations globally and makes the generated relational graph more robust.

The construction of enhanced relational graph is described as follows.

The construction of enhanced relational graph
---
1. Input: relational matrix W and each data point's k-nearest neighboring set $N^k(x_i)$.
2. For $i = 1,...N$
    For $j = 1,...N$
        IF $\quad x_z \in X \quad$ and $\quad x_z \in N^k(x_i), x_z \in N^k(x_j) \quad$ THEN
        $\sum_z W_{iz} W_{jz} \leftarrow W_{ij}^*$.
        ENDIF
    EndFor
  EndFor
3. Output: W^*

Finally we generate the relational matrix W^*. From the above procedure, we can see that if two points have the same neighboring points, the relation between those two points will then be enhanced. The weight of those two points is calculated as the sum of the dot product of their common neighbors' weights.

3.2 ESDA

We then propose an enhanced SDA algorithm, ESDA, by applying the generated enhanced relational matrix to SDA. Similarly, it is easy to applied the enhanced relational graph to any other dimension reduction algorithms based on graph embedding technique, such as ARE and MMP. Here we just take SDA as an example.

The improvement of ESDA is that, it changes the Laplacian matrix L used in SDA to a different Laplacian matrix L^*, which corresponds to W^* shown in Eq. (14). The objective function of ESDA is then given as:

$$XW_{SDA}X^T a = \lambda X(\tilde{I} + \alpha L^*)X^T a \tag{16}$$

After dimension reduction, the dimension of the data space of ESDA is 2, similar to SDA. In feedback-based image retrieval, we use the feedback information to modify Eq. (15) to improve the retrieval performance iteratively.

Generally speaking, there are two ways to solve the above objective function. One is singular value decomposition [15] and the other is regularization [16]. In this paper, we adopt the regularization method to solve it. Regularization is simple and efficient to change a singular matrix to the non-singular one. Take Eq. (8) as an example. We just change the right part of the equation from XLX^T to $XLX^T + \alpha I$, where I is a unit matrix and a full rank matrix will be generated after this transformation.

4 Experimental Results

To evaluate the performance of the proposed algorithm, we apply it to the relevance feedback-based image retrieval. To simplify, we just choose ESDA algorithm as an example to carry out our experiment and prove the effectiveness of the proposed enhanced relational graph algorithm.

4.1 Experiment Settings

We used the Corel5K [17] as the data set in our experiments, in which 30 categories are tested. Each category has different semantic meanings and has one hundred images. Therefore, the whole data set consists of 3,000 images in total. Examples of the tested images are shown in Figure 1(b)-(d). The query image was randomly selected from each category, and the rest of images served as the searched database. To evaluate algorithm performance, we used Precision-Scope Curve as the benchmark [6][18], which is more preferable to measure the performance of image retrieval than Precision-Recall Curve. Because the searched results may contain a large number of images, and usually a user won't examine the entire result set to evaluate the retrieval performance. Therefore, checking the accuracy in a certain scope is more reasonable in performance evaluation.

For image features, we took color moments, Tamura texture features, Gabor texture features and color histogram as image features. The image features and their dimensions are shown in Table 1. Each image was described by a vector with dimension of 112.

Table 1. Image Features

Feature Type	Dimension
Color moments(RGB)[19]	9
Color moments(LUV)[19]	9
Tamura texture features[20]	6
Gabor texture features[21]	24
Color histogram(HSV)	64

For image retrieval, we used automatic relevance feedback image mechanism as follows. We took the top 10 resulted images as the feedback information. The feedback information contained a group of positive and negative samples. This is identical to the real scenarios. Therefore, users do not need to take a long time to interact with the image retrieval system. The positive images selected in the first feedback iteration were not allowed to be selected again in the next time. In order to improve the efficiency of the retrieval procedure, we took the top 400 images, instead of all of the 3,000 images as the global data set in the construction of relational graph.

4.2 Image Retrieval Experiments

In the experiment, we used the optimal values for all parameters, as indicated in the original literature. For example, we set $\gamma = 50$ in ARE and MMP, and set the number of reduced dimension as 30 [6][7], the number of neighbors is 5, that is, $k = 5$. Both ESDA and SDA reduce the original data space into 2-dimension data space. According to the configuration, we firstly chose an image randomly to act as the query image, and used the top 20 resulted images to evaluate the performance. For each class of images, we ran an algorithm 50 times and for each algorithm only one feedback was taken. Finally, we calculated the average retrieval accuracy of all classes for each algorithm. The experimental results are shown in Figure 1(a). The "Baseline" means the accuracy without feedback. From the result, we can see that the accuracies of ESDA are greater than those of ARE and MMP on almost all categories with the exception of Category 9 and Category 19. Particularly, compared to SDA, ESDA outperforms it greatly. It can also show that relevance feedback can improve the accuracy globally, which proves that relevance feedback mechanism is an effective way to solve the semantic gap problem [1][2].

In order to further evaluate the retrieval performance of those algorithms, we conducted an experiment to test the feedback effectiveness. In this experiment, we ran an algorithm 50 times for each class of images, and for each algorithm we took feedback for 4 iterations. We then calculate the average retrieval accuracy of all classes for each algorithm, and also the Precision-Scope Curve shown in Figure 2(a). The X-axis shows the scope like top 10, 20 etc, and the Y-axis shows the accuracy under the scope. The result shows that ESDA achieved the highest accuracy in each feedback, particularly higher than SDA. It further improves the effectiveness of the enhanced relational graph algorithm. Figure 2(b)-(e) shows the top 10 retrieval results for the category of "tiger" with no feedback, 1-iteration feedback, 2-iteration feedback and 3-iteration feedback, respectively.

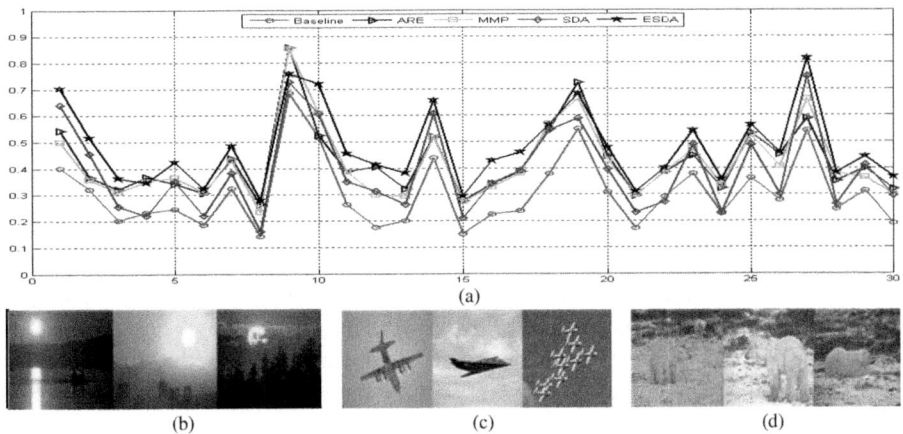

Fig. 1. The algorithms' accuracy curves after 1-iteration feedback, (b)-(d) Sample images from category 1, 2, and 3, respectively

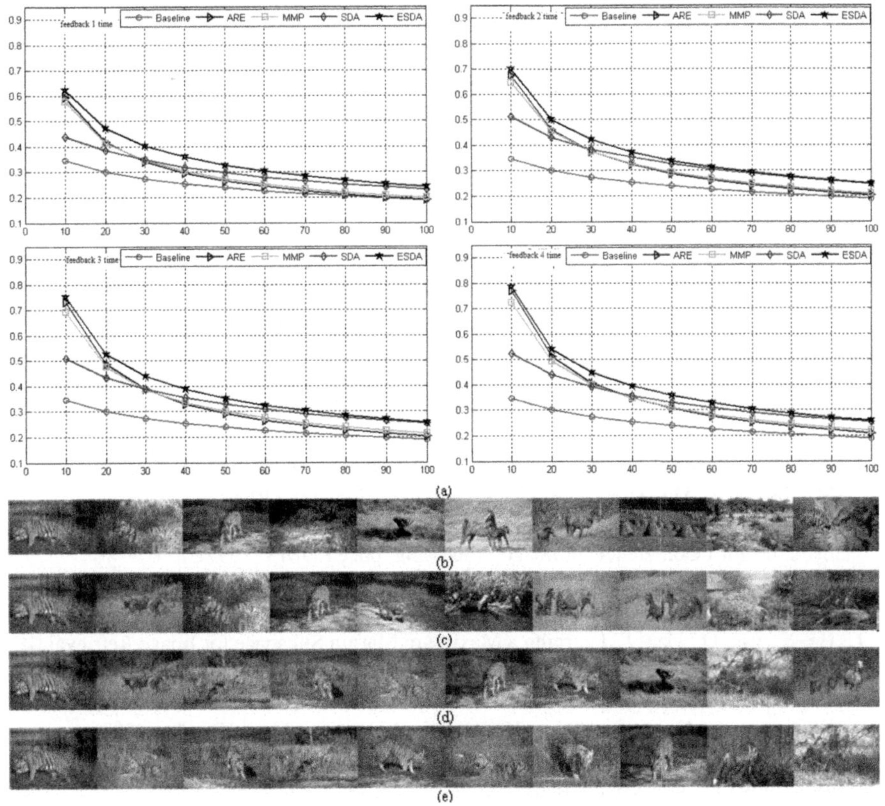

Fig. 2. (a) The algorithms' Precision-Scope Curves in a 4-iteration feedback; (b)-(e) Retrieval results with no feedback, 1-iteration feedback, 2-iteration feedback, and 3-itration feedback, respectively

5 Conclusions

A new algorithm for construct enhanced relational graph was proposed in this paper. Compared with traditional construction methods, the algorithm combines all of the class information in the data and effectively enhances the intra-class data relationship. It can be easily extended to the framework of semi-supervised learning. Based on the enhanced relational graph and SDA method, an algorithm called ESDA was developed. The experimental results in image retrieval show that ESDA outperforms the existed algorithms. The enhanced relational graph can also be used to improve the robustness and performance of MMP and ARE method. Therefore, it can be widely applied to the methods of data dimension reduction based on graph embedding. It can utilize the information from both labeled data and unlabeled data in the framework of semi-supervised learning to effectively improve the robustness of a semi-supervised learning algorithm.

Acknowledgements

We would like to acknowledge the supports from the National Science Foundation of China (Grant Nos. 60875011, 60975043, 60723003, 61021062), the National 973 Program of China (Grant No. 2010CB327903), and the Key Program of National Science Foundation of Jiangsu, China (Grant No. BK2010054).

References

1. Smeulders, W.M., et al.: Content-based image retrieval at the end of the early years. IEEE Trans. on Pattern Analysis and Machine Intelligence 22, 1349–1380 (2000)
2. Salton, G., McGill, M.: Introduction to Modern Information Retrieval. McGraw-Hill, New York (1982)
3. Ishikawa, Y., Subramanya, R., Faloutsos, C.: Mindreader: Querying databases through multiple examples. In: International Conference on Very Large Data Bases, pp. 218–227 (1998)
4. Porkaew, K., Chakrabarti, K.: Query refinement for multimedia similarity retrieval in MARS. In: ACM Conference on Multimedia, pp. 235–238 (1999)
5. Rui, Y., Huang, T., Mehrotra, S.: Content-based image retrieval with relevance feedback in mars. In: Int'l Conference on Image Processing, pp. 815–818 (1997)
6. Lin, Y.-Y., Liu, T.-L., Chen, H.-T.: Semantic manifold learning for image retrieval. In: Proceedings of the ACM Conference on Multimedia, Singapore (November 2005)
7. He, X., Cai, D., Han, J.: Learning a Maximum Margin Subspace for Image Retrieval. IEEE Transactions on Knowledge and Data Engineering 20(2), 189–201 (2008)
8. Cai, D., He, X., Han, J.: Semi-Supervised Discriminant Analysis. In: IEEE International Conference on Computer Vision (ICCV), Rio de Janeiro, Brazil (October 2007)
9. Tenenbaum, J., de Silva, V., Langford, J.: A global geometric framework for nonlinear dimensionality reduction. Science 290(5500), 2319–2323 (2000)
10. Roweis, S., Saul, L.: Nonlinear dimensionality reduction by locally linear embedding. Science 290(5500), 2323–2326 (2000)

11. Belkin, M., Niyogi, P.: Laplacian eigenmaps and spectral techniques for embedding and clustering. In: Advances in Neural Information Processing Systems, vol. 14, pp. 585–591. MIT Press, Cambridge (2001)
12. Yan, S., Xu, D., Zhang, B., Yang, Q., Zhang, H., Lin, S.: Graph embedding and extensions: A general framework for dimensionality reduction. TPAMI 29(1), 40–51 (2007)
13. Cai, D., He, X., Han, J.: Spectral Regression: A Unified Subspace Learning Framework for Content-Based Image Retrieval. In: ACM Multimedia, Augsburg, Germany (September 2007)
14. Von Luxburg, U.: A Tutorial on Spectral Clustering. Statistics and Computing 17(4), 395–416 (2007)
15. Stewart, G.W.: Matrix Algorithms: Eigensystems, vol. II. SIAM, Philadelphia (2001)
16. Friedman, J.H.: Regularized discriminant analysis. Journal of the American Statistical Association 84(405), 165–175 (1989)
17. Duygulu, P., Barnard, K., de Freitas, J.F.G., Forsyth, D.: Object recognition as machine translation: Learning a lexicon for a fixed image vocabulary. In: Heyden, A., Sparr, G., Nielsen, M., Johansen, P. (eds.) ECCV 2002. LNCS, vol. 2353, pp. 97–112. Springer, Heidelberg (2002)
18. Huijsmans, D.P., Sebe, N.: How to Complete Performance Graphs in Content-Based Image Retrieval: Add Generality and Normalize Scope. IEEE Trans. Pattern Analysis and Machine Intelligence 27(2), 245–251 (2005)
19. Stricker, M., Orengo, M.: Similarity of color images. In: SPIE Storage and Retrieval for Image and Video Databases III, vol. 2185, pp. 381–392 (February 1995)
20. Tamura, H., Mori, S., Yamawaki, T.: Texture features corresponding to visual perception. IEEE Trans. On Systems, Man, and Cybernetics Smc-8(6) (June 1978)
21. Bovic, A.C., Clark, M., Geisler, W.S.: Multichannel texture analysis using localized spatial filters. IEEE Trans. Pattern Analysis and Machine Intelligence 12, 55–73 (1990)

Prediction-Oriented Dimensionality Reduction of Industrial Data Sets

Maciej Grzenda

Warsaw University of Technology,
Faculty of Mathematics and Information Science,
00-661 Warszawa, Pl. Politechniki 1, Poland
M.Grzenda@mini.pw.edu.pl

Abstract. Soft computing techniques are frequently used to develop data-driven prediction models. When modelling of an industrial process is planned, experiments in a real production environment are frequently required to collect the data. As a consequence, in many cases the experimental data sets contain only limited number of valuable records acquired in expensive experiments. This is accompanied by a relatively high number of measured variables. Hence, the need for dimensionality reduction of many industrial data sets.

The primary objective of this study is to experimentally assess one of the most popular approaches based on the use of principal component analysis and multilayer perceptrons. The way the reduced dimension could be determined is investigated. A method aiming to control the dimensionality reduction process in view of model prediction error is evaluated. The proposed method is tested on two industrial data sets. The prediction improvement arising from the proposed technique is discussed.

1 Introduction

The development of data-based models of industrial processes requires a number of experiments to be performed. Such experiments may be difficult and costly [6,8]. Moreover, tests with a significant variation of the process parameters, in many cases outside optimal working limits, are frequently required [1]. In some cases, there is a very limited production capacity, available for experiment needs, too [6]. Hence, the resulting data sets used for process modelling are frequently limited in size. At the same time, the industrial processes can be described by many variables. As a consequence, the data sets arising from a limited set of experiments, but containing a significant number of attributes, may have to be used to set up process models with soft computing techniques. The high-dimensional data makes the training process much more difficult and error-prone. One of the main reasons is the "curse of dimensionality". As the number of dimensions grows, in the absence of simplifying assumptions, the number of data points required to estimate a function of several variables i.e. defined in a high-dimensional space, grows exponentially [5]. In the case of industrial processes, in spite of several dimensions being individual variables measured for a process of interest, frequently only a few hundred of records are available [1,8,9]. This clearly illustrates the "empty space phenomenon" i.e. the fact that in a highly-dimensional space only a very limited set of points is populated with the data. Not surprisingly, dimension reduction techniques are attempted to tackle the problem. These fall into two categories: linear and nonlinear transformations of the original space. Firstly, they contribute to the model development by reducing the complexity of it. At the same time, when a reduction to 1, 2 or 3 dimensions is made, the need for data visualisation and model interpretability is answered, too [7].

The primary objective of the work is to practically investigate the way dimensionality reduction (DR) techniques can be used with high-dimension industrial data sets. The key assumption is that the goal of the modelling is the prediction of process results under known conditions including, but not limited to, tool settings. Multilayer perceptrons (MLP) [2], being one of the most widely used neural network architectures, were applied as prediction models. As the MLP networks were trained with gradient-based methods, an important aspect of the investigation is the impact of finding local optima only on method evaluation. It is a well known issue that the training sessions based on MLP networks produce different results in subsequent runs, hence making the method evaluation more problematic. Moreover, in the case of the data sets of a limited size only, a division of a data set into a training and testing subset may largely affect the prediction error rates. Finally, when a limited number of records is available, the risk of model overfitting is largely increased. For these reasons, the selection of an optimal reduced dimension in terms of the predictive capabilities of a model built using the transformed data, requires a systematic approach.

Principal Component Analysis (PCA) [2,4,5] is one of the most frequently used DR techniques. Hence, the remainder of this paper is devoted to this method. In technical sciences, in some works, components representing 90% [4] or 95% [4,6,8] of total variance are preserved when preprocessing available data. It should be noted, however, that this and many other methods refrain from taking into account the number of records present in the data set.

Hence, the need to provide a meta algorithm aiming to simplify the decision on the optimal level of PCA-based DR, providing the best prediction capabilities of a resulting prediction model. Such an algorithm based on the use of PCA, is proposed in the work. The remainder of the work is organised as follows:

- the problem of dimensionality reduction and the main features of PCA are summarised in Sect. 2,
- the algorithm controlling the process of dimensionality reduction is proposed in Sect. 3,
- Next, the results of using an algorithm are discussed. The experiments performed with two different industrial data sets provide basis for this part.
- Finally, conclusions and the main directions of future research are outlined.

2 Dimensionality Reduction with PCA

2.1 The Role of Intrinsic Dimension

Many data sets used in machine learning problems contain numerous variables of partly unknown value and mutual impact. Some of these variables may be correlated, which may negatively affect model performance [4]. Among other problems, an empty space phenomenon discussed above is a major issue when multidimensional data sets are processed. Hence, the analysis of high-dimensional data aims to identify and eliminate redundancies among the observed variables [5]. The process of DR is expected to reveal underlying latent variables. More formally the DR of a data set $D \subset \mathbb{R}^S$ can be defined by two functions used to code and decode element $x \in D$ [5]:

$$c : \mathbb{R}^S \longrightarrow \mathbb{R}^R, \ x \longrightarrow \tilde{x} = c(x) \tag{1}$$

$$d : \mathbb{R}^R \longrightarrow \mathbb{R}^S, \ \tilde{x} \longrightarrow x = d(\tilde{x}) \tag{2}$$

The dimensionality reduction may be used for better understanding of the data including its visualisation and may contribute to model development process. The work concentrates on the latter problem of dimensionality reduction as a technique improving the quality of prediction models.

An important aspect of DR is the estimation of *intrinsic dimension*, which can be interpreted as the number of latent variables [5]. More precisely, the intrinsic dimension of a random vector y can be defined as the minimal number of parameters or latent variables needed to describe y [3,5]. Unfortunately, in many cases, it remains difficult or even impossible to precisely determine the intrinsic dimension for a data set of interest. Two main factors contribute to this problem. The first reason is noisiness of many data sets making the separation of noise and latent variables problematic. The other reason is the fact that many techniques such as PCA may overestimate the number of latent variables [5]. It is worth emphasizing that even if intrinsic dimension is correctly assessed, in the case of some prediction models built using the multidimensional data, some latent variables may have to be skipped to reduce model complexity and avoid model overfitting. In the case of MLP-based models, this can be due to the limited number of records comparing to the number of identified latent variables. The latter number defines the number of input signals and largely influences the number of connection weights that have to be set by a training algorithm. Nevertheless, since $R < S$, the pair of functions $c()$ and $d()$ usually does not define a reversible process i.e. for all or many $x \in D$ it may be observed that $d(c(x)) \neq x$. Still, the overall benefits of the transformation may justify the information loss caused by DR.

To sum up, for the purpose of this study an assumption is made that intrinsic dimension will not be assessed. Instead of it, the performance of prediction models built using transformed data sets $\tilde{D} \subset \mathbb{R}^R$ is analyzed.

2.2 Principal Component Analysis

One of the most frequently used DR techniques, being PCA, was used for the simulations performed in this study. The PCA model assumes that the S observed variables result from linear transformation of R unknown latent variables. An important requirement for the method is that the input variables are centred by subtracting the mean value and if necessary standardised. The standardisation process should take into account the knowledge on the data set and should not be applied automatically, especially in the presence of noise. Nevertheless, the basic initial transformation can be defined as $X_i \longrightarrow \frac{X_i - \mu_i}{\sigma_i}$, where μ_i and σ_i stand for mean and standard deviation of variable i, $i = 1, 2, \ldots, S$. In the remainder of the work, for the sake of simplicity, an input data set D denotes the data centred and standardised.

Let ϱ denote a $(S \times S)$ covariance matrix, $\lambda_1, \ldots, \lambda_S$ denote eigenvalues of the matrix sorted in descending order and $\mathbf{q}_1, \ldots, \mathbf{q}_S$ the associated eigenvectors i.e. $\varrho \mathbf{q}_i = \lambda_i \mathbf{q}_i, i = 1, 2, \ldots, S$. Finally, let $\mathbf{a} : a_j = \mathbf{q}_j^T \mathbf{x} = \mathbf{x}^T \mathbf{q}_j$, $j = 1, \ldots, S$. Then, the coding function $c^R(\mathbf{x})$ can be defined as follows: $c^R(\mathbf{x}) = [a_1, \ldots, a_R]$ [2]. Hence, $c^R(\mathbf{x}) : \mathbb{R}^S \longrightarrow \mathbb{R}^R$. Similarly, $d^R(\mathbf{a}) = \sum_{j=1}^{R} a_j \mathbf{q}_j$ [2].

One of the objective techniques of selecting the number of preserved components R and hence transforming $D \longrightarrow \tilde{D} \subset \mathbb{R}^R$ is based on *proportion of explained variation* (PEV) α [4,5]. In technical sciences, $\alpha = 0.95$ may be applied [4].

$$R = min_{l=1,2,\ldots,S} : \frac{\sum_{d=1}^{l} \lambda_d}{\sum_{d=1}^{S} \lambda_d} \geq \alpha \qquad (3)$$

According to this criterion, the reduced dimension R is set to the lowest number of largest eigenvalues satisfying the condition stated in formula 3.

DR, both in its linear and non-linear form is frequently used in engineering applications, especially in image processing problems. In the latter case, even mid-size resolution of images results in potentially thousands of dimensions. Hence, numerous applications of DR to problems such as shape recognition, face recognition, motion understanding or colour perception were proposed. Among other studies, [10] provides a survey of works in this field. Even more new applications of PCA are proposed, such as the detection of moving objects [11].

When modelling production processes, PCA is usually used to run a separate preprocessing process. In this case, frequently [6,8], a decision is made to leave the principal components representing $\alpha = 0.95$ of overall variance in the set, as stated in formula 3. In this way, a transformed data set $\tilde{D} \subset \mathbb{R}^R$ is obtained and used for model development. In the next stage, it is typically divided into the training and testing set [8] or a 10-fold cross-validation technique is applied to assess the merits of model construction algorithm [6]. In the latter stage, different prediction and classification models based on neural models, such as MLPs [8], Radial Basis Function (RBF) [8], or Support Vector Machines (SVM) [7] are frequently developed. It is important to note that usually a decision is made to apply an a priori reduction of a data set. Also, some other popular methods of selecting the reduced dimension R, such as *eigenvalue, scree plot* and *minimum communality criterion* [4] refrain from evaluating the impact of R on model creation and its generalisation capabilities.

3 Algorithm

Prediction models based on MLP networks can suffer from poor generalisation. When the data are scarce, and this is often the case, a model may not respond correctly on the data not used in the training process. In industrial applications, the main objective is the credibility of a model i.e. the minimisation of prediction error on new data. Therefore, the criterion for selecting the reduced dimension R might be expressed as follows:

$$R: E(M(\tilde{D}_T^R)) = min_{l=1,2,...,S} E(M(\tilde{D}_T^l)) \qquad (4)$$

where M denotes the prediction model trained on $\tilde{D}_L^l$ training set of dimension l, $\tilde{D}_T^l$ denotes the testing data set, and $E()$ the error observed on the testing set $\tilde{D}_T^l$, when using model M. In reality, the results of applying criterion 4 strongly depend on the division of $\tilde{D}^l$ into $\tilde{D}_T^l$ and $\tilde{D}_L^l$. Moreover, when a model M is an MLP network trained with gradient algorithms, diverse results of the training process are inevitable. Hence, the need for averaging the $E()$ error rates over different divisions of the $\tilde{D}^l$ data set and r algorithm runs. In addition, as individual gradient-based training sessions may result in extremaly large error rates, median error value can be calculated to provide a more representative error measure than mean value. Finally, validation data set $\tilde{D}_V^l$ can be created out of the training data to avoid model overfitting. Therefore, to address all these issues an algorithm 1 is proposed in this study.

The *CalculatePCAMatrix()* routine returns a PCA transformation matrix reducing the dimension from S to R i.e. implementing the coding function $c^R(x)$ and preserving $\frac{\sum_{d=1}^{R} \lambda_d}{\sum_{d=1}^{S} \lambda_d}$ of total variance. In addition, the following standard data mining procedures are vital to objectively estimate the merits of reduced dimension R when limited data sets D are involved:

- *elimination of outliers*. What should be emphasised is that when the size of the data set $card(D)$ is limited, the impact of individual records on final error rates is even higher. In our case $EliminateOutliers(P, \gamma, D)$ is a function eliminating any elements $a \in P$ and corresponding elements of D such that $a \notin [Q1 - \gamma(Q3 - Q1), Q3 + \gamma(Q3 - Q1)]$, where $Q1, Q3$ stand for the lower and upper quartiles of the data set.
- *cross-validation* (CV). CV is a standard procedure when working with data sets of limited size. Not only this procedure should be applied, but also CV folders should have possibly equal sizes i.e. $card(D_i) \geq \lfloor \frac{card(D)}{K} \rfloor, i = 1, \ldots, K$. Should some of the data sets D_i contain less than $\lfloor \frac{card(D)}{K} \rfloor$ elements, the gradient-based training process will rely on the validation error

Input: D - matrix of input attributes, $P \subset \mathbb{R}$ - vector of corresponding output features, $card(D)=card(P)$, R - reduced dimension, K - the number of CV folders, γ - outlier identification parameter, r - the number of training sessions

Data: F - $(R \times card(D))$ PCA transformation matrix, D_i - a family of K sets: $D_i \cap D_j = \emptyset, i \neq j$, $\cup_{i=1}^{K} D_i = D$; N - an average value of P; P_i, P_L, P_T - output features corresp. to D_i, D_L, D_T

Result: $E_L^R(D,P)$ - the average error rate of the models on the training sets; $E_T^R(D,P)$ - the average error rate of the models on the testing sets; $\beta^R(D,P)$ - the average proportion of testing error rates calculated against the naive baseline N.

begin
  $D,P = \text{EliminateOutliers}(P, \gamma, D)$;
  $F = \text{CalculatePCAMatrix}(D,R)$;
  for $i = 1, \ldots r$ **do**
    $\{D_j, P_j\}_{j=1,\ldots,K} = \text{DivideSet}(D,P,K)$;
    for $k = 1 \ldots K$ **do**
      $D_T = D_k$;
      $D_V = D_{(k+1) \bmod K}$;
      $D_L = \cup_{j \in \{1,\ldots,K\} - \{k\} - \{(k+1) \bmod K\}} D_j$;
      $\tilde{D}_L^R = F \times D_L$;
      $\tilde{D}_T^R = F \times D_T$;
      $\tilde{D}_V^R = F \times D_V$;
      $M = train(\tilde{D}_L^R, P_L, \tilde{D}_V^R, P_V)$;
      $E_T((i-1) * K + k) = MSE(M(\tilde{D}_T^R) - P_T)$;
      $E_L((i-1) * K + k) = MSE(M(\tilde{D}_L^R) - P_L)$;
      $\beta((i-1) * K + k) = \frac{MSE(M(\tilde{D}_T^R) - P_T)}{MSE(N - P_T)}$;
    end
  end
  $E_T^R(D,P) = median(E_T())$;
  $E_L^R(D,P) = median(E_L())$;
  $\beta^R(D,P) = median(\beta())$;
end

Algorithm 1: The evaluation of the impact of dimensionality reduction on model performance

rate calculated on just a few elements, which might be largely misleading. Formally, every training data set composed of $(K - 1)$ CV folders is subdivided into the actual training part D_T and a validation part D_V.
- The prediction error rates have to be compared to baseline rates. Hence, N denotes an average value of $P \subset \mathbb{R}$ used for method evaluation.

Thus, the optimal dimension $\tilde{R}$ in view of prediction model can be proposed to be defined as follows:

$$\tilde{R} : E_T^{\tilde{R}}(D,P) = min_{l=1,2,\ldots,S} \; E_T^l(D,P) \qquad (5)$$

Alternatively, the median of individual error proportions can be considered:

$$\tilde{R} : \beta^{\tilde{R}}(D,P) = min_{l=1,2,\ldots,S}\ \beta^l(D,P) \tag{6}$$

4 Experimental Results

4.1 Algorithm Settings

In all experiments, the number of CV folders was set to $K = 10$. Moreover, to ensure unbiased error rate calculations $r = 100$ runs were made. The MLP networks were trained with Levenberg-Marquardt algorithm, with the error on the validation set $\widetilde{D}_V^R$ used as a stopping criterion. The Alg. 1 was run for $R = S - 1, \ldots, 2$ potential reduced dimensions. The error rates reported below for $R = S$ were calculated by using the original data set D in Alg. 1, with no PCA transformation applied. The outliers were identified with $\gamma = 1.5$.

4.2 Drilling Process

The first data set comes from a series of deep drilling experiments [1]. The experimental test provided a dataset with 7 input variables: tool type, tool diameter, hole length, federate per revolution av, cutting speed Vc, type of lubrication system and axial cutting force. A data set of 219 records was used for the simulations in this study. Some of the records were originally incomplete. Hence, a data imputation algorithm was applied. The experimental procedure and the imputation of the data set were discussed in detail in [1]. The main modelling objective is the prediction of a roughness of a drill hole under different drilling conditions and tool settings.

Fig. 1 shows the results of the simulations. The lowest error rate $E_T^R(D,P)$ and reduction factor $\beta^R(D,P)$ is observed for $R = 6$. Further dimensionality reduction results in the raise of the error rates measured on training and testing data sets. In other words, from the prediction perspective, the reduction of dimensionality with PCA technique made to $R = 6$ dimensions, yields the best roughness prediction. When a reduction to $\tilde{D} \subset \mathbb{R}^2$ is made, the error rate of MLP networks is virtually identical to naive baseline N. Hence, PCA-based reduction can not be used to visualise the problem in $\mathbb{R}^2$.

4.3 Milling Process

The second data set used in the simulations was obtained from a milling process. Milling consists of removing the excess material from the workpiece in the form of small individual chips. A resulting surface consists of a series of elemental surfaces generated by the individual cutter edges. Surface roughness average (Ra) is commonly used to evaluate surface quality. Hence, the need for a model predicting the Ra value depending on the other experimentally measured factors. Further details on the process can be found in [9].

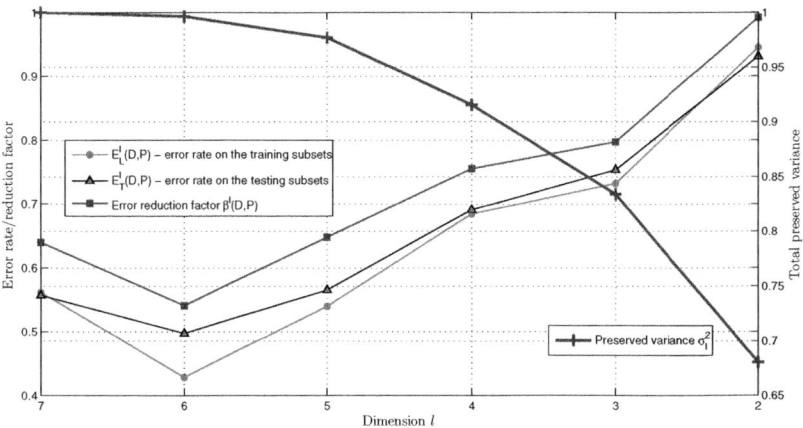

Fig. 1. The impact of DR on prediction error rates for the drilling process

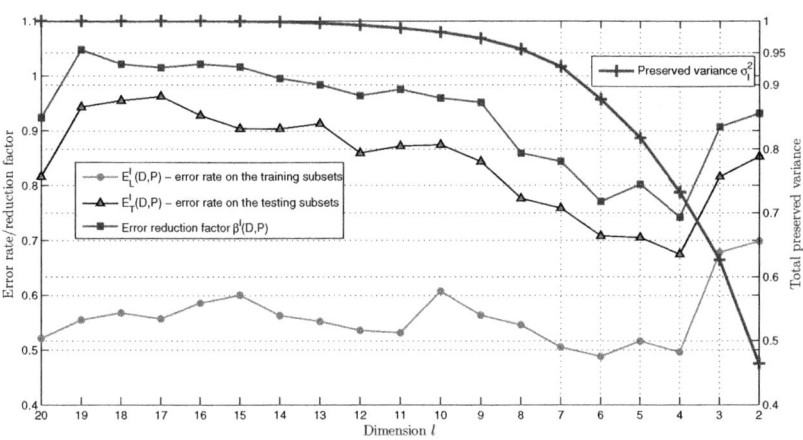

Fig. 2. The impact of DR on prediction error rates for the milling process

The experimental data used in the study consists of only $card(D) = 60$ records. Every record contains 20 variables, which include radial depth of cut, axial depth of cut, feed rate, spindle speed, radial depth of cut expressed in percentage of the tool radius, feed per tooth, cutting speed, tooth passing frequency, cutting section, material removal rate, and low, medium and high frequency vibration amplitudes in X and Y axes. As $D \subset \mathbb{R}^{20}$, the data are extremely scarce considering the number of dimensions. A decision was made to make an attempt to use a very compact MLP network with 2-1 architecture, hyperbolic tangent neurons in a hidden layer and a linear transfer function in the output layer.

The results of the simulations are shown in Fig. 2. When no PCA transformation is applied to the data set, $\beta^R(D,P)$ is close to 1 i.e. the MLP-based prediction is similar to naive guessing of mean value. However, as the dimension R is reduced, the number of connection weights of MLP network is reduced, too. Therefore, the $\beta^R(D,P)$ is gradually reduced and reaches its minimum at $R = 4$. In this case, even a reduction to $\mathbb{R}^2$ does not fully diminish predictive capabilities of MLP models.

4.4 Discussion

The error rates calculated on the testing sets and the reduction factors were summarised in Table 1. Reduced dimension, calculated according to PEV i.e. formula 3 is denoted by $\widehat{R}$. $\tilde{R}$ denotes the dimension resulting from formula 6.

Table 1. Results summary

Data set	Original dimension S	Reduced dimension $\widehat{R}$	$\tilde{R}$	Error reduction rates $\beta^{\widehat{R}}$	$\beta^{\tilde{R}}$
Drilling	7	5	6	0.6475	0.5405
Milling	20	8	4	0.8586	0.7418

Not only $\tilde{R} \neq \widehat{R}$, but also there is no straightforward relation between the two values. In the case of a drilling data set, the number of principal components preserved by PEV rule is lower, while in the case of milling data much higher than dimension R at which the error reduction β^R reaches its minimum. This can be easily explained, when the size of both data sets is considered. For the drilling data set, the best prediction models rely on components representing 0.9962 of the total variance. This is due to larger data set than in the case of a milling data set. In the latter case, due to just 60 records present in the data set, the best reduction in terms of predictive capabilities preserves only 0.7317 of total variance. Nevertheless, when a reduction is made to $\mathbb{R}^R, R \geq 5$, the overfitting of the MLP models diminishes the benefits arising from lower information loss.

5 Summary

When the prediction of industrial processes is based on experimental data sets, PCA-based dimensionality reduction is frequently applied. The simulations described in this work show that the PEV criterion may not always yield the best dimensionality reduction from prediction perspective. When the number of available records is sufficiently large, the number of preserved components may be set to a higher value. In the other cases, even significant information loss may contribute to the overall model performance. A formal framework for testing different DR settings was proposed in the work. It may be used with other DR methods, including non-linear transformations. Further research on the selection of the optimal combination of DR method, reduced dimension and neural network architecture using both unsupervised and supervised DR methods is planned.

Acknowledgements. This work has been made possible thanks to the support received from dr Andres Bustillo from University of Burgos, who provided both data sets and industrial problems description. The author would like to especially thank for his generous help and advice.

References

1. Grzenda, M., Bustillo, A., Zawistowski, P.: A Soft Computing System Using Intelligent Imputation Strategies for Roughness Prediction in Deep Drilling. Journal of Intelligent Manufacturing, 1–11 (2010),
   http://dx.doi.org/10.1007/s10845-010-0478-0
2. Haykin, S.: Neural Networks and Learning Machines. Person Education (2009)
3. Kegl, B.: Intrinsic Dimension Estimation Using Packing Numbers. In: Adv. In Neural Inform. Proc. Systems, Massachusetts Inst. of Technology, vol. 15, pp. 697–704 (2003)
4. Larose, D.T.: Data Mining Methods and Models. John Wiley & Sons, Chichester (2006)
5. Lee, J., Verleysen, M.: Nonlinear Dimensionality Reduction. Springer, Heidelberg (2010)
6. Li, D.-C., et al.: A Non-linear Quality Improvement Model Using SVR for Manufacturing TFT-LCDs. Journal of Intelligent Manufacturing, 1–10 (2010)
7. Maszczyk, T., Duch, W.: Support Vector Machines for Visualization and Dimensionality Reduction. In: Kůrková, V., Neruda, R., Koutník, J. (eds.) ICANN 2008, Part I. LNCS, vol. 5163, pp. 346–356. Springer, Heidelberg (2008)
8. Pal, S., et al.: Tool Wear Monitoring and Selection of Optimum Cutting Conditions with Progressive Tool Wear Effect and Input Uncertainties. Journal of Intelligent Manufacturing, 1–14 (2009)
9. Redondo, R., Santos, P., Bustillo, A., Sedano, J., Villar, J.R., Correa, M., Alique, J.R., Corchado, E.: A Soft Computing System to Perform Face Milling Operations. In: Omatu, S., Rocha, M.P., Bravo, J., Fernández, F., Corchado, E., Bustillo, A., Corchado, J.M. (eds.) IWANN 2009. LNCS, vol. 5518, pp. 1282–1291. Springer, Heidelberg (2009)
10. Rosman, G., et al.: Nonlinear Dimensionality Reduction by Topologically Constrained Isometric Embedding. Int. J. of Computer Vision 89(1), 56–68 (2010)
11. Verbeke, N., Vincent, N.: A PCA-Based Technique to Detect Moving Objects. In: Ersbøll, B.K., Pedersen, K.S. (eds.) SCIA 2007. LNCS, vol. 4522, pp. 641–650. Springer, Heidelberg (2007)

Informative Sentence Retrieval for Domain Specific Terminologies

Jia-Ling Koh and Chin-Wei Cho

Department of Information Science and Computer Engineering,
National Taiwan Normal University, Taipei, Taiwan, R.O.C.
jlkoh@csie.ntnu.edu.tw

Abstract. Domain specific terminologies represent important concepts when students study a subject. If the sentences which describe important concepts related to a terminology can be accessed easily, students will understand the semantics represented in the sentences which contain the terminology in depth. In this paper, an effective sentence retrieval system is provided to search informative sentences of a domain-specific terminology from the electrical books. A term weighting model is constructed in the proposed system by using web resources, including Wikipedia and FOLDOC, to measure the degree of a word relative to the query terminology. Then the relevance score of a sentence is estimated by summing the weights of the words in the sentence, which is used to rank the candidate answer sentences. By adopting the proposed method, the obtained answer sentences are not limited to certain sentence patterns. The results of experiment show that the ranked list of answer sentences retrieved by our proposed system have higher NDCG values than the typical IR approach and pattern-matching based approach.

Keywords: sentence retrieval, information retrieval, definitional question answering.

1 Introduction

When students study a course in specific domain, the learning materials usually make mention of many domain specific terminologies. For example, "supervised learning" or "decision tree" are important domain specific terminologies in the field of data mining. The domain specific terminologies represent important concepts in the learning process. If a student didn't know the implicit meaning of a domain specific terminology, it is difficult for the student to understand the complete semantics or concepts represented in the sentences which contain the terminology. For solving this problem, a student would like to look for some resources to understand the domain specific terminologies. Accordingly, it is very useful to provide an effective retrieval system for searching informative sentences of a domain-specific terminology.

Although various kinds of data on the Internet can be accessed easily by search engines, the quality and correctness of the data are not guaranteed. Books are still the main trustable learning resources of specific-domain knowledge. Furthermore, a book published in electrical form is a trend in the digital and information age. It makes the electrical books, especially the electrical textbooks, form a good resource for searching the semantic related sentences to a domain-specific terminology in a subject. For this reason, the goal of this paper is to design an effective sentence retrieval system for searching informative sentences of a domain-specific terminology X from the given electrical books.

A research topic related to this problem is automatic question answering. The goal of a question answering system is to provide an effective and efficient way for getting answers of a given natural language question. The most common types of queries in English are 5W1H (Who, When, Where, What, Why, and How). For processing different types of queries, various strategies were proposed to get proper answers. The question answering track of TREC 2003 first introduced definitional question answering (definitional QA) [14]. The task of definitional QA system is to answer the questions "What is X?" or "Who is X?" for a topic term X. The problem of searching informative sentences of a domain-specific term is similar to find answers of a definitional question. The typical definitional QA systems commonly used lexical patterns to identify sentences that contain information related to the query target. It has been shown that the lexical and syntactic patterns works well for identifying facet information of query targets such as the capital of a country or the birth day of a person. However, the lexical patterns are usually applicable to general topics or to certain types of entities. As described in [10], an information nugget is a sentence fragment that describes some factual information about the query term. Determined by the type and domain of the query term, an information nugget can include properties of the term, usage or application of the term, or relationship that the term has with related entities. In order to let users understand various aspects of the query term, it is better to cover diverse information nuggets. The pattern matching approach is in direction contrast to discover all interesting nuggets which are particular to a domain-specific term. Therefore, the informative sentences of the term could not be discovered if they didn't match the patterns.

In this paper, for retrieving as complete informative sentences of a domain-specific term as possible, we adopt a relevance-based approach. A term weighting model is constructed in our proposed system by using web resources, including Wikipedia and FOLDOC, to measure the degree of a word relative to the query term. Then the relevance score of a sentence is estimated by summing the weights of the words in the sentence, which is used to rank the candidate answer sentences. By adopting the proposed method, the retrieved answer sentences are not limited to certain sentence patterns. The results of experiment show that the ranked list of answer sentences retrieved by the proposed system have higher NDCG values than the typical IR approach and pattern-matching based approach.

The rest of this paper is organized as follows. The related works are discussed in Section 2. Section 3 introduces the proposed system for retrieving informative sentences of domain-specific terminologies from electrical books. The results of experiment for evaluating the effectiveness of the proposed system are reported in Section 4. Finally, Section 5 concludes this paper and discusses the future work.

2 Related Works

The traditional IR systems provide part of the solution for searching informative data of a specific term, which can only retrieve relevant documents for a query topic but not the relevant sentences. To aim at finding exact answers to natural language questions in a large collection of documents, open domain QA has become one of the most actively investigated topics over the last decade [13].

Among the research issues of QA, many works focused on constructing short answers for relatively limited types of questions, such as factoid questions, from a large document collection [13]. The problem of definitional question answering is a task of finding out conceptual facts or essential events about the question target [14], which is similar to the problem studied in this paper. Contract to the facet questions, a definitional question does not clearly imply an expected answer type but only specifies its question target. Moreover, the answers of definitional questions may consist of small segments of data with various conceptual information called information nuggets. Therefore, the challenge is how to find the information which is essential for the answers to a definitional question.

Most approaches used pattern matching for definition sentence retrieval. Many of the previously proposed systems created patterns manually [12]. To prevent the manually constructed rules from being too rigid, a sequence-mining algorithm was applied in [6] to discover definition-related lexicographic patterns from the Web. According to the discovered patterns, a collection of concept-description pairs is extracted from the document database. The maximal frequent word sequences in the set of extracted descriptions were selected as candidate answers to the given question. Finally, the candidate answers were evaluated according to the frequency of occurrence of their subsequences to determine the most adequate answers. In the joint predication model proposed in [9], not only the correctness of individual answers, but also the correlations of the extracted answers were estimated to get a list of accurate and comprehensive answers. For solving the problem of diversity in patterns, a soft pattern approach was proposed in [4]. However, the pattern matching approaches are usually applicable to general topics or to certain types of entities.

The relevance-based approaches explore another direction of solving definitional question answering. Chen et al. [3] used the bigram and bi-term language model to capture the term dependence, which was used to rerank candidate answers for definitional QA. The answer of a QA system is a smaller segment of data than in a document retrieval task. Therefore, the problems of data sparsity and exact matching become critical when constructing a language model for extracting relevant answers to a query. For solving these problems, after performing terms and n-grams clustering, a class-based language model was constructed in [11] for sentence retrieval. In [7], it was considered that an answer for the definitional question should not only contain the content relevant to the topic of the target, but also have a representative form of the definition style. Therefore, a probabilistic model was proposed to systematically combine the estimations of a sentence on topic language model, definition language model, and general language model to find retrieval essential sentences as answers for the definitional question. Furthermore, external knowledge from web was used in [10] to construct human interest model for extracting both informative and human-interested sentences with respect to the query topic.

From the early 2000s, rather than just made information consumption on the web, more and more users participated in content creation. Accordingly, the social media sites such as web forums, question/answering sites, photo and video sharing communities etc. are increasingly popular. For this reason, how to retrieve contents of social media to support question answering has became an important research topic in text mining. The problems of identifying question-related threads and their potential answers in forum were studied in [5] and [8]. A sequential patterns based classification method was proposed in [5] to detect questions in a forum thread; within the same thread, a graph-based propagation method was provided to detect the corresponding answers. Furthermore, it was shown in [8] that, in addition to the content features, the combination of several non-content features can improve the performance of questions and answers detection. The extracted question-answer pairs in forum can be applied to find potential solutions or suggestions when users ask similar questions. Consequently, the next problem is how to find good answers for a user's question from a question and answer archive. To solve the word mismatch problem when looking for similar questions in a question and answer achieve, the retrieval model proposed in [15] adopted a translation-based language model for the question part. Besides, after combining with the query likelihood language model for the answer part, it achieved further improvement on accuracy of the retrieved results. However, the main problem of the above tasks is that it is difficult to make sure the quality of content in social media [1].

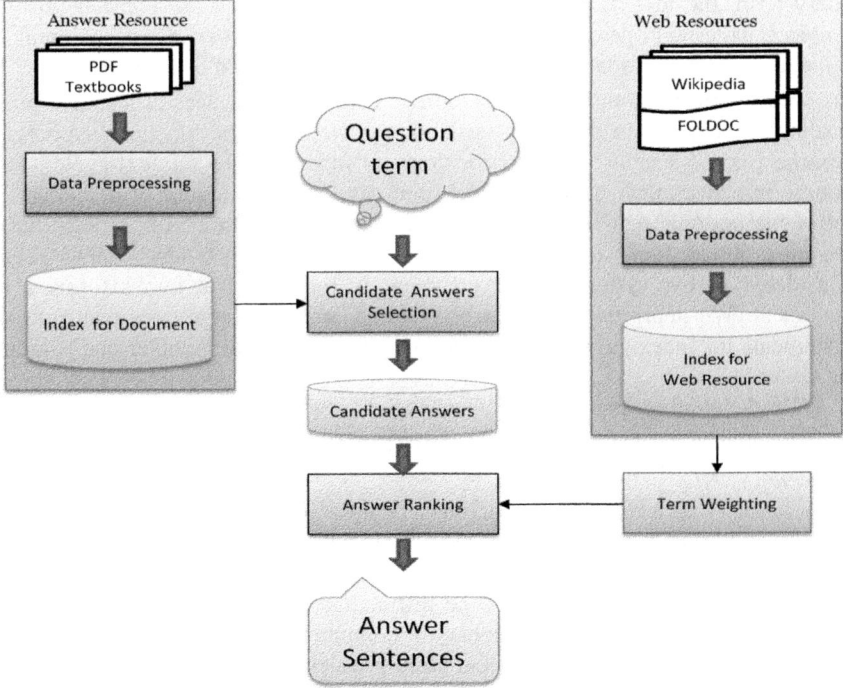

Fig. 1. The proposed system architecture for retrieving informative sentences of a query term

3 Proposed Methods

3.1 System Architecture

The overall architecture of our proposed system for retrieving informative sentences of a query term is shown as Fig. 1. The main processing components include Data Preprocessing, Candidate Answers Selection, Term Weighting, and Answer Ranking. In the following subsections, we will introduce the strategies used in the processing components in detail.

3.2 Data Preprocessing

<1> Text extraction: In our system, the electrical books in PDF format are selected as the answer resources. Therefore, the text content in the books is extracted and maintained in a text file per page by using a pdf-to-text translator. In addition, a HTML parser is implemented to extract the text content of the documents got from the web resources.

<2> Stemming and stop word removing: In this step, the English words are all transformed to its root form. The Poster's stemming algorithm is applied to do stemming. After that, we use a stop word list to filter out the common words which do not contribute significant semantics.

<3> Sentence separation: Because the goal is to retrieve informative sentences of the query term, the text has to be separated per sentence. We use some heuristics of sentence patterns to separate sentences, such as capital letter at the beginning of a sentence and the punctuation marks: '?', '!', and '.' at the end of a sentence.

<4> Index construction: In order to retrieve candidate sentences efficiently from the textbooks according to the query terminology, we apply the functions supported by Apache Lucene search engine to construct an index file for the text content of the books. In a document database which consists of large amount of documents, the index file not only maintains the ids of documents in which a term appears, but also the appearing locations and frequencies of the term in the documents. In the scenario considered in our system, the text content in the electrical books forms a large document. We apply two different ways to separate the large document into small documents for indexing: one way is indexing by pages and the other one is indexing by sentences.

Let B denote the electrical book used as the answer resource. The set of pages in B is denoted as $B.pages = \{p_1, p_2, p_3, ..., p_n\}$, where p_i denotes the document content of page i in B. The set of sentences in B is denoted as $B.sentences = \{s_1, s_2, s_3, ..., s_m\}$, where s_i denotes the document content of the ith sentence in B. The method of indexing by pages constructs an index file for $B.pages$; the indexing by sentences constructs an index file for $B.sentences$.

The training corpus consists of the documents got from the web resources: the Wikipedia and the free online dictionary of computing (FOLDOC). We also construct an index file for the training corpus in order to calculate the degree of a word relative to the query terminology efficiently.

3.3 Candidate Answers Selection

Whenever the system gets a query terminology X, at first, IndexSearcher method provided by Lucene is used to retrieve candidate sentences according to the constructed index file. As the Boolean model adopted popularly in information retrieval, a naive approach of getting candidate answer sentences is to retrieve the sentences which contain the query term. In addition to the sentences containing the query term, it is possible that the other sentences which are close to the query term in the document represent the related concepts of the query term. For this reason, three approaches are proposed for retrieving candidate answer sentences. The first approach uses the index file which is constructed according to the indexing by sentences. Only the sentences which contain the query term will become candidate sentences. The first retrieval method is denoted as "Sentence" method in the following. The second approach retrieves not only the sentences which contain the query term but also their previous two sentences and their following two sentences to be candidate sentences, which is denoted as "Sentence+-2" method. The last approach, denoted as "Page" method, uses the index file which is constructed according to the indexing by page. By adopting the last approach, the sentences in a page where the query term appears will all become candidate sentences.

3.4 Term Weighting and Answer Ranking

In order to perform ranking for the retrieved candidate sentences, the next step is to estimate the relevance scores of the candidate sentences to the query term. Therefore, we use the web resources, including the Wekipedia and the FOLDOC, as the training corpus for mining the relevance degrees of words with respect to the query term.

The term frequency-inverse document frequency (TF-IDF) of a term is a weight often used in information retrieval to evaluate the importance of a word in a document within a corpus. The Lucene system also applies a TF-IDF based formula to measure the relevance score of the indexing objects, the sentences here, to a query term. Therefore, in our experiments, we use the sort method supported by Lucene as one of the baseline methods for comparison.

In our approach, we consider the words which appear in a sentence as features of the sentence. A sentence should have higher score when it has more words with high ability to distinguish the sentences talking about the query term from the whole corpus. Therefore, we apply the Jensen-Shannon Divergence (JSD) distance measure to perform term weighting, which was described in [2] to extract important terms to represent the documents within the same cluster. The weight of a term (word) w with respect to the query terminology X is estimated by measuring the contribution of w to the Jensen-Shannon Divergence (JSD) between the set of documents containing the query term and the whole training corpus.

Let P denote the set of query related documents returned by Lucene, which are found out from the training corpus. Let the set of documents in the training corpus be denoted by Q. The JSD term weighting of a word w, denoted as $W_{JSD}(w)$ is computed according to the following formula:

$$w_{JSD}(w) = \frac{1}{2}\left(p(w|P) \cdot \log\frac{p(w|P)}{p(w|M)} + p(t|Q) \cdot \log\frac{p(w|Q)}{p(w|M)}\right) \quad (1)$$

The equations for getting the values of $p(w|P)$, $p(w|Q)$ and $p(w|M)$ are defined as the following:

$$p(w|P) = \frac{tf(w,P)}{|P.words|} \quad (2)$$

$$p(w|Q) = \frac{tf(w,Q)}{|Q.words|} \quad (3)$$

$$p(w|M) = \frac{1}{2}(p(w|P) + p(w|Q)) \quad (4)$$

where $tf(w,P)$ and $tf(w,Q)$ denote the frequencies of word w appearing in P and Q, respectively. Besides, $|P.words|$ and $|Q.words|$ denote the total word counts in P and Q, respectively.

According to the JSD term weighting method, the JSD weight of each word in the candidate sentences is evaluated. The relevance score of a candidate sentence s, denoted as $Score(s)$, is obtained by summarizing the JSD weights of the words in s as the following formula:

$$Score(s) = \sum_{w \in s} w_{JSD}(w) \quad (5)$$

where w denote a word in s. The top scored sentences are then selected as the informative sentences of the given domain-specific terminology.

4 Experiments and Results

4.1 Experiment Setup and Evaluation Metric

We used the electrical books "Introduction to Data Mining and Knowledge Discovery" and the first four chapters of "Web Data Mining: Exploring Hyperlinks, Contents, and Usage Data" as the answer resources. Both books cover the related techniques of data mining, which consist of 222 pages in total.

We implemented three versions of the proposed system according to the three methods of candidate sentences retrieving, which are denoted by "Sentence", "Sentence+-2", and "Page", respectively. The sort method supported in Lucene system was used as a baseline method. Furthermore, a pattern-based approach was also implemented as another baseline.

In the experiment, the following 6 important domain specific terminologies: "Web Mining", "Supervised Learning", "Neural Network", "Naïve Bayesian Classification", "Clustering" and "Decision Tree" were selected to be the test terms. Each test term was used to be a query inputted to the system, where the top 25 sentences in the ranking result provided by our system are returned to be the answers.

We invited 8 testers to participate the experiment for evaluating the quality of the returned answer sentences. All the testers are graduate students in the university, whose majors are computer science. Besides, they are familiar with the field of data mining. For each test term, the sets of the top 25 answer sentences returned by the 5 different methods are grouped together. An interface was developed to collect the satisfying levels of the testers for each returned answer, where the meanings of the 5 satisfying levels are defined as the following.

Level 5: the answer sentence explains the definition of the test term clearly.
Level 4: the answer sentence introduces the concepts relative to the test term.
Level 3: the answer sentence describes the test term and other related terms, but the content is not very helpful for understanding the test term.
Level 2: the content in the answer sentence is not relative to the test term.
Level 1: the semantics represented in the answer sentence is not clear.

4.2 Experimental Results and Discussion

We evaluated the proposed methods by computing Normalized Discounted Cumulative Gain (NDCG) of the ranked answer sentences. The equation for getting NDCG at a particular rank position n is defined as the following:

$$NDCG_n = \frac{DCG_n}{IDCG_n} \tag{6}$$

The DCG_n denotes the Discounted Cumulative Gain accumulated at a particular rank position n, which is defined as the following equation:

$$DCG_n = rel_1 + \sum_{i=2}^{n} \frac{rel_i}{\log_2 i} \tag{7}$$

The rel_i in the equation denotes the actual relevance score of the sentence at rank i. We estimated the actual relevance score of a sentence by averaging the satisfying scores given by the 8 testers. In addition, the $IDCG_n$ is an ideal DCG at position n, which occurs when the sentences are sorted according to the descending order of their actual relevance scores. In the experiment, the top 25 answer sentences are returned by each method to compute $NDCG_{25}$ for evaluating the quality of the returned sentences. The results of experiment are shown in Tab. 1.

As shown in Tab. 1, it is indicated that the answers retrieved by our proposed method are better than the ones retrieved by either the Lucene system or the pattern-based approach. The main reason is that a sentence usually consists of dozens of words at most. In the Lucene system, a TF-IDF based formula is used to measure the relevance scores of a sentence to the query. Accordingly, only the sentences which contain the query terms are considered. As shown in Fig. 2, the short sentences which contain all the query terms have high ranks in the Lucene system but do not have important information nuggets. On the other hand, most of the sentences retrieved by the JSD term weighting method proposed by this paper describe important concepts relative to the query terminology, which are not limited to specific pattern. In addition, it is not required that the answer sentences must contain the query term. The

Table 1. The NDCG$_{25}$ values of the returned answers of different methods

	Page	Sentence	Sentence+-2	Pattern	Lucene
"Web Mining"	0.95565	0.94099	0.93864	0.68087	0.75688
"Supervise Learning"	0.87728	0.89939	0.90548	0.73065	0.36578
"Neural Network"	0.95748	0.95172	0.95162	0.67896	0.77962
"Naive Bayesian Classification"	0.85521	0.85960	0.91597	0.76243	0.66769
"Decision Tree"	0.93926	0.92211	0.86469	0.73763	0.64161
"Clustering"	0.85544	0.85809	0.86029	0.63042	0.67952
Average	0.90672	0.90532	0.90611	0.70350	0.64852

---

1. Supervised Learning classification.
2. Partially Supervised Learning 5.
3. Partially Supervised Learning 5.
4. Partially Supervised Learning 1.
5. Partially Supervised Learning 5.

Fig. 2. The top 5 answers of "Supervised Learning" returned by Lucene

---

1. In supervised learning, the learning algorithm uses labeled training examples from every class to generate a classification function.
2. This type of learning has been the focus of the machine learning research and is perhaps also the most widely used learning paradigm in practice.
3. Supervised learning is also called classification or inductive learning in machine learning.
4. Supervised Learning computational learning theory shows that maximizing the margin minimizes the upper bound of classification errors.
5. Bibliographic Notes Supervised learning has been studied extensively by the machine learning community.

Fig. 3. The top 5 answers of "Supervised Learning" returned by "Sentence+-2"

top 5 answers returned by "Sentence+-2" are shown in Fig. 3 when querying "supervise learning". Furthermore, it is shown in Tab. 1 that none of the three methods for retrieving candidate sentences is superior to the other two methods in all cases.

5 Conclusion and Future Work

In this paper, an effective sentence retrieval system is provided to search informative sentences of a domain-specific terminology from electrical books. A term weighting model is constructed in the proposed system by using the web resources, including Wikipedia and FOLDOC, to measure the degree of a word relative to the query terminology. Then the relevance score of a sentence is estimated by summing the weights of the words in the sentence, which is used to rank the candidate answer sentences. By adopting the proposed method, the retrieved answer sentences are not limited to certain sentence patterns. Accordingly, informative sentences of a domain-specific term with various information nuggets can be retrieved as complete as possible. The results of experiment show that the ranked answer sentences retrieved by the proposed system have higher NDCG value than the typical IR approach and pattern-matching based approach.

Among the retrieved informative sentences of a query term, some of the sentences may represent similar concepts. How to cluster the semantics related sentences of a query terminology for providing an instructive summarization of the term is under our investigation currently.

References

1. Agichtein, E., Castillo, C., Donato, D., Gionis, A., Mishne, G.: Finding High-Quality Content in Social Media. In: Proc. the International Conference on Web Search and Data Mining, WSDM (2008)
2. Carmel, D., Roitman, H., Zwerdling, N.: Enhancing Cluster Labeling Using Wikipedia. In: Proc. the 32nd Annual International ACM SIGIR Conference on Research and Development in Information Retrieval, SIGIR (2009)
3. Chen, Y., Zhou, M., Wang, S.: Reranking Answers for Definitional QA Using Language Modeling. In: Proc. the 21st International Conference on Computational Linguistics and 44th Annual Meeting of the ACL (2006)
4. Chi, H., Kan, M.-Y., Chua, T.-S.: Generic Soft Pattern Models for Definitional Question Answering. In: Proc. the 28th International ACM SIGIR Conference on Research and Development in Information Retrieval, SIGIR (2005)
5. Cong, G., Wang, L., Lin, C.Y., Song, Y.I., Sun, Y.: Finding Question-Answer Pairs from Online Forums. In: Proc. the 31st international ACM SIGIR Conference on Research and Development in Information Retrieval, SIGIR (2008)
6. Denicia-carral, C., Montes-y-gómez, M., Villaseñor-pineda, L., Hernández, R.G.: A Text Mining Approach for Definition Question Answering. In: Proc. the 5th International Conference on Natural Language Processing, FinTal 2006 (2006)

7. Han, K.S., Song, Y.I., Rim, H.C.: Probabilistic Model for Definitional Question Answering. In: Proc. the 29th International ACM SIGIR Conference on Research and Development in Information Retrieval, SIGIR (2006)
8. Hong, L., Davison, B.D.: A Classification-based Approach to Question Answering in Discussion Boards. In: Proc. the 32nd International ACM SIGIR Conference on Research and Development in Information Retrieval, SIGIR (2009)
9. Ko, J., Nyberg, E., Si, L.: A Probabilistic Graphical Model for Joint Answer Ranking in Question Answering. In: Proc. the 30th International ACM SIGIR Conference on Research and Development in Information Retrieval, SIGIR (2007)
10. Kor, K.W., Chua, T.S.: Interesting Nuggets and Their Impact on Definitional Question Answering. In: Proc. the 30th International ACM SIGIR Conference on Research and Development in Information Retrieval, SIGIR (2007)
11. Momtazi, S., Klakow, D.: A Word Clustering Approach for Language Model-based Sentence Retrieval in Question Answering Systems. In: Proc. the 18th ACM International Conference on Information and Knowledge Management, CIKM (2009)
12. Sun, R., Jiang, J., Tan, Y.F., Cui, H., Chua, T.-S., Kan, M.-Y.: Using Syntactic and Semantic Relation Analysis in Question Answering. In: Proc. the 14th Text REtrieval Conference, TREC (2005)
13. Voorhees, E.M.: Overview of the TREC 2001 Question Answering Track. In: Proc. the 10th Text REtrieval Conference, TREC (2001)
14. Voorhees, E.M.: Overview of the TREC 2003 Question Answering Track. In: Proc. the 12th Text REtrieval Conference, TREC (2003)
15. Xue, X., Jeon, J., Croft, W.B.: Retrieval Models for Question and Answer Archives. In: Proc. the 31st International ACM SIGIR Conference on Research and Development in Information Retrieval, SIGIR (2008)

Factoring Web Tables

David W. Embley[1], Mukkai Krishnamoorthy[2],
George Nagy[2], and Sharad Seth[3]

[1] Brigham Young University, Provo, UT, USA
embley@cs.byu.edu
[2] Rensselaer Polytechnic Institute, Troy, NY, USA
mskmoorthy@gmail.com, nagy@ecse.rpi.edu
[3] University of Nebraska, Lincoln, Lincoln, NE, USA
seth@cse.unl.edu

Abstract. Automatic interpretation of web tables can enable database-like semantic search over the plethora of information stored in tables on the web. Our table interpretation method presented here converts the two-dimensional hierarchy of table headers, which provides a visual means of assimilating complex data, into a set of strings that is more amenable to algorithmic analysis of table structure. We show that Header Paths, a new purely syntactic representation of visual tables, can be readily transformed ("factored") into several existing representations of structured data, including category trees and relational tables. Detailed examination of over 100 tables reveals what table features require further work.

Keywords: table analysis, table headers, header paths, table syntax, category trees, relational tables, algebraic factorization.

1 Introduction

The objective of this work is to make sense of (analyze, interpret, transform) tables the best we can without resorting to any external semantics: we merely manipulate symbols. Remaining firmly on the near side of the semantic gap, we propose a canonical table representation based on *Header Paths* that relate column/row headers and data cells. We show that this "canonical" representation is adequate for subsequent transformations into three other representations that are more suitable for specific data/information retrieval tasks. The three targets for transformations are

Visual Table (VT). The VT provides the conventional two-dimensional table that humans are used to. Header and contents rows and columns may, however, be jointly permuted in the VT generated from Header Paths, and it does not preserve style attributes like typeface and italics.

Category Tree (CT). The CT is an "abstract data structure" proposed by X. Wang in 1996[1]. It is a convenient format for storing tables in an internal XML format and thus can also be used for information exchange.

Relational Table (RT). RT provides the link to standard relational databases and their extensive apparatus for storing and retrieving data from a collection of tables.

Header Paths are a general approach to tables and capture the essential features of two-dimensional indexing. A *Well Formed Table (WFT)* is a table that provides a unique string of Header Paths to each data cell. If the Header Paths of two distinct data cells are identical, then the table is ambiguous. Although the concept of Header Paths seems natural, we have not found any previous work that defines them or uses them explicitly for table analysis. Our contribution includes extracting Header Paths from web tables, analysis of Header Paths using open-source mathematical software, and a new way of transforming them into a standard relational representation.

We have surveyed earlier table processing methods--most directed at scanned printed tables rather than HTML tables--in [2]. An excellent survey from a different perspective was published by Zanibbi et al [3]. More recent approaches that share our objectives have been proposed in [4], [5], and [6]. Advanced approaches to document image analysis that can be potentially applied to tables are reported in [7] and [8].

After providing some definitions, we describe a simple method of extracting Header Paths from CSV versions of web tables from large statistical sites. Then we show how the structure of the table is revealed by a decomposition that incorporates algebraic factoring. The next section on analysis illustrates the application of Header Paths to relational tables. Our experiments to date demonstrate the extraction of header paths, the factorization, and the resulting Wang categories. Examination of the results suggests modifications of our algorithms that will most increase the fraction of automatically analyzed tables, and enhancements of the functionality of the interactive interface necessary to correct residual errors.

2 Header Paths

We extract Header Paths from CSV versions of web tables imported into Excel, which is considerably simpler than extracting them directly from html [9]. The transformation into the standard comma-separated-variables format for information exchange is not, however, entirely lossless. Color, typeface, type size, type style (bold, italics), and layout (e.g. indentations not specified explicitly in an HTML style sheet) within the cells are lost. Merged table cells are divided into elementary cells and the cell content appears only in the first element of the row. Anomalies include demoted superscripts that alter cell content. Nevertheless enough of the structure and content of the web tables is usually preserved for potentially complete understanding of the table. Some sites provide both HTML and CSV versions of the same table.

An example table is shown in Fig. 1a. The *(row-) stub* contains Year and Term, the *row header* is below the stub, the *column header* is to the right of the stub, and the 36 *delta* (or *data*, or *content*) *cells* are below the column header. Our Python routines convert the CSV file (e.g. Figs. 1b and 1c) to Header Paths as follows:

> 1. Identify the stub header, column header, row header, and content regions.
> 2. Eliminate blank rows and almost blank rows (that often designate units).
> 3. Copy into blank cells the contents of the cell above. ⎫
> 4. Copy into blank cells the contents of the cell to the left.⎭ *reverse for rows*

5. Underscore blanks within cells and add quote marks to cell contents.
6. Mark row headers roots in the stub with negative column coordinates.
7. Add cell identifiers (coordinates) to each cell.
8. Trace the column and row Header Paths.

(a)

(b)

Year,Term,Mark,,,,,EOL,,Assignments,,,Examinations,,GradeEOL,,Ass1,Ass2,Ass3,Midterm,Final,
EOL1991,Winter,85,80,75,60,75,75EOL,Spring,80,65,75,60,70,70,Fall,80,85,75,55,80,75EOL
1992,Winter,85,80,70,70,75,75EOL,Spring,80,80,70, 70,75,75EOL,Fall,75,70,65,60,80,70EOL
(c)

Fig. 1. (a) A table from [1] used as a running example. (b) Its CSV version. (c) CSV file string.

Step 1 can partition a table into its four constituent regions only if either the stub header is empty, or the content cells contain only digits. Otherwise the user must click on the top-left (*critical*) delta cell. Fig. 2 shows the paths for the example table.

The paths to the body of the table (delta cells) are traced in the same way. The combination of Header Paths uniquely identifies each delta cell. The three Wang Categories for this table are shown in Fig. 3. The algorithm yielded paths for 89 tables on a random sample of 107 tables from our collection of 1000 web tables. Most of the rejected tables had non-empty stub headers and blank separators or decimal commas.

colpaths =
 (("<0,2>Mark"*"<1,2>Assignments"*"<2,2>Ass1")
 +("<0,3>Mark"*"<1,3>Assignments"*"<2,3>Ass2")
 +("<0,4>Mark"*"<1,4>Assignments"*"<2,4>Ass3")
 +("<0,5>Mark"*"<1,5>Examinations"*"<2,5>Midterm")
 +("<0,6>Mark"*"<1,6>Examinations"*"<2,6>Final")
 +("<0,7>Mark"*"<1,7>Grade"*"<2,7>Grade"));

rowpaths =
 (("<-2,3>Year"*"<-1,3>1991"*"<0,3>Term"*"<1,3>Winter")
 +("<-2,4>Year"*"<-1,4>1991"*"<0,4>Term"*"<1,4>Spring")
 +("<-2,5>Year"*"<-1,5>1991"*"<0,5>Term"*"<1,5>Fall")
 +("<-2,6>Year"*"<-1,6>1992"*"<0,6>Term"*"<1,6>Winter")
 +("<-2,7>Year"*"<-1,7>1992"*"<0,7>Term"*"<1,7>Spring")
 +("<-2,8>Year"*"<-1,8>1992"*"<0,8>Term"*"<1,8>Fall"));

Fig. 2. Column Header and Row Header Paths for the table of Fig. 1

```
        Year                            Mark
                  1991                          Assignments
                  1992                                   Ass1
                                                         Ass2
        Term                                             Ass3
                  Winter                        Examinations
                  Spring                             Midterm
                  Fall                                  Final
                                              Grade*Grade
```

Fig. 3. Wang Categories for the table of Fig. 1

3 Algebraic Formulation and Factorization

The column and row Header Paths can be thought of as providing *indexing* relationships respectively, to the set of columns and rows in the table; their cross-product narrows these relationships to individual entries in the table. In this section, we develop an algebra of relations for Header Paths, which allows interpreting them in terms of categorical hierarchies using standard symbolic mathematical tools. As the essential formalism for row Header Paths is the same as it is for column Header Paths, we limit the exposition below to column Header Paths. We will use the example table shown in Fig. 1a with its column and row Header Paths shown in Fig. 2.

In our algebra of relations for column Header Paths, the domain of the relations is the set Γ of all columns in the table. Each cell c in the column header covers a subset of the columns in Γ. By identifying cells with their labels, the column-covering relation can be extended to cell labels.

In the example, the label Assignments covers columns 1, 2, and 3, and the label Mark covers all of Γ. These cell-label relations can be combined, using the union and intersection operations, to define new relations. We use the symbols + and *, respectively, for the union and intersection operations. Thus, the expression Assignments+Grade covers columns 1, 2, 3, and 6 and the expression Examinations*Midterm covers just column 4. We call the first a sum and the second a product relation. It will be seen that the collection of Header Paths, described in the previous section, can be represented as a sum of products (SOP) expression in our algebra. We will call this SOP an indexing relation on Γ because it covers all columns in Γ and each product term covers one column in Γ. This SOP is also an irredundant relation because each column is uniquely covered by one product term.

The notions of indexing and irredundancy can be extended to arbitrarily nested expressions by expanding the expression into its SOP-equivalent. For example, Mark*(Assignments*Ass1+Examinations*(Midterm+Final)) is an irredundant indexing relation on the column set {1,4,5}.

We will use the term *decomposition* to refer to converting a relational expression into a hierarchy of sub-expressions that are joined together by the union and intersection operations. Equivalent terms, *simplification* (in contrast to expansion) and *Horner nested representation* [10], are also used in the literature to denote the same concept. Now, we can formulate the problem of recovering a categorical structure of the column header as the decomposition of the column Header Paths expression into an equivalent expression E satisfying the following constraints:

(a) E is an indexing relation on Γ;
(b) E is irredundant ;
(c) E is minimal in the occurrence of the total number of labels in the expression.

A benefit of this formulation is that the decomposition problem for algebraic expressions has attracted attention for a long time from fields ranging from symbolic mathematics [11,12,13] to logic synthesis [14,15] and programs incorporating the proposed solutions are widely available [16,17]. In this work, we adopt Sis, a system for sequential circuit synthesis, for decomposition of header-path expressions [16]. Sis represents logic functions to be optimized in the form of a network with nodes representing the logic functions and directed edges (a,b) to denote the use of the function at node a as a sub-function at node b. An important step in network optimization is extracting, by means of a division operation, new nodes representing logic functions that are factors of other nodes. Because all good Boolean division algorithms are computationally expensive, Sis uses the ordinary algebraic division. The basic idea used is to look for expressions that are observed many times in the nodes of the network and extract such expressions.

Some caution is in order for this formulation because seemingly identical labels in two cells may carry different meaning. In our experience with many web tables, however, identical labels in the same row of a column header (or the same column of a row header) do not need to be differentiated for our Boolean-algebraic formulation. This is because either they carry the same meaning (e.g. repeated labels like Winter in the example table) or the differences in their meaning are preserved by other labels in their product terms. To illustrate the last point, assume ITEMS*TOTAL and AMOUNT*TOTAL are two product terms in a Header Paths expression, where the two occurrences of TOTAL refer to a quantity and a value, respectively. In the factored form, these two terms might appear as TOTAL*(ITEMS+AMOUNT), where the cell with label TOTAL now spans cells labeled ITEMS and AMOUNT, thus preserving the two terms in the Header Paths expression. On the other hand, suppose the column HEADER PATHS expression included TOTAL*MALES + TOTAL*FEMALES + TOTAL*TOTAL as a subexpression, where the last term spans the two rows corresponding to the first two terms. In Boolean algebra, the sub-expression could be simplified to TOTAL and the resulting expression would no longer cover the two columns covered by the first two terms. With the row indices attached to labels, the subexpression becomes: TOTAL1*MALES + TOTAL1*FEMALES + TOTAL1*TOTAL2, which cannot be simplified so as to eliminate one or more terms.

We illustrate the decomposition obtained by Sis for the column Header Paths of the example table:

> Input Order = Mark Assignments Ass1 Ass2 Ass3 Examinations Midterm Final Grade
> colpaths = Mark*[Examinations*[Final + Midterm]
>   + Assignments* [Ass3 + Ass2 + Ass1] + Grade]

Note that Sis does not preserve the input order in the output equations, but because the order is listed in the output, we can permute the terms of the output expression according to the left-to-right, top-to-bottom order of the labels occurring in the table:

colpaths=Mark*[Assignments*[Ass1+Ass2+Ass3] + [Examinations*[Midterm + Final]] + Grade]

Similarly, the rowpaths expression produces the following output:

rowpaths = Year*[1991+1992]*Term*[Winter+Spring+Fall]

We apply a set of Python regular expression functions to split each equation into a list of product terms and convert each product term into the form of [root, children]. A virtual header is inserted whenever a label for the category root is missing in the table, e.g. if the right-hand side of colpaths were missing "Mark*", we would insert a virtual header for the missing root. Then, the overall equation of the table can be printed out recursively in the form of Wang categories trees, as illustrated in Fig. 3. Further, we produce a canonical expression for the three categories in the form in Fig. 4, which is used in the generation of relational tables:

Mark*(Assignments*(Ass1+Ass2Ass3)+Examinations*(Midterm+Final)+Grade)
+ Year*(1991+1992)
+ Term*(Winter+Spring+Fall)

Fig. 4. Canonical expression for the table of Fig. 1

Currently, we don't have an automated way of verifying the visual table (VT) in CSV form against the category tree (CT) form obtained by factorization. Instead, we do the verification by comparing the following common characteristic features of the two forms: the number of row and column categories, and for each category: fanout at the root level, the total fanout, and whether the root category has a real or virtual label. The data for CT are unambiguous and readily derived. For VT, however, we derive it by labor-intensive and error-prone visual inspection, which limits the size and the objectiveness of the experiment. Still, we believe, the result demonstrates the current status of the proposed scheme for automated conversion of web tables.

Of 89 tables for which Header Paths were generated, 66 (74%) have correct row and column categories. Among the remaining 23 tables, 18 have either row *or* column category correct, including two where the original table contained mistakes (duplicate row entries). Both human and computer found it especially difficult to detect the subtle visual cues (indentation or change in type style) that indicate row categories. Since in principle the decomposition is error-free given the correct Header Paths, we are striving to improve the rowpath extraction routines.

4 Generation of Relational Tables

Given the headers of a table, factored into canonical form as explained in Section 3 along with their accompanying data (the delta cells), we can transform the table into a relational table for a relational database. We can then query the table with SQL or any other standard database query language. Given a collection of tables transformed into relations, we may also be able to join tables in the collection and otherwise manipulate the tables as we do in a standard relational database.

Our transformation of a factored table assumes that one of the Wang categories provides the attributes for the relational table while the remaining categories provide key values for objects represented in the original table. Without input from an oracle, we do not know which of the Wang categories would serve best for the attributes. We therefore transform a table with n Wang category trees into n complementary relational tables—one for each choice of a category to serve as the attributes.

The transformation from a factored table with its delta cells associated with one of its categories is straightforward. We illustrate with the example from Wang [1] in Fig. 1a This table has three categories and thus three associated relational tables (attributes can be permuted). We obtain the relational table in Fig. 5 using the factored expression in Fig. 4 and the data values from the original table in Fig. 1.

R(	Y	T	K_A_1	K_A_2	K_A_3	K_E_M	K_E_L	K_G	)
	91	W	85	80	75	60	75	75	
	91	S	80	65	75	60	70	70	
	91	F	80	85	75	55	80	75	
	92	W	85	80	70	70	75	75	
	92	S	80	80	70	70	75	75	
	92	F	75	70	65	60	80	70	

Fig. 5. Relational table for the table of Fig. 1

In Fig. 5, R is the title of the table ("The_average_marks_for_1991_to_1992"), Y is Year, T is Term, K is Mark, A is Assignment, 1–3 are Assignment numbers, E is Examinations, M is Midterm, L is Final, G is Grade, 91 and 92 are year values, and the terms are W for Winter, S for Spring, and F for Fall.

The key for the table is {Y, T}, a composite key with two attributes because we have two categories for key values. Since YT is the key, the objects for which we have attribute values are terms in a particular year, e.g., the Winter 1991 term. The attributes values are average marks for various assignments and examinations and for the final grade for the term.

In general, the algorithmic transformation to a relational table is as follows: (1) Create attributes from the first category-tree expression by concatenating labels along each path from root to leaf (e.g., Mark*(Assignments*(Ass1+ ... becomes the attribute Mark_Assignments_Ass1). (2) Take each root of the remaining category-tree expressions as additional attributes, indeed as (composite) primary-key attributes (e.g., Year and Term as attributes with the composite primary key {Year, Term}). (3) Form key values from the (cross product of) the labels below the root of the category-tree expressions (e.g., {1991, 1992} × {Winter, Spring Fall}). (4) Fill in the remaining table values from the delta-cell values in the original table as indexed by the header labels (e.g. for the attribute Mark_Assignment_Ass1 and the composite key value Year = 1991 and Term = Winter, the delta-cell delta-cell value 85).

Given the relational table in Fig. 5, we can now pose queries with SQL:

Query 1: What is the average grade for Fall, 1991?

>   select K_G
>   from R
>   where Y = 91 and T = "F"

With abbreviated names spelled out, the query takes on greater meaning:

>   select Mark_Grade
>   from The_average_marks_for_1991_to_1992
>   where Year = 1991 and Term = "Fall"

The answer for Query 1 is:

Mark_Grade
75

Query 2:

What is the overall average of the final for each year?

>   select Year, avg(Mark_Examinations_Final)
>   from The_average_marks_for_1991_to_1992
>   group by Year

The answer for Query 2 is:

Year	Avg(Mark_Examinations_Final)
1991	75.0
1992	76.7

Commuting the expression in Fig. 4 to let Term*(Winter+Spring+Fall) be first and stand for the attributes, and to let the Year expression be second and the Mark expression be third, this yields the relational table in Fig. 6a.

In this relational table the key is YK, the composite key Year-Mark. Thus, we have Year-Mark objects with Term attributes. Although less intuitive than our earlier choice, it certainly makes sense to say that the Winter term average mark for the 1991 Assignment #1 is 85.

Commuting the expression in Fig. 4 to let Year*(1991+1992) be first and stand for the attributes, and to let the Term expression be second and the Mark expression be third, yields the relational table in Fig. 6b. In this relational table the key is TK, the composite key Term-Mark. Thus, we have Term-Mark objects with Year attributes. Here, again, although less intuitive than our first choice, it makes sense to say that the 1991 average mark for the Winter term Assignment #1 is 85.

R (	Y	K	T_W	T_S	T_F	)
	91	A_1	85	80	80	
	91	A_2	80	65	85	
	91	A_3	75	75	75	
	91	E_M	60	60	55	
	91	E_L	75	70	80	
	91	G	75	70	75	
	92	A_1	85	80	75	
	92	A_2	80	80	70	
	92	A_3	70	70	65	
	92	E_M	70	70	60	
	92	E_L	75	75	80	
	92	G	75	75	70	

(a)

R (	T	K	Y_91	Y_92	)
	W	A_1	85	85	
	W	A_2	80	80	
	W	A_3	75	70	
	W	E_M	60	70	
	W	E_L	75	75	
	W	G	75	75	
	S	A_1	80	80	
	S	A_2	65	80	
	S	A_3	75	70	
	S	E_M	60	70	
	S	E_L	70	75	
	S	G	70	75	
	F	A_1	80	75	
	F	A_2	85	70	
	F	A_3	75	65	
	F	E_M	55	60	
	F	E_L	80	80	
	F	G	75	70	

(b)

Fig. 6. (a) A second relational table for the table of Fig. 1. (b) A third relational table

5 Discussion

We propose a natural approach for table analysis based on the indexing of data cells by the column and row header hierarchies. Each data cell is defined by paths through every header cell that spans that data cell. Automated extraction of paths from CSV versions of web tables must first locate the column and row headers via identification of an empty stub or by finding the boundaries of homogeneous rows and columns of content cells. It must also compensate for the splitting of spanning cells in both directions, and for spanning unit cells. We define a relational algebra in which the collection of row or column Header Paths is represented by a sum-of-products expression, and we show that the hierarchical structure of the row or column categories can be recovered by a decomposition process that can be carried out using widely available symbolic mathematical tools.

We demonstrate initial results on 107 randomly selected web tables from our collection of 1000 web tables from large sites. The experiments show that in 83% of our sample the header regions can be found using empty stubs or numerical delta cells. We estimate that the header regions can be found in at least a further 10%-15% with only modest improvements in our algorithms. The remainder will still require clicking on one or two cells on an interactive display of the table.

Most of the extracted column Header Paths are correct, but nearly 25% of the row Header Paths contain some mistake (not all fatal). The next step is more thorough analysis of indentations and of header roots in the stub. Interactive path correction is far more time consuming than interactive table segmentation. An important task is to develop adequate confidence measures to avoid having to inspect every table.

In addition to algorithmic extraction of an index, we show how to manipulate sum-of-products expressions along with a table's delta-cells to yield relational tables that can be processed by standard relational database engines. Thus, for tables our algorithms can interpret, we automate the process of turning human-readable tables into machine-readable tables that can be queried, searched, and combined in a relational database.

Acknowledgments. This work was supported by the National Science Foundation under Grants # 044114854 (at Rensselaer Polytechnic Institute) and 0414644 (at Brigham Young University) and by the Rensselaer Center for Open Software. We are thankful to students Landy Zhang, Douglas Mosher, Stephen Poplasky (RPI), Dongpu Jin (UNL), and Spencer Machado (BYU) for collecting and analyzing data and for coding the transformation algorithms.

References

1. Wang, X.: Tabular Abstraction, Editing, and Formatting, Ph.D Dissertation, University of Waterloo, Waterloo, ON, Canada (1996)
2. Embley, D.W., Hurst, M., Lopresti, D., Nagy, G.: Table Processing Paradigms: A Research Survey. Int. J. Doc. Anal. Recognit. 8(2-3), 66–86 (2006)
3. Zanibbi, R., Blostein, D., Cordy, J.R.: A survey of table recognition: Models, observations, transformations, and inferences. International Journal of Document Analysis and Recognition 7(1), 1–16 (2004)
4. Krüpl, B., Herzog, M., Gatterbauer, W.: Using visual cues for extraction of tabular data from arbitrary HTML documents. In: Proceedings. of the 14th Int'l Conf. on World Wide Web, pp. 1000–1001 (2005)
5. Pivk, A., Ciamiano, P., Sure, Y., Gams, M., Rahkovic, V., Studer, R.: Transforming arbitrary tables into logical form with TARTAR. Data and Knowledge Engineering 60(3), 567–595 (2007)
6. Silva, E.C., Jorge, A.M., Torgo, L.: Design of an end-to-end method to extract information from tables. Int. J. Doc. Anal. Recognit. 8(2), 144–171 (2006)
7. Esposito, F., Ferilli, S., Di Mauro, N., Basile, T.M.A.: Incremental Learning of First Order Logic Theories for the Automatic Annotations of Web Documents. In: Proceedings of the 9th International Conference on Document Analysis and Recognition (ICDAR-2007), Curitiba, Brazil, September 23-26, pp. 1093–1097. IEEE Computer Society, Los Alamitos (2007); ISBN 0-7695-2822-8, ISSN 1520-5363
8. Esposito, F., Ferilli, S., Basile, T.M.A., Di Mauro, N.: Machine Learning for Digital Document Processing: From Layout Analysis To Metadata Extraction. In: Marinai, S., Fujisawa, H. (eds.) Machine Learning in Document Analysis and Recognition. SCI, vol. 90, pp. 79–112. Springer, Berlin (2008); ISBN 978-3-540-76279-9
9. Jandhyala, R.C., Krishnamoorthy, M., Nagy, G., Padmanabhan, R., Seth, S., Silversmith, W.: From Tessellations to Table Interpretation. In: Carette, J., Dixon, L., Coen, C.S., Watt, S.M. (eds.) MKM 2009, Held as Part of CICM 2009. LNCS, vol. 5625, pp. 422–437. Springer, Heidelberg (2009)
10. `http://www.mathworks.com/help/toolbox/symbolic/horner.html`
11. Fateman, R. J.: Essays in Symbolic Simplification. MIT-LCS-TR-095, 4-1-1972, `http://publications.csail.mit.edu/lcs/pubs/pdf/MIT-LCS-TR-095.pdf` (downloaded November 10, 2010)

12. Knuth, D.E.: 4.6.2 Factorization of Polynomials". Seminumerical Algorithms. In: The Art of Computer Programming, 2nd edn., pp. 439–461, 678–691. Addison-Wesley, Reading (1997)
13. Kaltofen, E.: Polynomial factorization: a success story. In: ISSAC 2003 Proc. 2003 Internat. Symp. Symbolic Algebraic Comput. [-12], pp. 3–4 (2003)
14. Brayton, R.K., McMullen, C.: The Decomposition and Factorization of Boolean Expressions. In: Proceedings of the International Symposium on Circuits and Systems, pp. 49–54 (May1982)
15. Vasudevamurthy, J., Rajski, J.: A Method for Concurrent Decomposition and Factorization of Boolean Expressions. In: Proceedings of the International Conference on Computer-Aided Design, pp. 510–513 (November 1990)
16. Sentovich, E.M., Singh, K.J., Lavagno, L., Moon, C., Murgai, R., Saldanha, A., Savoj, H., Stephan, P.R., Brayton, R.K., Sangiovanni-Vincentelli, A.L.: SIS: A System for Sequential Circuit Synthesis. In: Memorandum No. UCB/ERL M92/41, Electronics Research Laboratory, University of California, Berkeley (May 1992), http://www.eecs.berkeley.edu/Pubs/TechRpts/1992/ERL-92-41.pdf (downloaded November 4, 2010)
17. (Quickmath-ref), http://www.quickmath.com/webMathematica3/quickmath/page.jsp?s1=algebra&s2=factor&s3=advanced (last accessed November 12, 2010)

Document Analysis Research in the Year 2021[*]

Daniel Lopresti[1] and Bart Lamiroy[2]

[1] Computer Science and Engineering, Lehigh University,
Bethlehem, PA 18015, USA
lopresti@cse.lehigh.edu
[2] Nancy Université, INPL, LORIA, Campus Scientifique,
BP 239, 54506 Vandoeuvre Cedex, France
Bart.Lamiroy@loria.fr

Abstract. Despite tremendous advances in computer software and hardware, certain key aspects of experimental research in document analysis, and pattern recognition in general, have not changed much over the past 50 years. This paper describes a vision of the future where community-created and managed resources make possible fundamental changes in the way science is conducted in such fields. We also discuss current developments that are helping to lead us in this direction.

1 Introduction: Setting the Stage

The field of document analysis research has had a long, rich history. Still, despite decades of advancement in computer software and hardware, not much has changed in how we conduct our experimental science, as emphasized in George Nagy's superb keynote retrospective at the DAS 2010 workshop [11].

In this paper, we present a vision for the future of experimental document analysis research. Here the availability of "cloud" resources consisting of data, algorithms, interpretations and full provenance, provides the foundation for a research paradigm that builds on collective intelligence (both machine and human) to instill new practices in a range of research areas. The reader should be aware that this paradigm is applicable to a much broader scope of machine perception and pattern recognition – we use document analysis as the topic area to illustrate the discussion as this is where our main research interests lie, and where we can legitimately back our claims. Currently under development, the platform we are building exploits important trends we see arising in a number of key areas, including the World Wide Web, database systems, and social and collaborative media.

The first part of this paper presents our view of this future as a fictional, yet realizable, "story" outlining what we believe to be a compelling view of community-created and managed resources that will fundamentally change the

---

[*] This work is a collaborative effort hosted by the Computer Science and Engineering Department at Lehigh University and funded by a Congressional appropriation administered through DARPA IPTO via Raytheon BBN Technologies.

way we do research. In the second part of the paper, we then turn to a more technical discussion of the status of our current platform and developments in this direction.

2 Document Analysis Research: A Vision of the Future

Sometime in the year 2021, Jane, a young researcher just getting started in the field, decides to turn her attention to a specific task in document analysis: given a page image, identify regions that contain handwritten notations.[1] Her intention is to develop a fully general method that should be able to take any page as input, although there is the implicit assumption that the handwriting, if present, covers only a relatively small portion of the page and the majority of the content is pre-printed text or graphics.

Note that in this version of the future, there is no such thing as "ground truth" – that term is no longer used. Rather, we talk about the *intent* of the author, the *product* of the author (*e.g.*, the physical page) [4], and the *interpretation* arrived at by a reader of the document (human or algorithm). There are no "right" or "wrong" answers – interpretations may naturally differ – although for some applications, we expect that users who are fluent in the language and the domain of the document will agree nearly all of the time.

The goal of document analysis researchers is to develop new methods that mimic, as much as possible, what a careful human expert would do when confronted with the same input, or at least to come closer than existing algorithms. Of course, some people are more "careful" or more "expert" than others when performing certain tasks. The notion of *reputation*, originally conceived in the early days of social networking, figures prominently in determining whose interpretations Jane will choose to use as the target when developing her new method. Members of the international research community – as well as algorithms – have always had informal reputations, even in the early days of the field. What is different today, in 2021, is that reputation has been formalized and is directly associated with interpretations that we use to judge the effectiveness of our algorithms, so that systems to support experimental research can take advantage of this valuable information. Users, algorithms, and even individual data items all have reputations that are automatically managed and updated by the system.

After Jane has determined the nature of the task, she turns to a well-known resource – a web server we shall call *DARE* (for "Document Analysis Research Engine") – to request a set of sample documents which she will use in developing and refining her algorithm. This server, which lives in the "cloud" and is not a single machine, has become the de facto standard in the field, just as certain

---

[1] Experience has taught us that we tend to be overly optimistic when we assume problems like this will be completely solved in the near future and we will have moved on to harder questions. Since we need a starting point for our story, we ask the reader to suspend skepticism on what is likely a minor quibble. Jane's problem can be replaced with any one that serves the purpose.

datasets were once considered a standard in the past. There are significant differences, however. In the early days, datasets were simply collections of page images along with a single interpretation for each item (which was called the "ground-truth" back then). In 2021, the DARE server supports a fundamentally different paradigm for doing experimental science in document image analysis. Jane queries the server to give her 1,000 random documents from the various collections it knows about. Through the query interface, she specifies that:

- Pages should be presented as a 300 dpi bitonal TIF image.
- Pages should be predominately printed material: text, line art, photographs, etc. This implies that the page regions have been classified somehow: perhaps by a human interpreter, or by another algorithm, or some combination. Jane indicates that she wants the classifications to have come from only the most "trustworthy" sources, as determined through publication record, citations to past work, contributions to the DARE server, *etc.*
- A reasonable number of pages in the set should contain at least one handwritten annotation. Jane requests that the server provide a set of documents falling within a given range.
- The handwritten annotations on the pages should be delimited in a way that is consistent with the intended output from Jane's algorithm. Jane is allowed to specify the requirements she would like the interpretations to satisfy, as well as the level of trustworthiness required of their source.

By now, the status of the DARE web server as the de facto standard for the community has caused most researchers to use compatible file formats for recording interpretations. Although there is no requirement to do so, it is just easier this way since so much data is now delivered to users from the server and no longer lives locally on their own machines. Rather than fight the system, people cooperate without having to be coerced.

The DARE server not only returns the set of 1,000 random pages along with their associated interpretations, it also makes a permanent record of her query and provides a URL that will return exactly same set of documents each time it is run. Any user who has possession of the URL can see the parameter settings Jane used. The server logs all accesses to its collections so that members of the research community can see the history for every page delivered by the server.

In the early days of document image analysis research, one of the major hurdles in creating and distributing datasets were the copyright concerns. In 2021, however, the quantity of image-based data available on the web is astounding. Digital libraries, both commercially motivated and non-profit, present billions of pages that have already been scanned and placed online. While simple OCR results and manual transcriptions allow for efficient keyword-based searching, opportunities remain for a vast range of more sophisticated analysis and retrieval techniques. Hence, contributing a new dataset to the DARE server is not a matter of scanning the pages and confronting the copyright issues one's self but, rather, the vast majority of new datasets are references (links) to collections of page images that already exist somewhere online. Access – whether free or through subscription services – is handled as though well-developed mechanisms (including user authentication, if it is needed) that are part of the much bigger web environment.

With dataset in hand, Jane proceeds to work on her new algorithm for detecting handwritten annotations. This part of the process is no different from the way researchers worked in the past. Jane may examine the pages in the dataset she was given by the DARE server. She uses some pages as a training set and others as her own "test" set, although this is just for development purposes and never for publication (since, of course, she cannot prove that the design of her algorithm was not biased by knowing what was contained in this set).

While working with the data, Jane notices a few problems. One of the page images was delivered to her upside down (rotated by 180 degrees). These sorts of errors, while rare, arise from time to time given the enormous size of the collections on the DARE server. In another case, the TIF file for a page was unreadable, at least by the version of the library Jane is using. Being a responsible member of the research community (and wanting her online reputation to reflect this), Jane logs onto the DARE server and, with a few mouse clicks, reports both problems – it just takes a minute. Everyone in the community works together to build and maintain the collections delivered via the web server. Jane's bug reports will be checked by other members of the community (whose reputations will likewise rise) and the problem images will be fixed in time.

In a few other cases, Jane disagrees with the interpretation that is provided for the page in question. In her opinion, the bounding polygons are drawn improperly and, on one page, there is an annotation that has been missed. Rather than just make changes locally to her own private copies of the annotation files (as would have happened in the past), Jane records her own interpretations on the DARE server and then refreshes her copies. No one has to agree with her, of course – the previous versions are still present on the server. But by adding her own interpretations, the entire collection is enriched. (At the same time Jane is doing her own work, dozens of other researchers are using the system.) The DARE server provides a wiki-like interface with text editing and graphical markup tools that run in any web browser. Unlike a traditional wiki, however, the different interpretations are maintained in parallel. The whole process is quite easy and natural. Once again, Jane's online reputation benefits when she contributes annotations that other users agree with and find helpful.

After Jane is done fine-tuning her algorithm, she prepares to write a paper for submission to a major conference. This will involve testing her claim that her technique will work for arbitrary real-world pages, not just for the data she has been using (and becoming very familiar with) for the past six months. She has two options for performing random, unbiased testing of her method, both of which turn back to the DARE server.[2] These are:

Option 1: Jane can "wrap" her code in a web-service framework provided by the DARE server. The code continues to run on Jane's machine, with the

---

[2] All top conferences and journals now require the sort of testing we describe here. This is a decision reached through the consensus of the research community, not dictated by some authority.

DARE server delivering a random page image that her algorithm has not seen before, but that satisfies certain properties she has specified in advance. Jane's algorithm performs its computations and returns its results to the DARE server within a few seconds. As the results are returned to the DARE server, they are compared to existing interpretations for the page in question. These could be human interpretations or the outputs from other algorithms that have been run previously on the same page.

Option 2: If she wishes, Jane can choose to upload her code to the DARE server, thereby contributing it to the community and raising her reputation. In this case, the server will run her algorithm locally on a variety of previously unseen test pages according to her specifications. It will also maintain her code in the system and use it in future comparisons when other researchers test their own new algorithms on the same task.

At the end of the evaluation, Jane is provided with:

– A set of summary results showing how well her algorithm matched human performance on the task.
– Another set of summary results showing how well her algorithm fared in comparison to other methods tested on the same pages.
– A *certificate* (*i.e.*, a unique URL) that guarantees the integrity of the results and which can be cited in the paper she is writing. Anyone who enters the certificate into a web browser can see the summary results of Jane's experiment delivered directly from the (trusted) DARE web server, so there can be no doubt what she reported in her paper is true and reproducible.

When Jane writes her paper, the automated analysis performed by the DARE server allows her to quantify her algorithms performance relative to that of a human, as well as to techniques that were previously registered on the system. Of course, given the specifics of our paradigm, performances can only be expressed in terms of statistical agreement and, perhaps, reputation, but perhaps not in terms of an absolute ranking of one algorithm with respect to another. Ranking and classification of algorithms and the data they were evaluated on will necessarily take more subtle and multi-valued forms. One may argue that having randomly selected evaluation documents for certification can be considered as marginally fair, since there is a factor of chance involved. While this is, in essence, true, the fact that the randomly generated dataset is available for reproduction (*i.e.* once generated, the certificate provides a link to the exact dataset used to certify the results), anyone arguing that the result was obtained on an unusually biased selection can access the very same data and use it in evaluating other algorithms.

It was perhaps a bit ambitious of Jane to believe that her method would handle all possible inputs and, in fact, she learns that her code crashes on two of the test pages. The DARE server allows Jane to download these pages to see what is wrong (it turns out that she failed to dimension a certain array to be big enough). If Jane is requesting a certificate, the DARE server will guarantee that her code never sees the same page twice. If she is not requesting a certificate, then this restriction does not apply and the server will be happy to deliver the same page as often as she wishes.

Unlike past researchers who had the ability to remove troublesome inputs from their test sets in advance, the DARE server prohibits such behavior. As a result, it is not uncommon for a paper's authors to report, with refreshing honesty, that their implementation of an algorithm matched the human interpretation 93% of the time, failed to match the human 5% of the time, and did not complete (*i.e.*, crashed) 2% of the time.

When other researchers read Jane's paper, they can use the URL she has published to retrieve exactly the same set of pages from the DARE server.[3] If they wish to perform an unbiased test of their own competing method, comparing it directly to Jane's – and receive a DARE certificate guaranteeing the integrity of their results – they must abide by the same rules she did.

In this future world, there is broad agreement that the new paradigm introduced (and enforced) by the DARE server has improved the quality of research. Results are now verifiable and reproducible. Beginning researchers no longer waste their time developing methods that are inferior to already-known techniques (since the DARE server will immediately tell you if another algorithm did a better job on the test set you were given). The natural (often innocent) tendency to bias an algorithm based on knowing the details of the test set have been eliminated. The overuse of relatively small "standard" collections that was so prevalent in the early days of the field is now no longer a problem.

The DARE server is not foolproof, of course – it provides many features to encourage and support good science, but it cannot completely eliminate the possibility of a malicious individual abusing the system. However, due to its community nature, all records are open and visible to every user of the system, which increases the risk of being discovered to the degree that legitimate researchers would never be willing to take that chance.

Looking back with appreciation at how this leap forward was accomplished, Jane realizes that it was not the result of a particular research project or any single individual. Rather, it was the collective effort and dedication of the entire document analysis research community.

3 Script and Screenplay for the Scenario

The scenario just presented raises a number of fundamental questions that must be addressed before document analysis research can realize its benefits. In this section we develop these questions and analyze the extent to which they already have (partial or complete) answers in the current state-of-the-art, those which are open but that can be answered with a reasonable amount of effort, and those which will require significant attention by the community before they are solved.

We also refer to a proof-of-concept prototype platform for Document Analysis and Exploitation (DAE – not to be confused with DARE), accessible at http://dae.cse.lehigh.edu, which is capable of storing data, meta-data and

---

[3] This form of access-via-URL is not limited to randomly generated datasets. Legacy datasets from the past are also available this way.

interpretations, interaction software, and complete provenance as more fully described elsewhere [8,9]. DAE is an important step in the direction of DARE, but still short of the grand vision described earlier.

3.1 Main Requirements

What the scenario describes, in essence, is the availability of a well identified, commonly available resource that offers storage and retrieval of document analysis data, complex querying of this data, collective yet personalized markup, representation, evaluation, organization and projection of data, as well as archival knowledge of uses and transformations of data, including certified interactions.

Rather than offer monolithic chunks of data and meta-data or interpretations as in the case of current standard datasets, the envisioned resource treats data on a finer-grained level. This level of detail is illustrated in the scenario by the complexity of the queries the system should be capable of answering.

This data need not conform to a predefined format, but can be polymorphic, originating both from individual initiatives as well as from collective contributions. Regardless of how it is stored, it can also be retrieved and re-projected into any format. Annotations need not be human-contributed but can be the result of complete document analysis pipelines and algorithms. As a result, the resource can hold apparently contradictory interpretations of identical documents, when these stem from different transformation and analysis processes [8].

Formats and representations cannot be rigidly dictated in advance for our scenario to have a chance of succeeding. Past experience has shown that attempts to "coerce" a community into using a single set of conventions does not work; at best, it contributes to locking it into a limited subset of possible uses, stifling creativity. This is contradictory to the standpoint we have taken with respect to *ground-truth* (or rather lack thereof [6,10,14,2]) and our preferring the term *interpretation*. This clearly advocates for as open as possible ways of representing data, keeping in mind, however, that abandoning any kind of imposed structure may make it impossible to realize our vision.

Turning now to the current DAE server, formats and representations are transparently handled by the system, since the user can define any format, naming, or association convention within our system. Data can be associated with image regions, image regions can be of any shape and format, there is no restriction on uniqueness or redundancy, so multiple interpretations are naturally supported. Because of its underlying data model and architecture, everything is queryable via SQL. The standard datasets that can be downloaded from the platform are no longer monolithic file collections, but potentially complex queries that generate these datasets on-the-fly [9].

Interactions with data are integrated in the DAE data model on the one hand (it represents algorithms as well as their inputs and outputs), but the model goes further by giving access to user-provided programs that can be executed on the stored data, thus producing new meta-data and interpretations. Queries like finding all results produced by a specified algorithm, class of algorithms, or user, can be used as an interpretation of a document, and can also serve as a benchmarking element for comparison with competitors. Since everything is hosted in the same context, it becomes possible to "certify" evaluations performed on the system.

3.2 Scientific (and Other) Challenges

While the DAE platform is a promising first step toward to the DARE paradigm, it still falls short in addressing some of the key concepts of the scenario we depicted in Section 2.

– Reputation is one suggestion we advanced to to treat disagreements between multiple interpretations. Not only can multiple interpretations arise from using the same data in different contexts, but there can be debate even in identical contexts when the data is "noisy." When such controversy arises, which would be the preferred interpretation to use? Here the notion of on-line reputation as practiced in Web 2.0 recommender systems may hold the key [12,13]. Researchers and algorithms already have informal reputations within the community. Extending this to interpretations can provide a mechanism for deciding which annotations to trust.

  How this actually needs to be implemented and made user-friendly is an interesting question. Success in doing so would likely also solve the problem of deliberately malicious interpretations.
– Semantic clutter is another major issue that will inevitably occur and that has no straightforward solution under the current state-of-the-art. Semantic clutter arises when different contributors are unaware of each other's interpretative contexts and label data with either identical labels even though their contexts are completely different, or, conversely, with different labels although they share the same interpretation context. In the case of Jane, for instance, some image region may be labeled as *handwriting* but is, in fact, the interpretation of a printed word spelled that way, rather than denoting actual handwritten text. On the other hand, she might miss regions that have been labeled by synonyms such as *manuscript, hand annotated, writing* ... There are probably very good opportunities to leverage work on *folksonomies* [7,3] to solve some of these issues, although our context is slightly different than the one usually studied in that community. Since, in our case, all data provenance is recorded and accessible, one can easily retrieve data produced by an algorithm using specific runtime parameters. This guarantees that the obtained data share the same interpretation context, and thus has some significant overlapping semantic value. Furthermore, since users are free to use and define their own, personal annotation tags, and since, again, provenance is recorded, one can assume that the semantics of a given user's tags will only evolve slowly over time, if at all. Exploring advanced formal learning techniques might yield a key to reducing the semantic cluttering mentioned

earlier, or at least provide tools for users to discover interpretations that are likely to be compatible with their own context.
– Query Interfaces, and especially those capable of handling the complex expressions used in this paper, are still open research domains [1]. Their application is also directly related to the previously mentioned semantic issues. Semantic Web [5] and ontology-folksonomy combinations [7,3] are therefore also also probable actors in our scenario. To make the querying really semantic in an as automated a way as possible, and by correctly capturing the power of expressiveness as people contribute to the resource pool, the DAE platform will need to integrate adequate knowledge representations. This goes beyond the current storage of attributes and links between data. Since individual research contexts and problems usually require specific representations and concepts, contributions to the system will initially focus on their own formats. However, as the need for new *interpretations* arises, users will want to combine different representations of similar concepts to expand their experiment base. To allow them to do that, formal representations and semantic web tools will need to be developed. Although their seems to be intuitively obvious inter-domain synergies between all cited domains (*e.g.* data-mining query languages need application contexts and data to validate their models, while our document analysis targeted platform needs query languages to validate the scalability and generality of its underlying research) only widespread adoption of the paradigm described in this paper will reveal potentially intricate research interactions.
– Archiving and Temporal Consistency, concern a fundamentally crucial part of this research. Numerous problems arise relating to the comprehensive usage and storage of all the information mentioned in this paper. In short, and given that our scenario is built on a distributed architecture, how shall availability of data be handled? Since our concept relies on complete traceability and inter-dependence of data, annotations and algorithms, how can we guarantee long term conservation of all these resources when parts of them are third-party provided? Simple replication and redundancy may rapidly run into copyright, intellectual property, or trade secret issues. Even data that was initially considered public domain may suddenly turn out to be owned by someone and need to be removed. What about all derived annotations, interpretations, and results? What about reliability and availability from a purely operational point of view, if the global resource becomes so widely used that it becomes of vital importance?

4 Conclusion

In this paper, we have presented a vision for the future of experimental research in document analysis and described how our current DAE platform [8,9] can exploit collective intelligence to instill new practices in the field. This forms a significant first step toward a crowd-sourced document resource platform that can contribute in many ways to more reproducible and sustainable machine perception research. Some of its features, such as its ability to host complex workflows, are currently being developed to support benchmarking contests.

We have no doubt that the paradigm we are proposing is largely feasible and we strongly believe that the future of experimental document analysis research will head in a new direction much like the one we are suggesting. When this compelling vision comes to pass, it will be through the combined efforts and ingenuity of the entire research community.

The DAE server prototype is open to community contributions. It is inherently cloud-ready and has the potential to evolve to support the grand vision of DARE. Because of this new paradigm's significance to the international research community, we encourage discussion, extensions and amendments through a constantly evolving Wiki: http://dae.cse.lehigh.edu/WIKI. This Wiki also hosts a constantly updated chart of DAE platform features realizing the broader goals discussed in this paper.

References

1. Boulicaut, J.F., Masson, C.: Data mining query languages. In: Maimon, O., Rokach, L. (eds.) Data Mining and Knowledge Discovery Handbook, pp. 655–664. Springer, US (2010), doi:10.1007/978-0-387-09823-4_33
2. Clavelli, A., Karatzas, D., Lladós, J.: A framework for the assessment of text extraction algorithms on complex colour images. In: DAS 2010: Proceedings of the 8th IAPR International Workshop on Document Analysis Systems, pp. 19–26. ACM, New York (2010)
3. Dotsika, F.: Uniting formal and informal descriptive power: Reconciling ontologies with folksonomies. International Journal of Information Management 29(5), 407–415 (2009)
4. Eco, U.: The limits of interpretation. Indiana University Press (1990)
5. Feigenbaum, L., Herman, I., Hongsermeier, T., Neumann, E., Stephens, S.: The semantic web in action. Scientific American (December 2007)
6. Hu, J., Kashi, R., Lopresti, D., Nagy, G., Wilfong, G.: Why table ground-truthing is hard. In: 6th International Conference on Document Analysis and Recognition, pp. 129–133. IEEE Computer Society, Los Alamitos (2001)
7. Kim, H.L., Decker, S., Breslin, J.G.: Representing and sharing folksonomies with semantics. Journal of Information Science 36(1), 57–72 (2010)
8. Lamiroy, B., Lopresti, D.: A platform for storing, visualizing, and interpreting collections of noisy documents. In: Fourth Workshop on Analytics for Noisy Unstructured Text Data - AND 2010. ACM International Conference Proceeding Series, IAPR. ACM, Toronto (2010)
9. Lamiroy, B., Lopresti, D., Korth, H., Jeff, H.: How carefully designed open resource sharing can help and expand document analysis research. In: Agam, G., Viard-Gaudin, C. (eds.) Document Recognition and Retrieval XVIII. SPIE Proceedings, vol. 7874. SPIE, San Francisco (2011)
10. Lopresti, D., Nagy, G., Smith, E.B.: Document analysis issues in reading optical scan ballots. In: DAS 2010: Proceedings of the 8th IAPR International Workshop on Document Analysis Systems, pp. 105–112. ACM, New York (2010)

11. Nagy, G.: Document systems analysis: Testing, testing, testing. In: Doerman, D., Govindaraju, V., Lopresti, D., Natarajan, P. (eds.) DAS 2010, Proceedings of the Ninth IAPR International Workshop on Document Analysis Systems, p. 1 (2010), http://cubs.buffalo.edu/DAS2010/GN_testing_DAS_10.pdf
12. Raub, W., Weesie, J.: Reputation and efficiency in social interactions: An example of network effects. American Journal of Sociology 96(3), 626–654 (1990)
13. Sabater, J., Sierra, C.: Review on computational trust and reputation models. Artificial Intelligence Review 24(1), 33–60 (2005)
14. Smith, E.H.B.: An analysis of binarization ground truthing. In: DAS 2010: Proceedings of the 8th IAPR International Workshop on Document Analysis Systems, pp. 27–34. ACM, New York (2010)

Markov Logic Networks for Document Layout Correction

Stefano Ferilli, Teresa M.A. Basile, and Nicola Di Mauro

Department of Computer Science, LACAM laboratory
University of Bari "Aldo Moro", via Orabona 4, 70125, Bari
{ferilli,basile,ndm}@di.uniba.it

Abstract. The huge amount of documents in digital formats raised the need of effective content-based retrieval techniques. Since manual indexing is infeasible and subjective, automatic techniques are the obvious solution. In particular, the ability of properly identifying and understanding a document's structure is crucial, in order to focus on the most significant components only. Thus, the quality of the layout analysis outcome biases the next understanding steps. Unfortunately, due to the variety of document styles and formats, the automatically found structure often needs to be manually adjusted. In this work we present a tool based on Markov Logic Networks to infer corrections rules to be applied to forthcoming documents. The proposed tool, embedded in a prototypical version of the document processing system DOMINUS, revealed good performance in real-world experiments.

1 Introduction

The task aimed at identifying the geometrical structure of a document is known as Layout Analysis, and represents a wide area of research in document processing, for which several solutions have been proposed in literature. The quality of the layout analysis outcome is crucial, because it determines and biases the quality of the next understanding steps. Unfortunately, the variety of document styles and formats to be processed makes the layout analysis task a non-trivial one, so that the automatically found structure often needs to be manually fixed by domain experts.

The geometric layout analysis phase involves several processes, among which page decomposition. Several works concerning the page decomposition step are present in the literature, exploiting different approaches and having different objectives. Basic operators of all these approaches are split and merge: they exploit the features extracted from an elementary block to decide whether splitting or merging two or more of the identified basic blocks in a top-down [8], bottom-up [15] or hybrid approach [12] to the page decomposition step. Since all methods split or merge blocks/components based on certain parameters, parameter estimation is crucial in layout analysis. All these methods exploit parameters that are able to model the split or merge operations in specific classes of the document domain. Few adaptive methods, in the sense that split or merge operations are performed using estimated parameter values, are present in the literature [2,10]. A step forward is represented by the exploitation of Machine Learning techniques in order to automatically assess the parameters/rules able to perform the document page decomposition, and hence the eventual correction of the performed split/merge operations, without requiring an empirical evaluation on the specific document domain at hand. To this regard, learning methods have been used to separate textual areas from graphical areas [5] and to classify text regions as headline, main text, etc. [3,9] or even to learn split/merge rules in order to carry out the corresponding operations and/or correction [11,16].

However, a common limit of the above reported methods regards the consideration that they are all designed with the aim of working on scanned documents, and in some cases on documents of a specified typology, thus lacking any generality of the proposal with respect to the online available documents that can be of different digital formats. On the other hand, methods that work on natively digital documents assume that the segmentation phase can be carried out by simply performing a matching of the document itself with a standard template, even in this case, of a specified format. In this work we propose the application of a Statistical Relational Learning [7] (SRL) technique to infer a probabilistic logical model recognising wrong document layouts from sample corrections performed by expert users in order to automatically apply them to future incoming documents. Corrections are codified in a first-order language and the learned correction model is represented as a Markov Logic Network [14] (MLN). Experiments in a real-world task confirmed the good performance of the solution.

2 Preliminaries

In this section we briefly describe DOC (Document Organization Composer) [4], a tool for discovering a full layout hierarchy in digital documents based primarily on layout information. The layout analysis process starts with a preprocessing step performed by a module that takes as input a generic digital document and extracts the set of its elementary layout components (*basic-blocks*), that will be exploited to identify increasingly complex aggregations of basic components.

The first step in the document layout analysis concerns the identification of rules to automatically shift from the basic digital document description to a higher level one. Indeed, the basic-blocks often correspond just to fragments of words (e.g., in PS/PDF documents), thus a preliminary aggregation based on their overlapping or adjacency is needed in order to obtain blocks surrounding whole words (*word-blocks*). Successively, a further aggregation of word-blocks could be performed to identify text lines (*line-blocks*). As to the grouping of blocks into lines, since techniques based on the mean distance between blocks proved unable to correctly handle cases of multi-column documents, Machine Learning approaches were applied in order to automatically infer rewriting rules that could suggest how to set some parameters in order to group together rectangles (words) to obtain lines. To do this, a kernel-based method was exploited to learn rewriting rules able to perform the bottom-up construction of the whole document starting from the basic/word blocks up to the lines.

The next step towards the discovery of the high-level layout structure of a document page consists in applying an improvement of the algorithm reported in [1]. To this aim, DOC analyzes the whitespace and background structure of each page in the document in terms of rectangular covers and identifies the white rectangles that are present in the page by decreasing area, thus reducing to the Maximal White Rectangle problem as follows: given a set of rectangular content blocks (*obstacles*) $C = \{r_0, \ldots, r_n\}$, all placed inside the page rectangular contour r_b, find a rectangle r contained in r_b whose area is maximal and that does not overlap any $r_i \in C$. The algorithm exploits a priority queue of pairs (r, O), where r is a rectangle and O is a set of obstacles overlapping r. The priority of the pair in the queue corresponds to the area of its rectangle. Pairs are iteratively extracted from the queue and if the set of obstacles corresponding to its rectangle is empty, then it represents the maximum white rectangle still to be discovered. Otherwise, one of its obstacles is chosen as a *pivot* and the rectangle is consequently split into four regions (above, below, to the right and to the left of the pivot). Each such region, along with the obstacles that fall in it, represents a new pair to be inserted in the queue. Complement of the found maximal white rectangles yield the document content blocks.

However, taking the algorithm to its natural end and then computing the complement would result again in the original basic blocks, while the layout analysis process aims at returning higher-level layout aggregates. This raised the problem of identifying a stop criterion to end this process. An empirical study carried out on a set of 100 documents of three different categories revealed that the best moment to stop the algorithm is when the ratio of the last white area retrieved with respect to the total white area in the current page of the document decreases up to 0, since before it the layout is not sufficiently detailed, while after it useless white spaces are found.

3 Learning Layout Correction Theories

The proposed strategy for automatically assessing a threshold to decide when stopping the background retrieval loop allows to immediately reach a layout that is already good for many further tasks of document image understanding. Nevertheless, such a threshold is clearly a trade off between several document types and shapes, and hence in some cases the layout needs to be slightly improved through a fine-tuning step that must be specific for each single document. To handle this problem, a tool was provided that allows the user to directly point out useful background fragments that were not yet retrieved and add them explicitly (*white forcing*), or, conversely, to select useless ones that were erroneously retrieved and remove them from the final layout (*black forcing*).

The forcing functionality allows the user to interact with the layout analysis algorithm and suggest which specific blocks are to be considered as background or content. To see how it can be obtained, let us recall that the algorithm, at each step, extracts from the queue a new area to be examined and can take three actions correspondingly: if the contour is not empty, it is split and the resulting fragments are enqueued; if the contour is empty and fulfils the constraints, it is added to the list of white areas; if the contour is empty but does not fulfil the constraints, it is discarded.

Allowing the user to interact with the algorithm means modifying the algorithm behaviour as a consequence of his choices. It turns out that the relevance of a (white or black) block to the overall layout can be assessed based on its position inside the document page and its relationships with the other layout components. According to this assumption, each time the user applies a manual correction, the information on his actions and on their effect can be stored in a trace file for subsequent analysis. In particular, each manual correction (user intervention) can be exploited as an example from which learning a model on how to classify blocks as meaningful or meaningless for the overall layout. Applying the learned models in subsequent incoming documents, it would be possible to automatically decide whether or not any white (resp., black) block is to be included as background (resp., content) in the final layout, this way reducing the need for user intervention.

3.1 Description Language

Now let us turn to the way in which the trace of manual corrections are codified. The assumption is that the user changes the document layout when he considers that the proposed layout is wrong, then it forces a specific block because he knows the resulting effect on the document and considers it as satisfactory. Thus, to properly learn rules that can help in automatically fixing and improving the document layout analysis outcome, one must consider what is available before the correction takes place, and what will be obtained after it is carried out. For this reason, each example, representing a correction, will include a description of the blocks' layout both before and after the correction. However, the modification is typically *local*, i.e. it does not affect the whole document layout, but involves just a limited area surrounding the forced block. This allows to limit the description to just such an area. To sum up, the log file of the manual corrections, applied by the user after the execution of the layout analysis algorithm, will include both the white and the black blocks he forced, and will record, for each correction, information about the blocks and frames surrounding the forced block, both before and after the correction.

In the following b stands for block, f for frame and r for rectangle. Each log example is represented as a set of first-order predicates and it is labelled as positive for forcing black (merge(b)) or white block (split(b)). Negative examples are denoted by negating the predicate with the not predicate.

A set of facts describing the other blocks in the area of interest before and after the correction are reported. It contains information on the document page in which the correction took place, the horizontal/vertical size and position of a block in the overall document, whether it is at the beginning, in the middle or at the end of the document, furthermore the forced block and the layout situation both before and after the correction are represented in the log example. The description of each of the two situations (before and after the correction) is based on literals expressing the page layout and describing the blocks and frames surrounding the forced block, and, among them, only those touching or overlapping the forced block. Each involved frame `frame(r)` or block `block(r)` is considered as a rectangular area of the page, and described according to the following parameters: horizontal and vertical position of its centroid with respect to the top-left corner of the page (`posX(r,x)` and `posY(r,y)`), height and width (`width(r)` and `height(r)`), and its content type (`type(r,t)`).

The relationships between blocks/frames are described by means of a set of predicates representing the spatial relationships existing among all considered frames and among all blocks belonging to the same frame (`belongs(b,f)`), that touch or overlap the forced block; furthermore for each frame or block that touches the forced block a literal specifying that they touch (`touches(b1,b2)`); finally, for each block of a frame that overlaps the forced block the percentage of overlapping (`overlaps(b1,b2,perc)`). It is fundamental to completely describe the mutual spatial relationships among all involved elements. All, and only, the relationships between each block/frame and the forced blocks are expressed, but not their inverses (i.e., the relationships between the forced block and the block/frame in question). To this aim, the model proposed in [13] for representing the spatial relationships among the blocks/frames was considered. Specifically, according to such a model, fixed a rectangle, its plane is partitioned in 25 parts and its spatial relationships to any other rectangle in the plane can be specified by simply listing the parts to which the other rectangle overlaps (`overlapPart(r1,r2,part)`).

3.2 Markov Logic Networks Background

Here, we briefly introduce the notions of a SRL approach combining first-order logic and probabilistic graphical models in a single representation. SRL seeks to combine the power of statistical learning to deal with the presence of uncertainty in many real-world application domains, and the power of relational learning in dealing the complex structure of such domains. We provide the background on Markov Logic Networks [14] (MLNs) representing first-order knowledge base with a weight attached to each formula.

A Markov Network (MN) models the joint distribution of a set of variables $X = (X_1, \ldots, X_n)$ made up of an undirected graph G and a set of potential functions ϕ_k. Each variable is represented with a node in the graph, and the model has a potential function for each clique in the graph. The joint distribution represented by a MN is given by

$$P(X = x) = \frac{1}{Z} \prod_k \phi_k(x_{\{k\}}),$$

where $x_{\{k\}}$ is the state of the k-th clique, and the partition function Z is given by $Z = \sum_{x \in \mathcal{X}} \prod_k \phi_k(x_{\{k\}})$. MNs may be represented as log-linear models, where each potential clique is replaced by an exponentiated weighted sum of features of the state, as in the following formula

$$P(X = x) = \frac{1}{Z} \exp\left(\sum_j w_j f_j(x)\right).$$

A first-order knowledge base (KB) is a set of formulas in first-order logic constructed using constants, variables, functions and predicates. A formula is satisfiable iff there exists at least one world (Herbrand interpretation) in which it is true. A first-order KB can be seen as a set of hard constraints on the set of possible worlds: if a world violates even one formula, it has zero probability, and MLNs soften these constraints. When a world violates one formula it is less probable, but not impossible. The fewer formulas a world violates, the more probable it is. In MLNs each formula has an associated weight representing how strong a constraint is. An high weight corresponds to a great difference in log probability between a world that satisfies the formula and one that does not.

More formally, A MLN L is a set of pairs (F_i, w_i), where F_i is a formula in first-order logic and w_i is a real number. Together with a finite set of constants $C = \{c_1, \ldots, c_n\}$, it defines a MN $M_{L,C}$ that a) contains one binary node[1] for each possible grounding of each predicate appearing in L, and one feature[2] for each possible grounding of each formula F_i in L, whose weight is the w_i associated with F_i in L. A MLN can be viewed as a template for constructing MNs, named ground Markov networks. The probability distribution over possible worlds x specified by the ground MN $M_{L,C}$ is given by

$$P(X = x) = \frac{1}{Z} \exp\left(\sum_i w_i n_i(s)\right) = \frac{1}{Z} \prod_i \phi_i(x_{\{i\}})^{n_i(x)}, \quad (1)$$

where $n_i(x)$ is the number of true groundings of F_i in x, $x_{\{i\}}$ is the state (truth values) of the atoms appearing in F_i, and $\phi_i(x_{\{i\}}) = e^{w_i}$.

Reasoning with MLNs can be classified as either learning or inference. Inference in SRL is the problem of computing probabilities to answer specific queries after having defined a probability distribution. Learning corresponds to infer both the structure and the parameters of the true unknown model. An inference task is computing the probability that a formula holds, given an MLN and set of constants, that, by definition, is the sum of the probabilities of the worlds where it holds. MLN weights can be learned generatively by maximizing the likelihood of a relational database consisting of one or more *possible worlds* that form our training examples. The inference and learning algorithms for MLNs are publicly available in the open-source Alchemy system[3]. Given a relational database and a set of clauses in the KB, many weights learning and inference procedures are implemented in the Alchemy system. For weight learning, we used

---
[1] The value of the node is 1 if the ground atom is true, and 0 otherwise.
[2] The value of the feature is 1 if the ground formula is true, and 0 otherwise.
[3] http://alchemy.cs.washington.edu/

the generative approach tha maximise the pseudo-likelihood of the data with standard Alchemy parameters (./learnwts -g), while for inference, we used the MaxWalkSAT procedure (./infer) with standard Alchemy parameters.

Predicates split(b) and merge(b) represent the query in our MLN, where b is the forced block. The goal is to assign a black (merge) or white (split) forcing to unlabelled blocks. The MLN clauses used in our system are reported in the following. There is one of the following rule for each of the 25 planes capturing the spatial relationships among the blocks:

```
overlapPart(b1,b,part) => split(b), merge(b)
```

Other relations are represented by the MLN rules:

```
belongs(b1,b) => split(b), merge(b)
touches(b1,b) => split(b), merge(b)
overlaps(b1,b,perc) => split(b), merge(b)
belongs(b1,b) => split(b1), merge(b1)
touches(b1,b) => split(b1), merge(b1)
overlaps(b1,b,perc) => split(b1), merge(b1)
```

Running weight learning with Alchemy, we will learn a weight for every clause representing how good a relation is for predicting the label. Then, whit this classifier, each test instance can be classified using inference.

4 Experimental Evaluation

The proposed description language was used to run an experiment aimed at checking whether it is possible to learn a theory that can profitably automatize, at least partially, the layout correction process. Two target concepts were considered: *split* (corresponding to the fact that a block discarded or not yet retrieved by the layout analysis algorithm must be forced to belong to the background) and *merge* (corresponding to the fact that a white rectangle found by the layout analysis algorithm must, instead, be discarded). A 10-fold cross-validation technique was exploited to obtain the training and test sets.

The experimental dataset concerned the corrections applied to obtain the correct layout on about one hundred documents, evenly distributed in four categories. According to the strategy described above, the examples concern significant background blocks that were not retrieved (split) or useless white blocks erroneously considered as background (merge) by the basic layout analysis algorithm. For the first dataset this activity resulted in a set of 786 examples of block correction, specifically 263 for split and 523 for merge. Positive examples for split were considered as negative for merge and *vice versa*, this way exploiting the whole dataset. Thus, each single correction was interpreted from two perspectives: as a positive example for the kind of forcing actually carried out by the user, and additionally as a negative example for the other kind of forcing.

The performances of the algorithm is evaluated by computing the area under the Receiver Operating Characteristic (ROC) curve, which shows how the

Table 1. Area under the ROC and PR curves for split (S) and merge (M)

		F1	F2	F3	F4	F5	F6	F7	F8	F9	F10	Mean ± St.Dev.
S	ROC	0.992	0.961	0.913	0.910	0.964	0.917	0.916	0.984	0.968	0.940	0.946 ± 0.031
	PR	0.989	0.968	0.810	0.868	0.949	0.916	0.762	0.967	0.946	0.852	0.903 ± 0.076
M	ROC	0.941	0.971	0.964	0.950	0.966	0.950	0.978	0.952	0.952	0.920	0.954 ± 0.017
	PR	0.966	0.983	0.933	0.970	0.985	0.972	0.989	0.934	0.963	0.930	0.962 ± 0.023

number of correctly classified positive examples varies with the number of incorrectly classified negative examples, and the Precision-Recall (PR) curve.

Table 1 reports the results for the queries split and merge in this first experiment. The outcomes of the experiment suggest that the description language proposed and the way in which the forcings are described are effective to let the system learn clause weights that can be successfully used for automatic layout correction. This suggested to try another experiment to simulate the actual behavior of such an automatic system, working on the basic layout analysis algorithm. After finishing the execution of the layout analysis algorithm according to the required stop threshold, three queues are produced (the queued areas still to be processed, the white areas discarded because not satisfying the constraints and the white blocks selected as useful background). Among these, the last one contains whites that can be forced to black, while the other two contain rectangles that might be forced white.

Since the rules needed by DOC to automatize the layout correction process must be able to evaluate each block in order to decide whether forcing it or not, it is not sufficient any more to consider each white block forcing as a counterexample for black forcing and *vice versa*, but to ensure that the learned MLN is correct, also all blocks in the document that have not been forced must be exploited as negative examples for the corresponding concepts. The adopted solution was to still express forcings as discussed above, including additional negative examples obtained from the layout configuration finally accepted by the user. Indeed, when the layout is considered correct, all actual white blocks that were not forced become negative examples for concept merge (because they could be forced as black, but weren't), while all white blocks, discarded or still to be processed become negative examples for the concept split (because they weren't forced). The dataset for this experiment was obtained by running the layout analysis algorithm until the predefined threshold was reached, and then applying the necessary corrections to fix the final layout. The 36 documents considered were a subset of the former dataset, evenly distributed among the four categories. Specifically, the new dataset included 113 positive and 840 negative examples for merge, and resulted in the performance reported in Table 2.

As to the concept split, the dataset was made up of 101 positive and 10046 negative examples. The large number of negative examples is due to the number of white blocks discarded or still to be processed being typically much greater than that of white blocks found. Since exploiting such a large number of negative examples might have significantly unbalanced the learning process, only

Table 2. Area under the ROC and PR curves for split (S) and merge (M)

		F1	F2	F3	F4	F5	F6	F7	F8	F9	F10	Mean ± St.Dev.
S	ROC	0.920	0.987	0.959	0.969	0.977	0.925	0.989	0.982	0.915	0.977	0.960 ± 0.029
	PR	0.841	0.917	0.699	0.909	0.692	0.914	0.778	0.774	0.807	0.913	0.824 ± 0.088
M	ROC	1.000	0.998	1.000	0.964	1.000	0.973	1.000	0.996	0.994	0.996	0.992 ± 0.013
	PR	1.000	0.986	1.000	0.804	1.000	0.874	1.000	0.971	0.966	0.967	0.957 ± 0.066

a random subset of 843 such examples was selected, in order to keep the same ratio between positive and negative examples as for the merge concept. The experiment run on such a subset provided the results shown in Table 2.

Figure 1 reports the plot obtained averaging ROC and PR curves for the ten folds. As reported in [6], a technique to evaluate a classifier over the results obtained with a cross validation method is to merge together the test instances belonging to each fold with their assigned scores into one large test set.

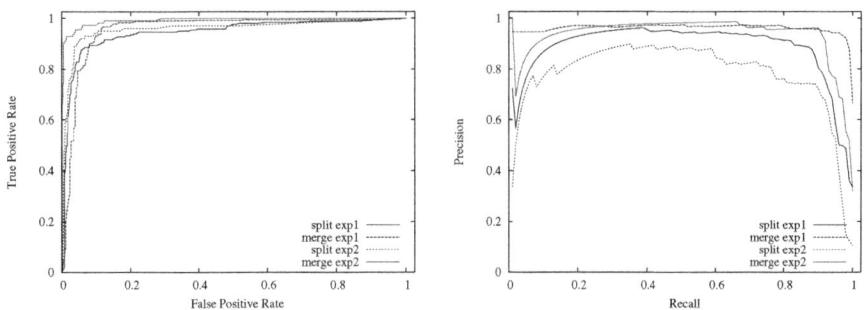

Fig. 1. ROC (left) and PR (right) curve by merging the 10-fold curves

5 Conclusions

The variety of document styles and formats to be processed makes the layout analysis task a non-trivial one and often the automatically found structure often needs to be manually fixed by domain experts. In this work we proposed a tool able to use the steps carried out by the domain expert, with the aim of correcting the outcome of the layout analysis phase, in order to infer models to be applied to future incoming documents. Specifically, the tool makes use of a first-order logic representation of the document structure as it is not fixed and a correction often depends on the relationships of the wrong components with the surrounding ones. Moreover, the tool exploits the statistical relational learning system Alchemy. Experiments in a real-world domain made up of scientific documents have been presented and discussed, showing the validity of the proposed approach.

References

1. Breuel, T.M.: Two geometric algorithms for layout analysis. In: Lopresti, D.P., Hu, J., Kashi, R.S. (eds.) DAS 2002. LNCS, vol. 2423, pp. 188–199. Springer, Heidelberg (2002)
2. Chang, F., Chu, S.Y., Chen, C.Y.: Chinese document layout analysis using adaptive regrouping strategy. Pattern Recognition 38(2), 261–271 (2005)
3. Dengel, A., Dubiel, F.: Computer understanding of document structure. International Journal of Imaging Systems and Technology 7, 271–278 (1996)
4. Esposito, F., Ferilli, S., Basile, T.M.A., Di Mauro, N.: Machine Learning for digital document processing: from layout analysis to metadata extraction. In: Marinai, S., Fujisawa, H. (eds.) Machine Learning in Document Analysis and Recognition. SCI, vol. 90, pp. 105–138. Springer, Heidelberg (2008)
5. Etemad, K., Doermann, D., Chellappa, R.: Multiscale segmentation of unstructured document pages using soft decision integration. IEEE Transactions on Pattern Analysis and Machine Intelligence 19(1), 92–96 (1997)
6. Fawcett, T.: Roc graphs: Notes and practical considerations for researchers. Tech. rep., HP Laboratories (2004),
   http://www.hpl.hp.com/techreports/2003/HPL-2003-4.pdf
7. Getoor, L., Taskar, B.: Introduction to Statistical Relational Learning (Adaptive Computation and Machine Learning). MIT Press, Cambridge (2007)
8. Krishnamoorthy, M., Nagy, G., Seth, S., Viswanathan, M.: Syntactic segmentation and labeling of digitized pages from technical journals. IEEE Transactions on Pattern Analysis and Machine Intelligence 15(7), 737–747 (1993)
9. Laven, K., Leishman, S., Roweis, S.: A statistical learning approach to document image analysis. In: Proceedings of the Eighth International Conference on Document Analysis and Recognition, pp. 357–361. IEEE Computer Society, Los Alamitos (2005)
10. Liu, J., Tang, Y.Y., Suen, C.Y.: Chinese document layout analysis based on adaptive split-and-merge and qualitative spatial reasoning. Pattern Recognition 30(8), 1265–1278 (1997)
11. Malerba, D., Esposito, F., Altamura, O., Ceci, M., Berardi, M.: Correcting the document layout: A machine learning approach. In: Proceedings of the 7th Intern. Conf. on Document Analysis and Recognition, pp. 97–103. IEEE Comp. Soc., Los Alamitos (2003)
12. Okamoto, M., Takahashi, M.: A hybrid page segmentation method. In: Proceedings of the Second International Conference on Document Analysis and Recognition, pp. 743–748. IEEE Computer Society, Los Alamitos (1993)
13. Papadias, D., Theodoridis, Y.: Spatial relations, minimum bounding rectangles, and spatial data structures. International Journal of Geographical Information Science 11(2), 111–138 (1997)
14. Richardson, M., Domingos, P.: Markov logic networks. Machine Learning 62, 107–136 (2006)
15. Simon, A., Pret, J.-C., Johnson, A.P.: A fast algorithm for bottom-up document layout analysis. IEEE Transactions on PAMI 19(3), 273–277 (1997)
16. Wu, C.C., Chou, C.H., Chang, F.: A machine-learning approach for analyzing document layout structures with two reading orders. Pattern Recogn. 41(10), 3200–3213 (2008)

Extracting General Lists from Web Documents: A Hybrid Approach

Fabio Fumarola[1], Tim Weninger[2], Rick Barber[2],
Donato Malerba[1], and Jiawei Han[2]

[1] Dipartimento di Informatica, Università degli Studi di Bari
"Aldo Moro", Bari, Italy
{ffumarola,malerba}@di.uniba.it
[2] Computer Science Department, University of Illinois at Urbana-Champaign,
Urbana-Champaign, IL
{weninge1,hanj,barber5}@uiuc.edu

Abstract. The problem of extracting structured data (*i.e.* lists, record sets, tables, etc.) from the Web has been traditionally approached by taking into account either the underlying markup structure of a Web page or the visual structure of the Web page. However, empirical results show that considering the HTML structure and visual cues of a Web page independently do not generalize well. We propose a new hybrid method to extract general lists from the Web. It employs both general assumptions on the visual rendering of lists, and the structural representation of items contained in them. We show that our method significantly outperforms existing methods across a varied Web corpus.

Keywords: Web lists, Web mining, Web information integration.

1 Introduction

The extraction of lists from the Web is useful in a variety of Web mining tasks, such as annotating relationships on the Web, discovering parallel hyperlinks, enhancing named entity recognition, disambiguation, and reconciliation. The many potential applications have also attracted large companies, such as Google, which has made publicly available the service Google Sets to generate lists from a small number of examples by using the Web as a big pool of data [13].

Several methods have been proposed for the task of extracting information embedded in lists on the Web. Most of them rely on the underlying HTML markup and corresponding DOM structure of a Web page [13,1,9,12,5,3,14,7,16,17]. Unfortunately, HTML was initially designed for rendering purposes and not for information structuring (like XML). As a result, a list can be rendered in several ways in HTML, and it is difficult to find an HTML-only tool that is sufficiently robust to extract *general* lists from the Web.

Another class of methods is based on the rendering of an HTML page [2,4,6,10,11]. These methods are likewise inadequate for general list extraction, since they tend to focus on specific aspects, such as extracting tables where each data record contains a link to a detail page [6], or discovering tables rendered from Web databases [10] (deep web pages) like Amazon.com. Due to the restricted notion of what constitutes a table on the web, these visual-based methods are not likely to effectively extract lists from the Web in the general case.

This work aims to overcome the limitations of previous works for what concerns the generality of extracted lists. This is obtained by combining several visual and structural features of Web lists. We start from the observation that lists usually contain items which are similar in type or in content. For example, the Web page shown in Figure 1a) shows eight separate lists. Looking closely at it, we can infer that the individual items in each list: 1) are visually aligned (horizontally or vertically), and 2) share a similar structure.

The proposed method, called HyLiEn (**Hy**brid approach for automatic **Li**st discovery and **En**traction on the Web), *automatically* discovers and extracts *general* lists on the Web, by using both information on the visual alignment of list items, and non-visual information such as the DOM structure of visually aligned items. HyLiEn uses the CSS2 visual box model to segment a Web page into a number of *boxes*, each of which has a position and size, and can either contain content (*i.e.*, text or images) or more boxes. Starting from the box representing the entire Web page, HyLiEn recursively considers inner boxes, and then extracts list boxes which are visually aligned and structurally similar to other boxes. A few intuitive, descriptive, visual cues in the Web page are used to generate candidate lists, which are subsequently pruned with a test for structural similarity in the DOM tree. As shown in this paper, HyLiEn significantly outperforms existing extraction approaches in specific and general cases.

The paper is organized as follows. Section 2 presents the Web page layout model. Sections 3 and 4 explain the methodology used by our hybrid approach for visual candidate generation and Dom-Tree pruning, respectively. Section 5 provides the methodology of our hybrid approach. The experimental results are

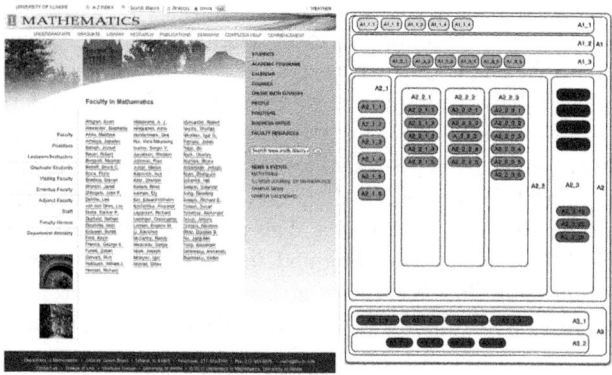

Fig. 1. a) Web page of Mathematics Department of University of Illinois. b) The boxes structure of the Web page.

reported in Section 6. Section 7 summarizes our contributions and concludes the paper.

2 Web Page Layout

When an HTML document is rendered in a Web browser, the CSS2 visual formatting model [8] represents the elements of the document by rectangular boxes that are laid out one after the other or nested inside each other. By associating the document with a coordinate system whose origin is at the top-left corner, the spatial position of each text/image element on the Web page is fully determined by both the coordinates (x, y) of the top-left corner of its corresponding box, and the box's height and width. The spatial positions of all text/image elements in a Web page define the *Web page layout*.

Each Web page layout has a tree structure, called *rendered box tree*, which reflects the hierarchical organization of HTML tags in the Web page. More precisely, let $\mathcal{H}$ be the set of occurrences of HTML tags in a Web page p, $\mathcal{B}$ the set of the rendered boxes in p, and $map : \mathcal{H} \to \mathcal{B}$ a bijective function which associates each $h \in \mathcal{H}$ to a box $b \in \mathcal{B}$. The markup text in p is expressed by a rooted ordered tree, where the root is the unique node which denotes the whole document, the other internal nodes are labeled by tags, and the leaves are labeled by the contents of the document or the attributes of the tags. The bijective map defines an isomorphic tree structure on $\mathcal{B}$, so that each box $b \in \mathcal{B}$ can be associated with a parent box $u \in \mathcal{B}$ and a set $CB = \{b_1, b_2, \cdots, b_n\}$ of child boxes. Henceforth, a box b will be formally defined as a 4-tuple $\langle n, u, CB, P \rangle$, where $n = map^{-1}(b)$ and $P = (x, y)$ is the spatial position of b, while a rendered box tree T will be formally defined as a directed acyclic graph $T = \{\mathcal{B}, \mathcal{V}, r\}$,

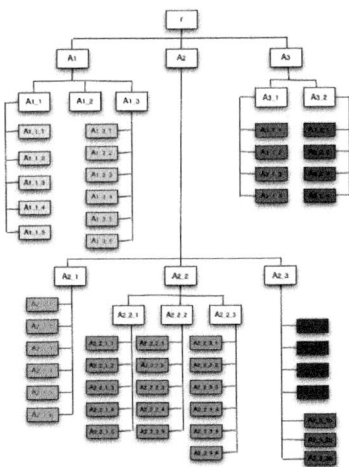

Fig. 2. The box tree structure of Illinois Mathematics Web Page

where $\mathcal{V} \subset \mathcal{B} \times \mathcal{B}$ is the set of directed edges between the *boxes*, and $r \in \mathcal{B}$ is the root box representing the whole page. An example of a rendered box tree is shown in Figure 2. The leaves of a rendered box tree are the non-breakable boxes that do not include other boxes, and they represent the minimum units of the Web page. Two properties can be reported for the rendered box trees.

Property 1. If box a is contained in box b, then b is an ancestor of a in the rendered box tree.

Property 2. If a and b are not related under property 1, then they do not overlap visually on the page.

3 Candidate Generation with Visual Features

Even though there is no strictly followed set of best practices for Web design, different information-carrying structures within Web page often have distinct visual characteristics which hold in general. In this work, we make the following basic assumption about the visual characteristics of a list:

Definition 1. *A list candidate $l = \{l_1, l_2, \ldots, l_n\}$ on a rendered Web page consists of a set of vertically and/or horizontally aligned boxes.*

We describe in [15] that this assumption alone is sufficient to outperform all existing list extraction methods. Indeed, it seems that a human user might find these alignment features to be most important in identifying lists manually. Therefore, with this assumption we can generate list candidates by comparing the boxes of a rendered Web page. Unfortunately, this assumption by itself does not cover Web pages such as the one in Figure 1 where the list elements are not necessarily all mutually aligned. For instance, all of the orange boxes inside Box A_{2_2} correspond to a single list in the page, but there are many pairs of elements in this list which are not visually aligned. Therefore, inside the region of A_{2_2}, the first step of our method will find three vertical list candidates and many horizontal list candidates based on our definition, and there will be some degree of overlap between these lists.

Definition 2. *Two lists l and l' are related ($l \sim l'$) if they have an element in common. A set of lists S is a tiled structure if for every list $l \in S$ there exists at least one other list $l' \in S$ such that $l \sim l'$ and $l \neq l'$. Lists in a tiled structure are called* tiled lists.

Three tiled lists, namely $A_{2\_2\_1}$, $A_{2\_2\_2}$ and $A_{2\_2\_3}$, are shown in Figure 1b). As explained below, the notion of tiled list is useful to handle more problematic cases by merging the individual lists of a tiled structure into a single tiled list.

4 Pruning with DOM-Tree Features

Several papers exploit the DOM-structure of Web pages to generate wrappers, identify data records, and discover lists and tables [13,1,9,12,5,3,14,7,16,17]. We

Algorithm 1: HybridListExtractor
input : Web site S, level of similarity α, max DOM-nodes β
output: set of lists L

RenderedBoxTree $T(S)$;
Queue Q;
Q.add(T.getRootBox());
while *!Q.isEmpty()* **do**
    Box $b = Q$.top();
    list *candidates* = b.getChildren();
    list *aligned* = **getVisAligned**(*candidates*);
    Q.addAll(**getNotAligned** (*candidates*));
    Q.addAll(**getStructNotAligned** (*candidates*, α, β));
    aligned = **getStructAligned** (*aligned*, α, β);
    L.add(*aligned*);
return L;

notice that, even if purely DOM-centric approaches fail in the general list finding problem, the DOM-tree could still be a valuable resource for the comparison of visually aligned boxes. In a list on a Web page, we hypothesize that the DOM-subtrees corresponding to the elements of the list must satisfy a structural similarity measure (*structSim*) to within a certain threshold α and that the subtrees not have a number of DOM-nodes (*numNodes*) greater than β.

Definition 3. *A candidate list* $l = \{l_1, l_2, \ldots, l_n\}$ *is a genuine list if and only if for each pair* (l_i, l_j), $i \neq j$, *structSim* $(l_i, l_j) \leq \alpha$, *numNodes*$(l_i) \leq \beta$ *and numNodes*$(l_j) \leq \beta$.

This DOM-structural assumption serves to prune false positives from the candidate list set. The assumption we make here is shared with most other DOM-centric structure mining algorithms, and we use it to determine whether the visual alignment of a certain boxes can be regarded as a real list or whether the candidate list should be discarded. Specifically, the α and β parameters are essentially the same as the K and T thresholds from MDR [9] and DEPTA [16,17], and the α and C thresholds from Tag Path Clustering [12].

5 Visual-Structural Method

The input to our algorithm is a set of unlabeled Web pages containing lists. For each Web page, Algorithm 1 is called to extract Web lists. We use the open source library *CSSBox*[1] to render the pages.

The actual rendered box tree of a Web page could contain hundreds or thousands of boxes. Enumerating and matching all of the boxes in search of lists of arbitrary size would take time exponential in the number of boxes if we used a

---
[1] http://cssbox.sourceforge.net

Algorithm 2: getVisAligned
  input : list *candidates*
  output: list *visAligned*

  Box *head* = *candidates*[0];
  list *visAligned*;
  for *(i = 1; i < candidates.length; i++)* **do**
  $\quad$ Box *tail* = *candidates*[i];
  $\quad$ **if** *(head.x == tail.x || head.y == tail.y)* **then**
  $\quad\quad$ *visAligned*.add(*tail*);

  if *visAligned.length* > 1 **then**
  $\quad$ *visAligned*.add(*head*);
  return *visAligned*;

brute force approach. To avoid this, our proposed method explores the space of the boxes in a top-down manner (*i.e.*, from the root to boxes that represent the elements of a list) using the edges of the rendered box tree ($\mathcal{V}$). This makes our method more efficient with respect to those reported in the literature.

Starting from the root box r, the *rendered box tree* is explored. Using a *Queue*, a breadth first search over children boxes is implemented. Each time a box b is retrieved from the *Queue*, all the children boxes of b are tested for visual and structural alignment using Algorithms 2 and 3, thereby generating candidate and genuine lists among the children of b. All the boxes which are not found to be visually or structurally aligned are enqueued in the *Queue*, while the tested boxes are added the resultset of the Web page. This process ensures that the search does not need to explore each atomic element of a Web page, and thus makes the search bounded on the complexity of the actual lists in the Web page.

Algorithm 2 uses the visual information of boxes to generate candidate lists which are horizontally or vertically aligned. To facilitate comprehension of the approach, we present a generalized version of the method where the vertical and horizontal alignments are evaluated together. However, in the actual implementation of the method these features are considered separately; this enables the method to discover both the horizontal, vertical and tiled lists on the Web page.

Algorithm 3 prunes false positive candidate lists. The first element of the visually aligned candidate list is used as an element for the structural test. Each time the *tail* candidate is found to be structurally similar to the *head*, the *tail* is added to the result list. At the end, if the length of the result list is greater than one, the head is added. If none of the boxes are found to be structurally similar, an empty list is returned.

To check if two boxes are structurally similar, Algorithm 4 exploits the DOM-tree assumption described in Def. 3. It works with any sensible tree similarity measure. In our experiments we use a simple string representation of the corresponding tag subtree of the boxes being compared for our similarity measurement.

Algorithm 3: getStructAligned

input : list *candidates*, min. similarity α, max. tag size β
output: list *structAligned*

Box *head* = *candidates*[0];
list *structAligned*;
for *(i = 1; i < candidates.length; i + +)* **do**
  Box *tail* = *candidates*[*i*];
  if getStructSim *(head, tail, β)* $\leq \alpha$ **then**
    $\lfloor$ *structAligned*.add(*tail*);

if *structAligned.length* > 1 **then**
  $\lfloor$ *structAligned*.add(*head*);
return *structAligned*;

Algorithm 4: getStructSim

input : Box a, b, max. tag size β
output: double *simValue*

TagTree *tA* = *a*.getTagTree();
TagTree *tB* = *b*.getTagTree();
double *simValue*;
if *(tA.length $\geq \beta$ || tB.length $\geq \beta$)* **then**
  $\lfloor$ **return** MAXDouble;
simValue = *Distance*(*tA, tB*);
return *simValue*;

At the end of the computation, Algorithm 1 returns the collection of all the lists extracted from the Web page. A post-processing step is finally applied to deal with tiled structures. Tiled lists are not directly extracted by this algorithm. Based on Section 2, each list discovered is contained in a box b. Considering the position P of these boxes, we can recognize tiled lists. We do this as a post-process by: 1) identifying the boxes which are vertically aligned, and 2) checking if, the element lists contained in that boxes are visually and structurally aligned. Using this simple heuristic we are able to identify tiled lists and update the result set accordingly.

6 Experiments

We showed in [15] that implementing a method that uses assumption in Def. 1 is sufficient to outperform all existing list extraction methods. Thus we tested HiLiEn on a dataset used to validate the Web Tables discovery in VENTex [4]. This dataset contains Web pages saved by WebPageDump[2] including the Web page after the "x-tagging", the coordinates of all relevant Visualized Words (VENs), and the manually determined ground truth. From the first 100 pages of

---
[2] http://www.dbai.tuwien.ac.at/user/pollak/webpagedump/

the original dataset we manually extracted and verified 224 tables, with a total number of 6146 data records. We can use this dataset as test set for our method because we regard tables on the Web to be in the set of lists on the Web, that is, a table is a special type of list. This dataset was created by asking students taking a class in Web information extraction at Vienna University of Technology to provide a random selection of Web tables. This, according to Gatterbauer et al.[4], was done to eliminate the possible influence of the Web page selection on the results. We use the generality advantage of this dataset to show that also our method is robust and the results are not biased from the selected test set.

HyLiEn returns a text and visual representation of the results. The former consists of a collection of all the discovered lists, where each element is represented by its HTML tag structure and its inner text. The latter is a png image, where all the discovered lists are highlighted with random colors.

We compared HyLiEn to VENTex, which returns an XML representation of the frames discovered. Because of the differences in output of the two methods, we erred on the side of leniency in most questionable cases. In the experiment, the two parameters α and β required by HeLiEn are empirically set to 0.6 and 50, respectively.

Table 1 shows that VENTex extracted 82.6% of the tables and 85.7% of the data records, and HyLiEn extracted 79.5% of the tables and 99.7% of the data records. We remind readers that HyLiEn was not initially created to extract tables, but we find that our method can work because we consider tables to be a type of list.

We see that VENTex did extract 8 more tables than HyLiEn. We believe this is because HyLiEn does not have any notion of element distance that could be used to separate aligned but separated lists. On the contrary, in HyLiEn, if elements across separate lists are aligned and structurally similar they are merged into one list. Despite the similar table extraction performance, HyLiEn extracted many more records (i.e., rows) from these tables than VENTex.

We did judge the precision score here for comparison sake. We find that, from among the 100 Web pages only 2 results contained false positives (i.e., incorrect list items) resulting in 99.9% precision. VENTex remained competitive with a precision of 85.7%. Table 2 shows the full set of results on the VENTex data set. We see that HyLiEn consistently and convincingly outperforms VENTex.

Interestingly, the recall and precision values that we obtained for VENTex were actually higher than the results presented in Gatterbauer et al. [4] (they show precision: 81%, and recall: 68%). We are confident this difference is because we use only the first 100 Web pages of the original dataset.

Table 1. Recall for table and record extraction on the VENTex data set

	Ground truth	VENTex	HyLiEn
# tables	224	**82.6%**	79.5%
# records	6146	85.7%	**99.7%**

Table 2. Precision and Recall for record extraction on the VENTex data set

	Recall	Precision	F-Measure
VENTex	85.7%	78.0%	81.1%
HyLiEn	**99.7%**	**99.9%**	**99.4%**

7 Discussion and Concluding Remarks

We have empirically shown that by exploiting the visual regularities in Web page rendering and structural properties of pertinent elements, it is possible to accurately extract general lists from Web pages. Our approach does not require the enumeration a large set of structural or visual features, nor does it need to segment a Web page into atomic elements and use a computationally demanding process to fully discover lists.

In contrast, the computation cost of HyLiEn list extraction is actually bounded on the structural complexity of lists in a Web page. Considering the assumption that the number of lists in a Web page is many orders of magnitude smaller than the number of all the HTML tags, the computation time for HyliEn is quite small: only 4.2 seconds on average.

As a matter of future work we plan on finding a better HTML rendering engine, which can be used to speed up the execution of our method. When such an engine is found we will perform a more rigorous computational performance evaluation. With our new list finding algorithm we plan on using extracted lists to annotate and discover relationships between entities on the Web.

Part of this future work is the theory that entities (*e.g.*, people) which are commonly found together in lists are more similar than those which are not frequently found together in lists. Other interesting avenues for future work involve tasks such as indexing the Web based on lists and tables, answering queries from lists, and entity discovery and disambiguation using lists.

Acknowledgments

This research is funded by an NDSEG Fellowship to the second author. The first and fourth authors are supported by the projects "Multi-relational approach to spatial data mining" funded by the University of Bari "Aldo Moro," and the Strategic Project DIPIS (Distributed Production as Innovative Systems) funded by Apulia Region. The third and fifth authors are supported by NSF IIS-09-05215, U.S. Air Force Office of Scientific Research MURI award FA9550-08-1-0265, and by the U.S. Army Research Laboratory under Cooperative Agreement Number W911NF-09-2-0053 (NS-CTA).

References

1. Cafarella, M.J., Halevy, A., Wang, D.Z., Wu, E., Zhang, Y.: Webtables: exploring the power of tables on the web. Proc. VLDB Endow. 1(1), 538–549 (2008)
2. Cai, D., Yu, S., Rong Wen, J., Ying Ma, W.: Extracting content structure for web pages based on visual representation. In: Zhou, X., Zhang, Y., Orlowska, M.E. (eds.) APWeb 2003. LNCS, vol. 2642, pp. 406–417. Springer, Heidelberg (2003)
3. Crescenzi, V., Mecca, G., Merialdo, P.: Roadrunner: automatic data extraction from data-intensive web sites. SIGMOD, 624–624 (2002)
4. Gatterbauer, W., Bohunsky, P., Herzog, M., Krüpl, B., Pollak, B.: Towards domain-independent information extraction from web tables. In: WWW, pp. 71–80. ACM, New York (2007)
5. Gupta, R., Sarawagi, S.: Answering table augmentation queries from unstructured lists on the web. Proc. VLDB Endow. 2(1), 289–300 (2009)
6. Lerman, K., Getoor, L., Minton, S., Knoblock, C.: Using the structure of web sites for automatic segmentation of tables. SIGMOD, 119–130 (2004)
7. Lerman, K., Knoblock, C., Minton, S.: Automatic data extraction from lists and tables in web sources. In: IJCAI. AAAI Press, Menlo Park (2001)
8. Lie, H.W., Bos, B.: Cascading Style Sheets:Designing for the Web, 2nd edn. Addison-Wesley Professional, Reading (1999)
9. Liu, B., Grossman, R., Zhai, Y.: Mining data records in web pages. In: KDD, pp. 601–606. ACM Press, New York (2003)
10. Liu, W., Meng, X., Meng, W.: Vide: A vision-based approach for deep web data extraction. IEEE Trans. on Knowl. and Data Eng. 22(3), 447–460 (2010)
11. Mehta, R.R., Mitra, P., Karnick, H.: Extracting semantic structure of web documents using content and visual information. In: WWW, pp. 928–929. ACM, New York (2005)
12. Miao, G., Tatemura, J., Hsiung, W.-P., Sawires, A., Moser, L.E.: Extracting data records from the web using tag path clustering. In: WWW, pp. 981–990. ACM, New York (2009)
13. Tong, S., Dean, J.: System and methods for automatically creating lists. In: US Patent: 7350187 (March 2008)
14. Wang, R.C., Cohen, W.W.: Language-independent set expansion of named entities using the web. In: ICDM, pp. 342–350. IEEE, Washington, DC, USA (2007)
15. Weninger, T., Fumarola, F., Barber, R., Han, J., Malerba, D.: Unexpected results in automatic list extraction on the web. SIGKDD Explorations 12(2), 26–30 (2010)
16. Zhai, Y., Liu, B.: Web data extraction based on partial tree alignment. In: WWW, pp. 76–85. ACM, New York (2005)
17. Zhai, Y., Liu, B.: Structured data extraction from the web based on partial tree alignment. IEEE Trans. on Knowl. and Data Eng. 18(12), 1614–1628 (2006)

Towards a Computational Model of the Self-attribution of Agency

Koen Hindriks[1], Pascal Wiggers[1], Catholijn Jonker[1], and Willem Haselager[2]

[1] Delft University of Technology
[2] Radboud University Nijmegen

Abstract. In this paper, a first step towards a computational model of the self-attribution of agency is presented, based on Wegner's theory of apparent mental causation. A model to compute a *feeling of doing* based on first-order Bayesian network theory is introduced that incorporates the main contributing factors to the formation of such a feeling. The main contribution of this paper is the presentation of a formal and precise model that can be used to further test Wegner's theory against quantitative experimental data.

1 Introduction

The difference between falling and jumping from a cliff is a significant one. Traditionally, this difference is characterized in terms of the contrast between something happening to us and doing something. This contrast, in turn, is cashed out by indicating that the person involved had mental states (desires, motives, reasons, intentions, etc.) that produced the action of jumping, and that such factors were absent or ineffective in the case of falling. Within philosophy, major debates have taken place about a proper identification of the relevant mental states and an accurate portrayal of the relation between these mental states and the ensuing behavior (e.g. [1,2]). In this paper, however, we will focus on a psychological question: how does one decide that oneself is the originator of one's behavior? Where does the feeling of agency come from? Regarding this question we start with the assumption that an agent generates explanatory hypotheses about events in the environment, a.o. regarding physical events, the behavior of others and of him/herself. In line with this assumption, in [3] Wegner has singled out three factors involved in the self-attribution of agency; the principles of priority, consistency and exclusivity. Although his account is detailed, both historically and psychologically, Wegner does not provide a formal model of his theory, nor a computational mechanism. In this paper, we will provide a review of the basic aspects of Wegner's theory, and sketch the outlines of a computational model implementing it, with a focus on the priority principle. Such a model of self-attribution can be usefully applied in interaction design to establish whether a human attributes the effects of the interaction to itself or to a machine.

The paper is organized as follows: Section 2 provides an outline of Wegner's theory and introduces the main contributing factors in the formation of an experience of will. In section 3, it is argued that first-order Bayesian network theory is the appropriate modeling tool for modeling the theory of apparent mental causation and a model of this theory is presented. In section 4, the model is instantiated with the parameters of the *I Spy* experiment as performed by Wegner and the results are evaluated. Finally, section 5 concludes the paper.

2 Apparent Mental Causation

Part of a theory of mind is the link between an agent's state and its actions. That is, agents describe, explain and predict actions in terms of underlying mental states that cause the behavior. In particular, human agents perceive their intentions as causes of their behavior. Moreover, intentions to do something that occur prior to the corresponding act are interpreted as reasons for doing the action. This understanding is not fully present yet in very young children.

It is not always clear-cut whether or not an action was caused by ones own prior intentions. For example, when one finds someone else on the line after making a phone call to a friend using voice dialing, various explanations may come to mind. The name may have been pronounced incorrectly making it hard to recognize it for the phone, the phone's speech recognition unit may have mixed up the name somehow, or, alternatively, one may have more or less unconsciously mentioned the name of someone else only recognizing this fact when the person is on the line. The perception of agency thus may vary depending on the perception of one's own mind and the surrounding environment.

In the self-attribution of agency, intentions play a crucial role, but the conscious experience of a feeling that an action was performed by the agent itself still may vary quite extensively. We want to gain a better understanding of the perception of agency, in particular of the attribution of agency to oneself. We believe that the attribution of agency plays an important role in the interaction and the progression of interaction between agents, whether they are human or computer-based agents. As the example of the previous paragraph illustrates, in order to understand human interaction with a computer-based agent it is also important to understand the factors that play a role in human self-attribution of agency. Such factors will enhance our understanding of the level of control that people feel when they find themselves in particular environments. One of our objectives is to build a computational model to address this question which may also be useful in the assessment by a computer-based agent of the level of control of one of its human counterparts in an interaction.

As our starting point for building such a model, we use Wegner's theory of apparent mental causation [4]. Wegner argues that there is more to intentional action than forming an intention to act and performing the act itself. A causal relation between intention and action may not always be present in a specific case, despite the fact that it is perceived as such. This may result in an illusion of control. Vice versa, in other cases, humans that perform an act do not perceive themselves as the author of those acts, resulting in more or less automatic behavior (automatisms). As Wegner shows, the causal link between intention and action cannot be taken for granted.

Wegner interprets the self-attribution of agency as an experience that is generated by an interpretive process that is fundamentally separate from the mechanistic process of real mental causation [3]. He calls this experience the *feeling of doing* or the *experience of will*. The fact that Wegner's theory explains the feeling of doing as the result of an interpretive process is especially interesting for our purposes. It means that this theory introduces the main factors that play a role in interpreting action as caused by the agent itself retrospectively. It thus provides a good starting point for constructing a computational model that is able to correctly attribute agency to a human agent.

Wegner identifies three main factors that contribute to the experience of a feeling of doing: (i) An intention to act should have been formed just before the action was performed. That is, the intention must appear within an appropriately small window of time before the action is actually performed. Wegner calls this the *priority principle*. (ii) The intention to act should be consistent with the action performed. This is called the *consistency principle*. (iii) The intention should exclusively explain the action. There should not be any other prevailing explanations available that would explain the action and discount any intention, if present, as a cause of the action. This is called the *exclusivity principle*.

A crucial factor in assessing the contribution of the priority principle to the feeling of doing is the timing of the occurrence of the intention. In [5] it is experimentally established that the experience of will typically is greatest when the intention is formed about 1 second before the action is performed. As Wegner argues, the priority principle does not necessarily need to be satisfied in order to have a feeling of doing. *People may sometimes claim their acts were willful even if they could only have known what they were doing after the fact* [3]. Presumably, however, an agent that makes up an intention after the fact to explain an event will (falsely) *believe* that it occurred prior to that event.

The contribution of the consistency principle to the experience of will *depends [...] on a cognitive process whereby the thoughts occurring prior to the act are compared to the act as subsequently perceived. When people do what they think they were going to do, there exists consistency between thought and act, and the experience of will is enhanced* [3]. The comparison of thought and action is based on a semantic relation that exists between the content of the thought and the action as perceived. The thought may, for example, name the act, or contain a reference to its execution or outcome. The mechanism that determines the contribution of the consistency principle to a feeling of doing thus relies on a measure of how strongly the thought and action are semantically related. Presumably, the contribution of the consistency principle is dependent on the priority principle. Only thoughts consistent with the act that occurred prior to the perceived act, within a short window of time, contribute to a feeling of doing.

The contribution of the exclusivity principle to the experience of will consists in the weighing of various possible causes that are available as explanations for an action. The principle predicts that when the own thoughts of agents do not appear to be the exclusive cause of their action, they experience less conscious will; and, when other plausible causes are less salient, in turn, they experience more conscious will [3]. People discount the causal influence of one potential cause if there are others available [6]. Wegner distinguishes between two types of competing causes: (i) internal ones such as: emotions, habits, reflexes, traits, and (ii) external ones such as external agents (people, groups), imagined agents (spirits, etc.), and the agent's environment. In the cognitive process which evaluates self-agency these alternative causes may discount an intention as the cause of action. Presumably, an agent has background knowledge about possible alternative causes that can explain a particular event in order for such discounting to happen. Wegner illustrates this principle by habitual and compulsive behavior like eating a large bag of potato chips. In case we know we do this because of compulsive habits, any intentions to eat the chips are discounted as causes.

3 Computational Model

One of our aims is to provide a computational model in order to validate and explicate Wegner's theory of apparent mental causation. This theory defines the starting point for the computational model. But the theory does not describe the functioning of the affective-cognitive mechanisms that lead to a feeling of doing at the level of detail which is required for achieving this goal. We thus have to make some modeling choices in order to specify *how* a feeling of doing is created. Here a computational model is introduced that provides a tool for simulating the feeling of doing. In the next section the model is instantiated with an experiment performed by Wegner as a means to validate that the model also fits some of the empirical evidence that Wegner presents to support his theory.

It is clear that any model of the theory of apparent mental causation must be able to account for the varying degrees or levels in the experience of a feeling of doing, the variation in timing of intention and action, the match that exists between those, and the competition that may exist between various alternative causes. Neither one of these factors nor the feeling of doing itself can be represented as a two-valued, binary state, since humans can experience more or less control over particular events. As observed in [3], even *our conscious intentions are vague, inchoate, unstudied, or just plain absent. We just don't think consciously in advance about everything we do, although we try to maintain appearances that this is the case.*

Given the considerations above, it seems natural to use a probabilistic approach to model the degrees of priority, and consistency and to weigh the various competing alternative explanations. Moreover, the cognitive process itself that results in an experience of will is an interpretive or inferential process. Given the various inputs relating to time and perceived action, a cause that explains the action is inferred which may or may not induce a feeling of doing. A natural choice to model such dependencies is to use Bayesian networks. Bayesian networks [7] have been used extensively to model causal inference based on probabilistic assessments of various sorts of evidence (see for examples of this in research on a *theory of mind* e.g. [8,9]). Bayesian networks also allow us to use symbolic representations of the thoughts formed and the actions performed by an agent, which need to be compared in order to compute a feeling of doing in the theory of apparent mental causation.

In this paper, Multi-Entity Bayesian Network (MEBN) Theory is used [10]. MEBN is *a knowledge representation formalism that combines the expressive power of first-order logic with a sound and logically consistent treatment of uncertainty*. An MEBN Theory consists of several MEBN fragments that together define a joint probability distribution over a set of first order logic predicates. Figure 1 shows two MEBN fragments, each depicted as a rounded rectangle, that model the priority principle. A fragment contains a number of nodes that represent random variables. In accordance with the mathematical definition, random variables are seen as functions (predicates) of (ordinary) variables.

The gray nodes in the top section of a fragment are called *context nodes*; they function as a *filter* that constrains the values that the variables in the fragment can take. In contrast to the nodes in the bottom section of a fragment, context nodes do not have an associated probability distribution but are simply evaluated as true or false. Another perspective on these nodes is that they define what the network is about. The context nodes labeled with the $IsA(t, v)$ predicate define the type t of each of the variables v used. In our model, we distinguish intentions, events, opportunities, and time intervals in which the former may occur. Intentions are *mental states* which are to be distinguished from events, which are temporally extended and may change the state of the world. Opportunities are states which enable the performance of an action. In the model, the probabilities associated with each of these nodes should be interpreted as the likelihood that the agent attaches to the occurrence of a particular state, event or other property (e.g. causal relationship) given the available evidence.

Dark nodes in the bottom section of a fragment are called *input nodes* and are references to nodes that are defined in one of the other fragments. In Figure 1, the node in the right fragment labeled $Exists(a, t_a)$ is an input node. To ensure that the model defines a proper probability distribution, a node can be defined in a single fragment only, in which it is said to be *resident*. The node labeled $Exists(a, t_a)$ is resident in the left fragment in Figure 1.

As usual, the links between nodes represent dependencies. Every resident node has a conditional probability table attached that gives a probability for every state of the node given the states of its parent nodes. Prior distributions are attached to resident nodes without parents. Essentially, every fragment defines a parameterized Bayesian network that can be instantiated for all combinations of its variables that satisfy the constraints imposed by its context nodes.

In order to be able to compute a feeling of doing, the prior probability distributions are assumed to be given in this paper. The computational model presented does not explain how explanatory hypotheses about perceived events are generated, nor does it include an account of the perception of these events. Even though the model assumes this information somehow has already been made available, it is setup in such a way that it already anticipates an account for computing at least part of this information. In particular, the mechanism approach of [6] to explain causal attribution has played a guiding role in

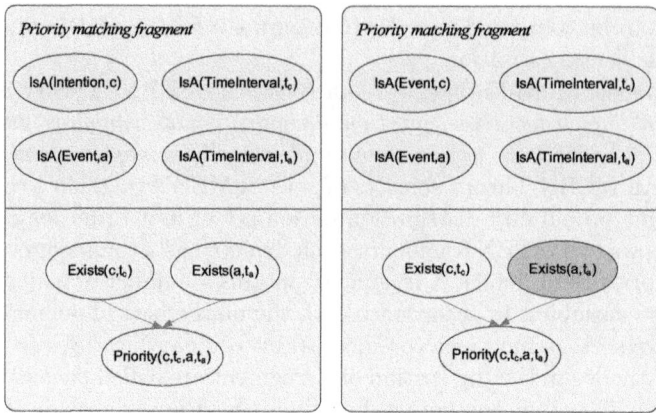

Fig. 1. Priority Fragments

defining the model. The basic idea of this approach is that *causal attribution involves searching for underlying mechanism information (i.e. the processes underlying the relationship between the cause and the effect)*, given evidence made available through perception and introspection. Assuming that each mechanism defines a particular covariation (or joint probability distribution) of the contributing factors with the resulting outcome, the introduction of separate probability distributions for each particular event that is to be explained can be avoided. As a result, the number of priority and causality fragments needed is a function linear in the number of mechanisms instead of the number of events.

3.1 Priority Fragments

The priority principle is implemented by the Priority fragments in Figure 1. Though these fragments are structurally similar, two fragments are introduced in line with the idea that different causal mechanisms may associate different time frames with a cause and its effect. For reasons of space and simplicity, Figure 1 only depicts two fragments, one associated with intentional mechanisms leading to action and a second one for other causal events. The exact time differences depend on the mechanism involved. For example, when moving the steering wheel of a car one expects the car to respond immediately, but a ship will react to steering with some delay.

The *Exists* random variables model that an agent may be uncertain whether a particular state or event has actually taken place at a particular time (also called the *existence condition* in [11]). If there is no uncertainty these nodes will have value true with probability one. The probability associated with the *Priority* random variable is non-zero if the potential cause occurs more or less in the right time frame before the event that is explained by it and the associated probability that the relevant events actually occurred is non-zero. In line with [5], the probability associated with the intentional mechanism increases as the time difference decreases to about one second. As one typically needs some time to perform an action, the probability starts to decrease again for time intervals less than one second. Each fragment may be instantiated multiple times, illustrated in Section 4, depending on the number of generated explanatory hypotheses.

3.2 Causality Fragments

Figure 2 depicts two fragments corresponding respectively with the intentional mechanism (left) and another type of mechanism (right) that may explain an event. In this case, the fragments are structurally different in two ways. First, even though both fragments require that cause c and effect a are consistent with the mechanism associated with the fragment, the consistency nodes are different. The type of consistency associated with the intentional fragment, called *intentional consistency*, is fundamentally different in nature from that associated with other mechanisms as it is based on the degree of *semantic* relatedness of the content of intention c and the event a (represented as a probability associated with the node). This reflects the fact that one of Wegner's principles, the consistency principle, is particular to intentional explanations. Second, an additional context node representing an opportunity o to act on the intention is included in the fragment corresponding with the intentional mechanism. An intention by itself does not result in action if no opportunity to act is perceived. In line with common sense and philosophical theory [2], the intentional mechanism leads to action given an intention and the right opportunity as input. The model entails that the presence of multiple opportunities increases the probability that a relevant intention is the cause of an event. Additional detail is required to model this relation precisely, but for reasons of space we refer to [12] for a formal model.

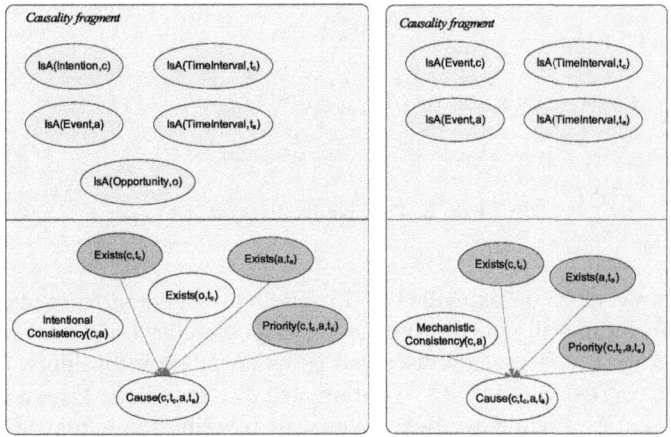

Fig. 2. Causality Fragments

The node labeled $Cause(c, t_c, a, t_a)$ in the intentional fragment models the *feeling of doing*. The associated probability of this node represents the probability that the intention c of an agent has caused event a. In other words, it represents the level of self-attribution of agency for that agent. The probability associated with the node depends on the priority and consistency as well as on the presence (i.e. existence) of both c and a. Obviously, if either c or a is not present, $Cause(c, t_c, a, t_a)$ will be false with probability 1. Additionally, in the intentional fragment an opportunity o must exist.

3.3 Exclusivity Fragment

In order to model the exclusivity principle, an exclusivity fragment is introduced as depicted in Figure 3. In general, if there are multiple plausible causes for an event, exclusivity will be low. Technically, this is modeled as an exclusive-or relation between the competing causes. The value of the random variable *Exclusivity* is set to true to enforce exclusivity. As a result, given two causes of which only one is very likely, the posterior probability of the unlikely cause is reduced. This effect is known as the *discounting effect*, also called *explaining away* [7], and has been studied extensively (e.g. [6]).

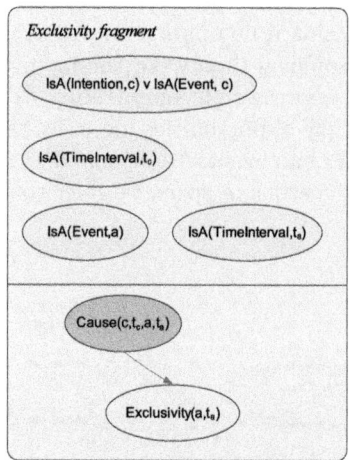

Fig. 3. Exclusivity Fragment

Given an event to be explained and a number of generated explanatory hypotheses (including all contributing factors associated with a particular mechanism), each of the fragments discussed is instantiated accordingly, taking into account the context conditions. To obtain a single, connected Bayesian network, all of the resulting fragments are connected by merging the reference nodes with their resident counterparts. Using this network, the *feeling of doing* can be computed by performing probabilistic inference and querying the $Cause(c, t_c, a, t_a)$

variable in the intentional fragment given the values of the other nodes in the network. By querying other *Cause* variables we can find by means of comparison which of the potential causes is the most plausible one. As a result, only when the node representing the feeling of doing has a high associated probability an agent would explain the occurrence of an event as caused by itself.

4 Simulation of the *I SPY* Experiment

In this section, an instantiation of the model that corresponds with an experiment performed by Wegner is presented. In [5] the results of the *I Spy* experiment are presented that tested whether participants report an experience of agency for something that is most likely the result of someone else's action. In the experiment two participants are seated on opposite sides of a table. On the table a square board that is attached to a computer mouse is located and both participants are asked to put their fingertips on the board and to move the mouse by means of the board in slow sweeping circles. By doing so, a cursor is moved over a computer screen showing a photo from the book *I Spy* [13], hence the name of the experiment, picturing about 50 small objects. The subjects had to move the mouse for about 30 seconds after which they would hear a 10 second clip of music through headphones and within this period they had to stop moving the mouse and then rate on a continuous scale whether they allowed the stop to happen or intended the stop to happen. In addition to the music, subjects would occasionally hear words over the headphones. Participants were told that they would hear different bits of music and different words. One of the participants however did not hear music at all, but was a confederate who received instructions from the experimenter to stop on a particular picture or to let the other participant determine the stop. The forced stops were timed to occur at specific intervals from when the participant heard a corresponding word that was intended to prime a thought about items on the screen. By varying timing, priority was manipulated. For unforced stops the words heard by the participant corresponded about half of the time to an object on the screen.

It turned out that in initial experiments in which the confederate did not force stops the mean distance between stops and the pictures that were primed by words was not significantly different from the mean distance in trials in which the prime word did not refer to an object on the screen. These initial experiments were performed to confirm that participants would not stop the cursor on an object simply because of hearing the word. In consecutive experiments, however, where the stops were forced by the confederator, participants tended to perceive the stops as more or less intended, dependent on the time interval between the hearing of the prime word and the actual stop. If the word occurred between 5 and 1 seconds before the stop, a significant increase in self-attribution was observed.

Based on the description of the *I Spy* experiment and the results presented in [5], an instantiation of the computational model has been derived. Due to space limitations we cannot provide all details.

The resulting model gives the same results as those reported in [5]: If the a priori probability associated with the *Priority* variables is higher (corresponding to the time interval between 5 to 1 seconds), then a significantly higher feeling of doing is produced than otherwise. The second column of Table 1 shows the posterior probability of the $Cause(I_p, t_p, S, t_s)$ node that models the feeling of doing for several a priori probabilities of the $Priority$ variable. For a probability of 0.85 for priority the probability of $Cause$ corresponds to the feeling of doing for a time difference of about 1 second as described in [5]. Similarily, the values obtained with a probability for priority of 0.8 and 0.35 correspond to the feeling of doing reported in [5] for respectively 5 seconds and 30 seconds time difference between the prime word and the stop of the cursor.

In [5], also the variance in feeling of doing observed in the experiment is reported. One would expect that a person's personality influences his feeling of doing. Various people, for example, might be more or less sensitive to priming or might have a strong or weak tendency to claim agency in a setup such as in the *I Spy* experiment. We tested the model with different values of priority with a moderated a priori probability for the existence of intention of 0.45 and with a high a priori probability of 0.65 for the existence of an intention. The corresponding posterior probabilities of the cause node are shown in Table 1. These probabilities adequately correspond with the variance reported by Wegner, which gives some additional support for the proposed computational model.

Table 1. Posterior probability of $Cause(I_p, t_p, S, t_s)$ for different a priori probabilities of $Priority(I_p, t_p, S, t_s)$ and $Exists(I_p, t_p)$

	$P(Exists(I_p, t_p))$		
$P(Priority)$	0.55	0.45	0.65
0.3	0.41	0.36	0.45
0.35	0.44	0.39	0.48
0.5	0.51	0.46	0.56
0.8	0.62	0.56	0.66
0.85	0.63	0.58	0.67

5 Conclusion and Future Work

In this paper, a first step towards a computational model of the self-attribution of agency is presented, based on Wegner's theory of apparent mental causation [3]. A model to compute a *feeling of doing* based on first-order Bayesian network theory is introduced that incorporates the main contributing factors (according to Wegner's theory) to the formation of such a feeling. The main contribution of this paper is the presentation of a formal and precise model that provides detailed predictions with respect to the self-attribution of agency and that can be used to further test such predictions against other quantitative experimental data. An additional benefit of the model is that given empirical, quantitative data parameters of the network can be learned, using an algorithm as in [10].

A number of choices had to be made in order to obtain a computational model of Wegner's theory of apparent mental causation. Not all of these choices are explicitly supported by Wegner's theory. In particular, it has been hard to obtain quantitative values to define the probability distributions in our model. The report of the *I Spy* experiment in [5] does detailed information, but did not provide sufficient information to construct the probability distributions we need. Certain values had to be guessed in order to obtain outcomes corresponding with the results in [5]. The only validation of these guesses we could perform was to verify whether variation of some of the input values of our model could be said to reasonably correspond with the reported variations in the experiment in [5]. It is clear that more work needs to be done to validate the model. In future work, we want to design and conduct actual experiments to validate and/or refine the model of self-attribution.

To conclude, we want to remark that there are interesting relations here with other work. As is argued in [9], Bayesian networks are not sufficient as cognitive models of how humans infer causes. These networks are very efficient for computing causes, but are themselves instantiations from more general, higher-level theories. In a sense, this is also the case in our model since both the consistency fragment as well as the causality fragment in our first-order Bayesian theory of apparent mental causation need to be instantiated by other domain-specific theories in order to derive the right semantic relations between thoughts and actions, and to identify potential other causes of events. Additional work has to fill in these gaps in the model, starting from e.g. ideas presented in [6,9].

References

1. Anscombe, G.E.M.: Intention. Harvard University Press, Cambridge (1958)
2. Dretske, F.: Explaining behavior. MIT Press, Cambridge (1988)
3. Wegner, D.M.: The Illusion of Conscious Will. MIT Press, Cambridge (2002)
4. Wegner, D.M.: The mind's best trick: How we experience conscious will. Trends in Cognitive Science 7, 65–69 (2003)
5. Wegner, D.M., Wheatley, T.: Apparent mental causation: Sources of the experience of will. American Psychologist 54 (1999)
6. Ahn, W.K., Bailenson, J.: Causal attribution as a search for underlying mechanisms. Cognitive Psychology 31, 82–123 (1996)
7. Pearl, J.: Probabilistic Reasoning in Intelligent Systems - Networks of Plausible Inference. Morgan Kaufmann Publishers, Inc., San Francisco (1988)
8. Gopnik, A., Schulz, L.: Mechanisms of theory formation in young children. Trends in Cognitive Science 8, 371–377 (2004)
9. Tenenbaum, J., Griffiths, T., Niyogi, S.: Intuitive Theories as Grammars for Causal Inference. In: Gopnik, A., Schulz, L. (eds.) Causal Learning: Psychology, Philosophy, and Computation. Oxford University Press, Oxford (in press)
10. Laskey, K.B.: MEBN: A Logic for Open-World Probabilistic Reasoning. Technical Report C4I-06-01, George Mason University Department of Systems Engineering and Operations Research (2006)
11. Kim, J.: Supervenience and Mind. Cambridge University Press, Cambridge (1993)
12. Jonker, C., Treur, J., Wijngaards, W.: Temporal modelling of intentional dynamics. In: ICCS, pp. 344–349 (2001)
13. Marzollo, J., Wick, W.: I Spy. Scholastic, New York (1992)

An Agent Model for Computational Analysis of Mirroring Dysfunctioning in Autism Spectrum Disorders

Yara van der Laan and Jan Treur

VU University Amsterdam, Department of Artificial Intelligence
De Boelelaan 1081, 1081 HV Amsterdam, The Netherlands
vanderlaan.yara@gmail.com, treur@cs.vu.nl

Abstract. Persons with an Autism Spectrum Disorder (ASD) may show certain types of deviations in social functioning. Since the discovery of mirror neuron systems and their role in social functioning, it has been suggested that ASD-related behaviours may be caused by certain types of malfunctioning of mirror neuron systems. This paper presents an approach to explore such possible relationships more systematically. As a basis it takes an agent model incorporating a mirror neuron system. This is used for functional computational analysis of the different types of malfunctioning of a mirror neuron system that can be distinguished, and to which types of deviations in social functioning these types of malfunctioning can be related.

Keywords: agent model, analysis, mirror neuron systems, autism spectrum disorder.

1 Introduction

Autism is a developmental disorder showing a wide range of (gradual) differences in functioning across patient populations. The major features that characterize autistic persons are deficits in social interaction, but also abnormalities in language within a social context and repetitive and stereotyped patterns of behavior are seen. Some characteristics can be observed in all autistic persons, but it is very rare that in one person all kinds of autism-related behavior are found. Different persons with some form of autism may feature different ranges of behaviors. Therefore nowadays the term Autism Spectrum Disorder (ASD) is used rather than autism. There seem to be no strict boundaries in the spectrum of autism. It is assumed that ASD may be caused by a number of neurological factors responsible for the specific type of ASD-related behaviours shown by a person. Since the discovery of mirror neuron systems and their role in social functioning, it has been assumed that in some way ASD relates to malfunctioning of the person's mirror neuron system; e.g. [16], [20]. Given the wide variety of deviant behaviours observed in different persons with ASD, a natural step then is to analyse functionally what types of malfunctioning of a mirror neuron system can be distinguished, and to explore how each of these types of malfunctioning relates to which types of deviant behaviours. To describe an agent-based model for such a computational analysis is the main focus of the current paper.

A large amount of publications is available describing the behavior of persons with ASD; e.g., [10], [16], [19], [20]. There is no unique physical appearance that shows that someone has ASD. It is seen in persons of all ages and in different ranges within a broad spectrum; it is universal and timeless. The major characteristics concern deficits in social interaction. Also difficulties and a shortage in language acquisition, the tendency towards repetition of actions and narrowed focus may occur. In most cases, persons with ASD are not mentally disabled. Examples of possible characteristics found are reduced forms of self-other differentiation, empathy, imitation, eye contact, facial expression, gestures, shared intention and attention, or strong concentration on or occupation with a subject. Pioneers in research in autistic phenomena were Leo Kanner [17] and Hans Asperger [1]. They both came up with detailed case studies and made an attempt to give an explanation. The term 'autistic' was originally used by the psychiatrist Eugen Bleuler, to describe a particular case of schizophrenia, which narrows the view of the patient in an immense way; see also [3]. The main difference between the papers is that Asperger's descriptions and definitions of the disorder are broader than Kanner's view on it.

The presented approach is based on the assumption that to obtain a solid basis for a computational analysis: (1) an agent model incorporating the functionality of mirror neuron systems is designed, allowing modifications to model certain types of malfunctioning, and (2) this agent model is formally specified in an executable form to obtain the possibility to exploit computational formal analysis as a means to explore which behaviours may result from such an agent model and modified variants to model types of malfunctioning of the agent's mirroring system. With this agent model incorporating a mirror neuron system as a basis (presented in Section 2). In Section 3 different modifications of the agent model to model different types of malfunctioning are explored, and it is analysed which types of deviant social behaviour emerge from these types of malfunctioning. Section 4 is a discussion.

2 An Agent Model with a Mirror Neuron System as a Basis

Within a person's neurological processes, sensory representations of stimuli usually lead to preparations for responses. Recent neurological findings more and more reveal that socalled 'preparation' or premotor neurons have multiple functions; preparing for an action that actually is to be executed is only one of these functions. Sometimes, an action is prepared, but execution of the action is not taking place. For example, preparation of actions may play a role in interpreting an observation of somebody else performing an action, by internally simulating that action, or in imagining the action and its consequences; e.g., [15], [14], [16]. In these cases, actual execution of the prepared action is prevented. Without altering any body state, activation of preparation states can lead to further mental processing via an *as-if body loop* [7], [8] from preparation state to emotions felt by sensory representation of body states associated to the (expected) outcome of the prepared action. For the agent model, the following internal causal chain for a stimulus s is assumed; see [7], [8]:

Sensory representation of s → preparation for response → sensory representation of body state

This causal chain is extended to a recursive loop by assuming that the preparation for the response is also affected by the level of feeling the emotion associated to the expected outcome of the response:

sensory representation of body state → preparation for response

Thus the obtained agent model is based on reciprocal causation relations between emotions felt and preparations for responses. Within the agent model presented here, states are assigned a quantitative (activation) level or gradation. The positive feedback loops between preparation states for responses and their associated body states, and the sensory representations of expected outcomes are triggered by a sensory representation of a stimulus and converge to a certain level of feeling and preparation.

Apparently, activation of preparation neurons by itself has no unambiguous meaning; it is strongly context-dependent. Suitable forms of context can be represented at the neurological level based on what are called *supermirror neurons* [14, pp. 196-203], [18]; see also [5]. These are neurons which were found to have a function in control (allowing or suppressing) action execution after preparation has taken place. In single cell recording experiments with epileptic patients [18], cells were found that are active when the person prepares an own action to be executed, but shut down when the action is only observed, suggesting that these cells may be involved in the distinction between a preparation state to be used for execution, and a preparation state generated to interpret an observed action. In [14, pp. 201-202] it is also described that as part of this context representation, certain cells are sensitive to a specific person, so that in the case of an observed action, this action can be attributed to the person that was observed. Within the agent model presented in this section, the functions of super mirror neurons have been incorporated as focus states, generated by processing of available (sensory) context information. For the case modeled, this focus can refer to the person her or himself, or to an observed person.

To formalise the agent model in an executable manner, the hybrid dynamic modeling language LEADSTO has been used; cf. [4]. Within LEADSTO the dynamic property or temporal relation a $\rightarrow\!\!\!\rightarrow_D$ b denotes that when a state property a occurs, then after a certain time delay (which for each relation instance can be specified as any positive real number D), state property b will occur. Below, this D will be taken as the time step Δt, and usually not be mentioned explicitly. Both logical and quantitative calculations can be specified, and a software environment is available to support specification and simulation. The modeled agent receives input from the external world, for example, another agent is sensed (see also Fig. 1). Not all signals from the external world come in with the same level, modelled by having a sensor state of certain strength. The sensor states, in their turn, will lead to sensory representations. Sensory representations lead to a state called *supermirroring state* and to a specific motor *preparation state*. The supermirroring state provides a focus state for regulation and control, it also supports self-other distinction. In the scenario used as illustration, it is decisive in whether a prepared action is actually executed by the observing agent, or a communication to the observed agent is performed reflecting that it is understood what the other agent is feeling. Note that the internal process modelled is not a linear chain of events, but cyclic: the preparation state of the agent is updated constantly in a cyclic process involving both a body loop and an internal as-if body loop (via the connections labeled with w_6 and w_7). All updates of states take place in parallel.

Capitals in the agent model are variables (universally quantified), lower case letters specific instantiations. All strengths are represented by values between *0* and *1*. A capital *V* with or without subscripts indicates a *real number* between *0* and *1*. The variable S reflects that it is of the sort *signal* and B of the sort that concerns the agent's *body state*. What is outside the dotted lining is not a part of the internal process of the agent. The first two sensor states (sensor_state(A,V) and sensor_state(S,V)) are possibly coming from a single source in the external source, but are not further specified: their specific forms are not relevant for the processes captured in this agent model. A more detailed description will follow below. For each of the dynamic properties an informal and formal explanation is given.

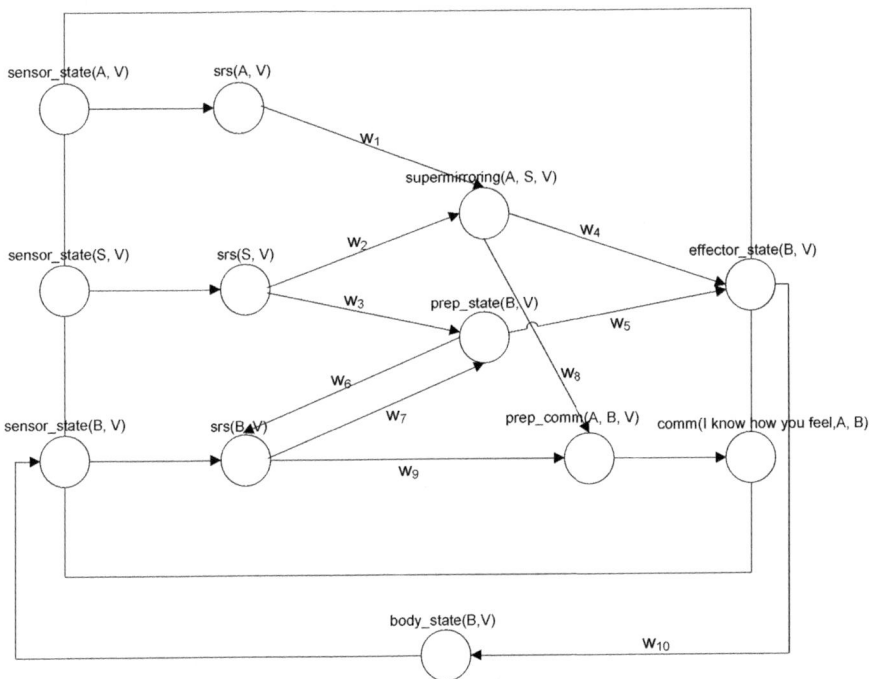

Fig. 1. Overview of the agent model

Dynamic Property 1 Generating a sensory representation for another agent

If   the presence of another agent A is sensed with a certain strength *V*,
then a sensory representation of A will occur with level *V*.
   sensor_state(A,V) →» srs(A,V)

Dynamic Property 2 Generating a sensory representation for a property of the world

If   a sensor state for S is present with a certain strength *V*,
then a sensory representation of S will occur with strength *V*.
   sensor_state(S,V) →» srs(S,V)

Super mirroring controls whether an action is to be performed (self-other distinction).

Dynamic Property 3 Super mirroring

If   a sensory representation of another agent A occurs with a certain strength V_1,
and   a sensory representation of any world state property S occurs with certain strength V_2,
then supermirroring for A and S will take place, with strength $f(w_1V_1, ow_2V_2)$.

srs(A,V_1) & srs(S,V) →» supermirroring(A, S, f(w₁V₁, ow₂V₂))

If   a sensory representation of state property instance s occurs with certain strength V,
then supermirroring for self will take place, with strength V.

srs(s,V) →» supermirroring(self, s, sw₂V)

Here w_1 and ow_2 are the strengths of the connections from, respectively, the sensory representation of the other agent and the sensory representation of the observed action (stimulus) to the super mirroring state, and sw_2 the strength of the connection from the sensory representation of the stimulus to the super mirroring state for the case of self. These weights are in the normal situation set to value *1*. By altering these connections, different output may show in the simulations. This is interesting when validating theories about dysfunctioning in some point of the process in persons with ASD. The function $f(w_1V_1, ow_2V_2)$ is a combination function, mapping values from the interval $[0, 1]$ onto values in the same interval. It can be calculated, for example, as $f(w_1V_1, ow_2V_2) = w_1V_1 + ow_2V_2 - w_1V_1 ow_2V_2$, or by a logistic threshold function $f(w_1V_1, ow_2V_2) = 1/(1 + e^{-\sigma(w_1V_1 + ow_2V_2 - \tau)})$, with σ a steepness and τ a threshold value.

Dynamic Property 4 Preparing for motor execution

If   sensory representation occurs of s_1 (movement/action) in the world with level V_1,
and   the preparation for body state b_1 has level V_2
then the preparation state for body state b1 will have level $f(V_1, w_3V_2)$.

srs(b₁,V₁) & srs(s₁,V₂) →» prep_state(b₁, f(V₁, w₃V₂))

Not every signal S that comes from the external world and which has generated a sensory representation will have a match in the sense that the body of the agent will prepare for an action. This specificity is seen in the rule, because signal s_1 will generate body state b_1, whereas any signal s_2 will not generate b_1, but maybe some b_2. These are no universal quantified variables, which are written with capital letters in this description. As earlier, also here a connection strength is given. The weight w_3 is the relation between the sensory representation of s_1 and the preparation state for b_1. Note that when a representation is not present, the value *0* is attributed to it.

Dynamic Property 5 Generating an updated sensory representation

If   the preparation state for body state B has level V_1
and   the sensor state for body state B has level V_2
then a sensory representation for body state B will be generated with level $f(V_1, V_2)$.

prep_state(B, V₁) & sensor_state(B, V₂) →» srs(B, f(V₁,V₂))

The state of the body and the preparation states are important in order to obtain a feeling; cf. Damasio (1999). The changes in the body will change the somatosensory system. In this way the altered body state produces the feeling. The earlier mentioned *body loop* (from preparation via execution to altered body state, and via sensing thia body state to sensory representation) and *as if body loop* (direct connection from preparation to sensory representation) are combined in this part of the agent model.

Dynamic Property 6 Generating action execution

If   supermirroring for self and s occurs with certain strength V_1.
and   preparation of motor execution of body state b_1 occurs with strength V_2
then   motor execution of body state b_1 will take place with level $f(w_4V_1, w_5V_2)$.
   supermirroring(self, s, V_1) & prep_state(b_1, V_2) $\twoheadrightarrow$ effector_state(b_1, $f(w_4V_1, w_5V_2)$)

The value for the effector state is based on connection strengths from super mirroring to effector state (w_4) and from preparation state to effector state (w_5).

Dynamic Property 7 From effector state to body state

If   the effector state for body state B occurs with level V,
then   body state B will have level V.
   effector_state(B, V) $\twoheadrightarrow$ body_state(B, V)

When the agent performs an action, this generates a new body state of the agent.

Dynamic Property 8 Generating a sensor state for a body state

If   a body state B is sensed with strength V,
then   a sensor state body state B with the same level V will be generated.
   body_state(B, V) $\twoheadrightarrow$ sensor_state(B, V)

Dynamic Property 9 Generating a preparation for communication

If   there is a sensory representation of body state B with level V_1,
   and   supermirroring indicating world state S and agent A occurs with level V_2,
then   preparation of communication to A about B will occur.with level $f(w_6V_1, w_7V_2)$
   srs(B, V_1) & supermirroring(A, S, V_2) $\twoheadrightarrow$ prep_communication(A, B, $f(w_6V_1, w_7V_2)$)

Also here, weights are used for connections between srs(B,V) and prep_communication(B, V) (w_6) and between supermirroring(S, V) and prep_communication(B, V) (w_7).

Dynamic Property 10 Communication

If   the preparation of a communication to A about B occurs with value V.
then   the agent will communicate 'I know how you feel B', to A with V as value.
   prep_communication(A, B, V) $\twoheadrightarrow$ communication(A, I know how you feel, B, V)

The communication that the agent knows what an observed agent feels is based upon feeling the same body state. After the observer gained the representation of the same body state, this can generate the feeling associated with it, and this is communicated.

3 Functional Analysis and Simulation Based on the Agent Model

To test the feasibility of the approach, the agent model described above was used as a basis for a functional analysis, also involving a number of simulations. The connection strengths w_1, sw_2, ow_2, w_3, w_4, w_5, w_6, w_7, w_8, w_9 and w_{10} which are parameters in the agent model, were systematically varied over the interval [0, 1], to inspect what the effects on the agent's functioning are. In the simulations, the state properties progress over time. Every delay in a temporal property corresponds to one time point in a simulation run. An example trace for normal functioning is shown in Fig. 2. The dark lines in the upper part of this picture indicate the time intervals for

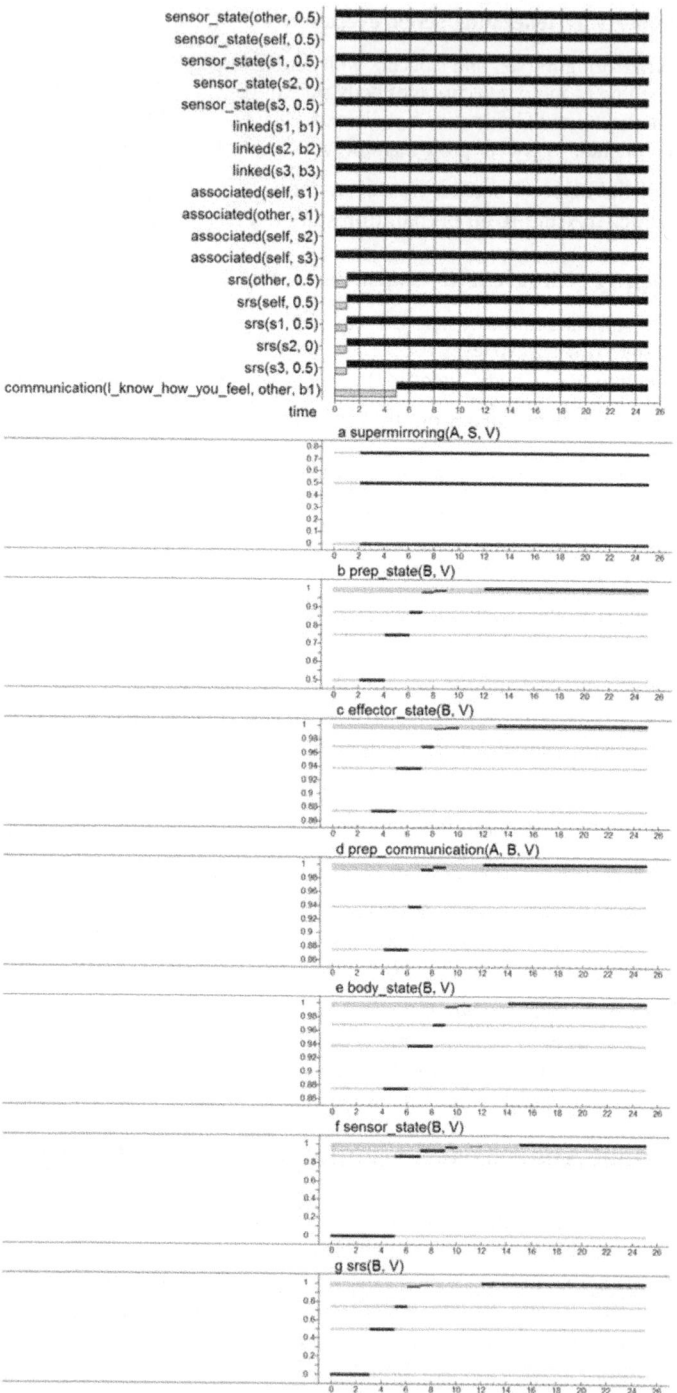

Fig. 2. Example simulation trace for normal functioning

which the indicated state properties hold. Extreme cases are when some of the connection strengths are *0*. Three of such types of malfunctioning are discussed.

Impaired basic mirroring. For example, when both w_3 and w_7 are *0*, from the agent model it can easily be deduced that the preparation state will never have a nonzero level. This indeed is confirmed in simulations (in this case the second graph in the lower part of Fig. 2 is just a flat line at level 0). This illustrates a situation of full lack of basic mirroring of an observed action or body state. Due to this type of malfunctioning no imitation is possible, nor empathic attribution to another agent.

Impaired supermirroring: self-other distinction, control. As another extreme case, when both w_1 and ow_2 are *0*, from the agent model it can be deduced that the super mirroring state for an observed action from another agent will never have a nonzero level (in this case the first graph in the lower part of Fig. 2 is just a flat line at level 0). Therefore, never a communication will be prepared, independent of the level of preparation for a body state. This covers cases in which basic mirroring is fully functional, but self-other distinction as represented in supermirroring is fully absent, and therefore the mirrored action or body state cannot be attributed to the other agent, although they still can be imitated. This is also shown in simulations.

Impaired emotion integration. Yet another extreme case occurs when w_6 and w_7 are *0*, in which case the emotion felt has no integration with the preparation for the action or body state. This covers persons who do not fully affect preparations by emotions felt. Here imitation still may be possible, and even attribution to another agent, but in general will be weaker, also depending on other connections (in this case some of the graphs in the lower part of Fig. 2 have a lower but nonzero level).

Further systematic exploration has been performed in the sense that one connection at a time was changed, from very low *(0.01* and *0.001)*, and low *(0.25)*, to medium *(0.5)*, and high *(0.75)* strengths. The connections w_4, w_5, w_7 and w_{10} showed more substantial deviations from the normal situation in comparison to the connections w_1, w_2, w_3, w_6, w_8 and w_9. As an example, a reduced connection w_5 (e.g., value *0.001*) from preparation to body state makes that it takes longer to reach an increased value for the body state. This corresponds to persons with low expressivity of prepared body states. However, when the other connections are fully functional, still empathy may occur, and even be expressed verbally.

4 Discussion

The presented approach to explore possible relationships between types of malfunctioning of a mirror neuron system and deviations in social behaviour occurring in persons with ASD is based on an executable agent model. Alterations in parameter values of this agent model are used to analyse from a functional perspective which types of malfunctioning of a mirror neuron system can be distinguished, and to which types of deviations in social functioning these types of malfunctioning lead. This approach requires a number of steps. First the normal process must be understood; for this inspiration was taken from the agent model described in [14]. In the case of autism spectrum disorder and the dysfunction of mirror neurons, there was no general description of the process in the sense of formalised causal relations. However, neurological evidence informally described what brain area would have an effect on the performance of certain tasks, resulting in (impaired) behavior. Modeling such causal relations as presented here does not take specific neurons into consideration but more abstract states, involving, for example, groups of neurons. At such an abstract level the proposed agent model summarizes the process in accordance with literature.

The agent model allows to distinguish three major types of malfunctioning, corresponding to impaired mirroring, impaired self-other distinction and control (supermirroring), and impaired emotion integration. Neurological evidence for *impaired mirroring* in persons with ASD is reported, for example, in [9], [16], [21]. This type of analysis fits well to the first case of malfunctioning discussed in Section 4. In [16] the role of super mirror neurons is also discussed, but not in relation to persons with ASD. In [5], [13] it is debated whether the social deviations seen in ASD could be related more to impaired self-other distinction and control (*impaired super mirroring*) than to the basic mirror neuron system; for example:

> 'Recent research has focused on the integrity of the mirror system in autistic patients and has related this to poor social abilities and deficits in imitative performance in ASD [21]. To date this account is still being debated. In contrast to this hypothesis, we would predict that autistic patients likely to have problems in the control of imitative behaviour rather then in imitation per se. Recent evidence has revealed no deficit in goal-directed imitation in autistic children, which speaks against a global failure in the mirror neuron system in ASD [13]. It is, therefore, possible that the mirror system is not deficient in ASD but that this system is not influenced by regions that distinguish between the self and other agents.' [4, p. 62]

The type of impaired mechanism suggested here fits well to the second case of malfunctioning discussed in Section 4.

In [11], [12] it is also debated whether the basic mirror neuron system is the source of the problem. Another explanation of ASD-related phenomena is suggested: *impaired emotion integration*:

> 'Three recent studies have shown, however, that, in high-functioning individuals with autism, the system matching observed actions onto representations of one's own action is intact in the presence of persistent difficulties in higher-level processing of social information (…). This raises doubts about the hypothesis that the motor contagion phenomenon – "mirror" system – plays a crucial role in the development of sociocognitive abilities. One possibility is that this mirror mechanism, while functional, may be dissociated from socio-affective capabilities. (…) A dissociation between these two mechanisms in autistic subjects seems plausible in the light of studies reporting problems in information processing at the level of the STS and the AMG (…) and problems in connectivity between these two regions.' [9, pp. 73-74]

This mechanism may fit to the third case of malfunctioning discussed in Section 4.

The agent-model-based computational analysis approach presented explains how a number of dysfunctioning connections cause certain impaired behaviors that are referred to as typical symptoms in the autism spectrum disorder. The agent model used, despite the fact that it was kept rather simple compared to the real life situation, seems to give a formal confirmation that different hypotheses relating to ASD, such as the ones put forward in [5], [11], [16] can be explained by different types of malfunctioning of the mirror neuron system in a wider sense (including super mirroring and emotion integration). An interesting question is whether the three types of explanation should be considered as in competition or not. Given the broad spectrum of phenomena brought under the label ASD, it might well be the case that these hypotheses are not in competion, but describe persons with different variants of characteristics. The computational analysis approach presented here provides a framework to both unify and differentiate the different variants and their underlying mechanisms and to further explore them. Further research will address computational analysis of different hypotheses about ASD which were left out of consideration in the current paper, for example, the role of enhanced sensory processing sensitivity in ASD; e.g., [6]. Moreover, the possibilities to integrate this model in human-robot interaction may be addressed in further research; see, e.g., [2].

References

1. Asperger, H.: Die 'autistischen psychopathen' im kindesalter. Arch. Psychiatr. Nervenkr 117, 76–136 (1944) ;Repr. as: 'Autistic psychopathy' in childhood. Transl. by Frith,U.: In: Frith U., (ed.) Autism and Asperger Syndrome. Cambridge University Press. Cambridge (1991)
2. Barakova, E.I., Lourens, T.: Mirror neuron framework yields representations for robot interaction. Neurocomputing 72, 895–900 (2009)
3. Bleuler, E.: Das autistische Denken. In: Jahrbuch für psychoanalytische und psychopathologische Forschungen 4 (Leipzig and Vienna: Deuticke), pp. 1–39 (1912)
4. Bosse, T., Jonker, C.M., Meij, L., van der Treur, J.: A Language and Environment for Analysis of Dynamics by Simulation. Intern.J. of AI Tools 16, 435–464 (2007)
5. Brass, M., Spengler, S.: The Inhibition of Imitative Behaviour and Attribution of Mental States. In: [20], pp. 52–66 (2009)
6. Corden, B., Chilvers, R., Skuse, D.: Avoidance of emotionally arousing stimuli predicts social-perceptual impairment in Asperger syndrome. Neuropsych. 46, 137–147 (2008)
7. Damasio, A.: The Feeling of What Happens. Body and Emotion in the Making of Consciousness. Harcourt Brace, New York (1999)
8. Damasio, A.: Looking for Spinoza: Joy, Sorrow, and the Feeling Brain. Vintage books, London (2003)
9. Dapretto, M., Davies, M.S., Pfeifer, J.H., Scott, A.A., Sigman, M., Bookheimer, S.Y., Iacoboni, M.: Understanding emotions in others: Mirroor neuron dysfunction in children with autism spectrum disorder. Nature Neuroscience 9, 28–30 (2006)
10. Frith, U.: Autism, Explaining the Enigma. Blackwell Publishing, Malden (2003)
11. Grèzes, J., de Gelder, B.: Social Perception: Understanding Other People's Intentions and Emotions through their Actions. In: [20], pp. 67–78 (2009)
12. Grèzes, J., Wicker, B., Berthoz, S., de Gelder, B.: A failure to grasp the affective meaning of actions in autism spectrum disorder subjects. Neuropsychologia 47, 1816–1825 (2009)
13. Hamilton, A.F.C., Brindley, R.M., Frith, U.: Imitation and action understanding in autistic spectrum disorders: How valid is the hypothesis of a deficit in the mirror neuron system? Neuropsychologia 45, 1859–1868 (2007)

14. Hendriks, M., Treur, J.: Modeling super mirroring functionality in action execution, imagination, mirroring, and imitation. In: Pan, J.-S., Chen, S.-M., Nguyen, N.T. (eds.) ICCCI 2010. LNCS, vol. 6421, pp. 330–342. Springer, Heidelberg (2010)
15. Hesslow, G.: Conscious thought as simulation of behavior and perception. Trends Cogn. Sci. 6, 242–247 (2002)
16. Iacoboni, M.: Mirroring People: the New Science of How We Connect with Others. Farrar, Straus & Giroux (2008)
17. Kanner, L.: Autistic disturbances of affective contact. Nervous Child 2, 217–250 (1943)
18. Mukamel, R., Ekstrom, A.D., Kaplan, J., Iacoboni, M., Fried, I.: Mirror properties of single cells in human medial frontal cortex. Soc. for Neuroscience (2007)
19. Richer, J., Coates, S. (eds.): Autism, The Search for Coherence. Jessica Kingsley Publishers, London (2001)
20. Striano, T., Reid, V.: Social Cognition: Development, Neuroscience, and Autism. Wiley-Blackwell (2009)
21. Williams, J.H., Whiten, A., Suddendorf, T., Perrtett, D.I.: Imitation, mirror neurons and autism. Neuroscience and Biobehavioral Reviews 25, 287–295 (2001)

Multi-modal Biometric Emotion Recognition Using Classifier Ensembles

Ludmila I. Kuncheva[1], Thomas Christy[1],
Iestyn Pierce[2], and Sa'ad P. Mansoor[1]

[1] School of Computer Science, Bangor University, LL57 1UT, UK
[2] School of Electronic Engineering, Bangor University, LL57 1UT, UK

Abstract. We introduce a system called AMBER (Advanced Multi-modal Biometric Emotion Recognition), which combines Electroencephalography (EEG) with Electro Dermal Activity (EDA) and pulse sensors to provide low cost, portable real-time emotion recognition. A single-subject pilot experiment was carried out to evaluate the ability of the system to distinguish between positive and negative states of mind provoked by audio stimuli. Eight single classifiers and six ensemble classifiers were compared using Weka. All ensemble classifiers outperformed the single classifiers, with Bagging, Rotation Forest and Random Subspace showing the highest overall accuracy.

1 Introduction

Affective computing covers the area of computing that relates to, arises from, or influences emotions [14]. Its application scope stretches from human-computer interaction for the creative industries sector to social networking and ubiquitous health care [13]. Real-time emotion recognition is expected to greatly advance and change the landscape of affective computing [15]. Brain-Computer Interface (BCI) is a rapidly expanding area, offering new, inexpensive, portable and accurate technologies to neuroscience [21]. However, measuring and recognising emotion as a brain pattern or detecting emotion from changes in physiological and behavioural parameters is still a major challenge.

Emotion is believed to be initiated within the limbic system, which lies deep inside the brain. Hardoon et al. [4] found that the brain patterns corresponding to basic positive and negative emotions are complex and spatially scattered. This suggests that in order to classify emotions accurately, the whole brain must be analysed.

Functional Magnetic Resonance Imaging (fMRI) and Electro Encephalography (EEG) have been the two most important driving technologies in modern neuroscience. No individual technique for measuring brain activity is perfect. fMRI has the spatial resolution needed for emotion recognition while EEG does not. fMRI, however, offers little scope for a low-cost, real-time, portable emotion classification system. In spite of the reservations, EEG has been applied for classification of emotions [1, 5, 19, 20]. Bos [1] argues that the projections of positive and negative emotions in the left and right frontal lobes of the brain make these two emotions distinguishable by EEG. He also warns that the granularity of the information collected from these regions through EEG may be insufficient for detecting more complex emotions. Different success rates of emotion recognition through EEG have been reported in the recent literature ranging from moderate [2] to excellent accuracy [10, 13]. The reason for the inconclusiveness of the results can be explained with the different experimental set-ups, different ways of eliciting and measuring emotion response, and the type and number of distinct emotions being recognised.

Chanel et al. [2] note that, until recently, there has been a lack of studies on combination of biometric modalities for recognising affective states. Some physiological signals can be used since they come as spontaneous reactions to emotions. Among other affective states, stress detection gives a perfect example of the importance of additional biometric modalities. It is known that stress induces physiological responses such as increased heart rate, rapid breathing, increased sweating, cool skin, tense muscles, etc. This gives stress detection systems good chances of success [9]. Considerable effort has been invested in designing low-power and high-performance readout circuits for the acquisition of biopotential signals such as EEG/EMG electrodes [16, 24, 25], skin conductance sensors [12], temperature sensors and muscle tightness gauges. Finally, combination of EEG and other biometric modalities has proved to be a successful route for affective state recognition [1, 2, 10, 20].

Here we present a system for multi-modal biometric emotion recognition (AMBER) consisting of a single-electrode headset, an EDA sensor and a pulse reader. These modalities were selected due to their low cost, commercial availability and simple design. We evaluate state-of-the art classifiers, including classifier ensembles, on data collected from the system. The goal is to assess the ability of the classification methodologies to recognise emotion from signals spanning several seconds.

2 Component Devices of AMBER

2.1 EEG Headset

We used a headset with single EEG electrode placed on the left of the forehead (The NeuroSky Mindset[1], Figure 1(a)). Mindset is a typical commercial EEG-headset of relatively low cost and good speed, suitable for real-time signal acquisition. It connects to a computer via a Bluetooth adapter, configured as a serial port. The data is received in variable sized packets and has to be reconstructed into readable form by a packet parser. A packet parser was written in Matlab to read and check the accuracy of the transmitted data.

2.2 EDA Sensor

Electro-Dermal Activity (EDA), also known as Galvanic Skin Response (GSR), is the measure of electrical resistance between two points across the skin. In its

---

[1] http:\www.neurosky.com

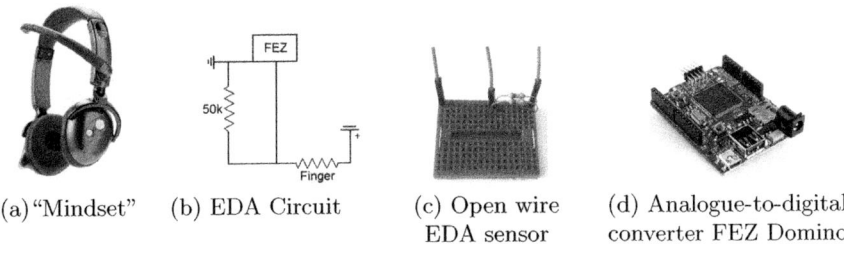

(a) "Mindset"   (b) EDA Circuit   (c) Open wire EDA sensor   (d) Analogue-to-digital converter FEZ Domino

Fig. 1. Components of AMBER

most basic form, human skin is used as an electrical resistor whose value changes when a small quantity of sweat is secreted. Figure 1(b) depicts the circuit, and 1(c) shows the electronic breadboard used in AMBER.

To feed the signal into the system we used a low-cost analogue to digital converter, FEZ Domino, shown in Figure 1(d). The FEZ Domino enables electrical and digital data to be controlled using the .NET programming language. The digital output was transmitted to a computer using a TTL Serial to USB converter cable.

2.3 Pulse Reader

Pulse sensors can determine levels of anxiety and stress, thereby contributing to the recognition of emotion. A commercially available pulse monitor kit was used for AMBER. The monitor uses a phototransistor to detect variances in blood flowing through a finger. An infra-red light is emitted through a finger and the level of light able to pass through to the other side is detected by the phototransistor. The signal is fed to the FEZ Domino and further transmitted to the computer.

The pulse sensor was attached to the middle finger on the right hand. The EDA sensor was connected to the index finger and the ring finger of the left hand. The EEG was positioned as recommended by NeuroSky. The sampling rate of all three input devices of AMBER was set at 330 readings per second.

3 Data

3.1 The Data Collection Experiment

The experiment involved presenting auditory stimuli to the subject in twenty 60-second runs. The stimuli were selected so as to provoke states of relaxation (positive emotion) or irritation (negative emotion). The positive audio stimuli were taken from an Apple iPhone application called Sleep Machine. The composition was a combination of wind, sea waves and sounds referred to as Reflection (a mixture of slow violins tinkling bells and oboes); this combination was considered by the subject to be the most relaxing. The negative audio stimuli were musical tracks taken from pop music, which the subject strongly disliked. The three biometric signals were recorded for 60 seconds for each of the 20 runs: 10 with the positive stimuli and 10 with the negative stimuli.

Typical examples of one-second run of the three signals is shown in Figure 2.

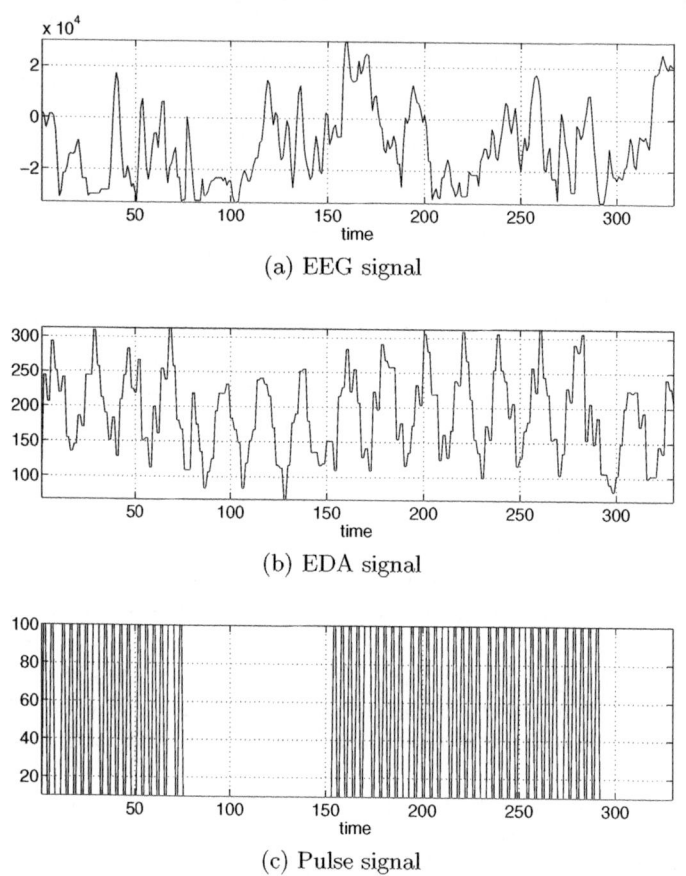

Fig. 2. Typical 1-second runs of the three input signals

3.2 Feature Extraction and Collating the Data Sets

Eight data sets were prepared by cutting the 60-second runs into sections of {3, 4, 5, 6, 10, 12, 15, and 20} seconds respectively. The sections were pooled to form a data set. All sections from a positive run were labelled as positive, and those from the negative runs were labelled as negative. For example, for the 5-second sections, there were $10 \times 60/5 = 120$ positive examples and 120 negative examples.

Ten features were extracted from each section. The power spectrum of the EEG signal was computed using the Welch method, and cut into 8 bands: delta

(1-3Hz), theta (4-7Hz), alpha 1 (8-9Hz), alpha 2 (10-12Hz), beta 1 (13-17Hz), beta 2 (18-30Hz), gamma 1 (31-40Hz) and gamma 2 (41-50Hz). The first 8 features for a particular section were the mean power within the respective frequency bands. Figure 3 shows the frequency powers for the 8 bands and the two classes, averaged across all examples from the respective class. The 4-second data set was used for this illustration.[2] The axes are omitted for clarity of the plot and error bars of the 95% confidence intervals are displayed. Significant differences between the curves for the two classes are observed in bands δ, γ_1 and γ_2.

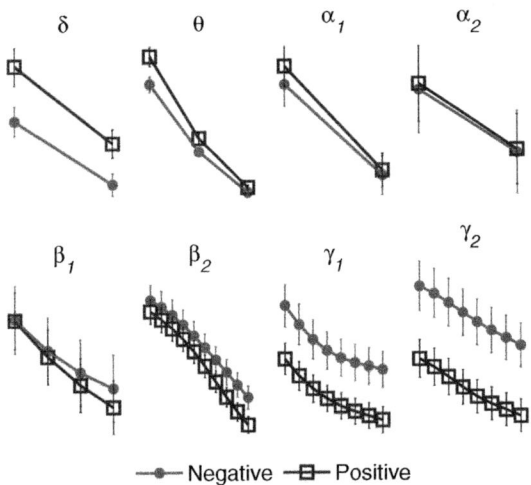

Fig. 3. Frequency powers for the 8 bands and the two classes

The remaining two features for the sections were the mean EDA signal and the mean pulse signal.

4 Classification Methods

The most widely used classification method in neuroscience analyses is the Support Vector Machine classifier (SVM) [3, 11, 18, 22]. Our previous research confirmed the usefulness of SVM but also highlighted the advantages of multiple classifier systems (classifier ensembles) [6, 7, 8].

All experiments were run within Weka [23] with the default parameter settings. The individual classifiers and the classifier ensemble methods chosen for this study are shown in Table 1.[3] Ten-fold cross-validation was used to estimate

---

[2] The curves for the remaining 7 data sets were a close match.
[3] We assume that the reader is familiar with the basic classifiers and ensemble methods. Further details and references can be found within the Weka software environment at http://www.cs.waikato.ac.nz/ml/weka/

Table 1. Classifiers and classifier ensembles used with the AMBER data

Single classifiers

1nn	Nearest neighbour
DT	Decision tree
RT	Random tree
NB	Naive Bayes
LOG	Logistic classifier
MLP	Multi-layer perceptron
SVM-L	Support vector machines with linear kernel
SVM-R	Support vector machines with Radial basis function (RBF) kernel

Ensembles

BAG	Bagging
RAF	Random Forest
ADA	AdaBoost.M1
LB	LogitBoost
RS	Random Subspace
ROF	Rotation Forest

the classification accuracy of the methods. All ensembles consisted of 10 single classifiers (the default value in Weka).

Since Rotation Forest (ROF) is a relatively recent ensemble method [17], we give a brief description here. ROF builds classifier ensembles using independently trained decision trees. Each tree uses a custom set of extracted features created in the following way. The original feature set is split randomly into K subsets (the default value in Weka is $K = 3$), principal component analysis (PCA) is run separately on each subset, and a new set of n linear extracted features is constructed by pooling all principal components. Different splits of the feature set will lead to different extracted features, thereby contributing to the diversity introduced by the bootstrap sampling. The average combination method is applied on the (continuous-valued) votes of the classifiers.

5 Results

Table 2 shows the correct classification (in %) for all methods and data sets. The highest accuracy for each data set is highlighted as a frame box, and the second highest is underlined. All highest accuracies are achieved by the ensemble methods. The individual classifiers reach only one of the second highest accuracies while the ensemble methods hold the remaining 7 second highest scores. This result confirms the advantage of using the classifier ensembles compared to using single classifiers, even the current favourite SVM. In fact, SVM-R was outperformed by all classifiers and ensembles, and SVM-L managed to beat only the logistic classifier. A series of pilot experiments revealed that none of the modalities alone were as accurate as the combination.

Table 2. Classification accuracy from the 10-fold cross-validation

Method	Data sets and number of instances							
	3s	4s	5s	6s	10s	12s	15s	20s
	400	300	240	200	120	100	80	60
1nn	62.84	64.89	63.44	62.04	61.11	60.87	56.48	59.93
DT	64.16	58.57	67.37	65.92	58.49	62.78	69.96	58.93
RT	60.02	63.02	61.9	62.63	57.11	66.66	66.75	57.10
NB	64.69	63.81	64.45	64.48	65.02	67.82	65.43	61.07
LOG	62.04	60.37	62.59	63.27	59.26	59.16	57.59	57.53
MLP	62.46	59.37	63.28	63.36	63.43	64.22	57.05	58.47
SVM-L	62.09	61.41	63.52	62.38	62.32	59.13	58.70	56.83
SVM-R	50.81	51.16	50.56	50.52	50.18	51.19	51.66	51.33
BAG	65.56	65.62	68.25	67.09	67.37	68.79	66.46	64.37
RAF	64.51	64.65	66.08	65.27	65.86	69.58	67.29	61.57
ADA	63.41	62.21	70.00	67.59	61.07	66.28	73.80	63.30
LB	65.34	62.92	68.78	68.05	62.04	64.02	68.27	60.70
RS	64.96	64.78	66.25	68.21	64.61	67.43	68.95	61.77
ROF	66.90	65.41	66.86	67.23	67.36	69.30	65.46	62.27

To visualise the results, Figure 4 shows the 14 ensemble methods sorted by their overall ranks. Each method receives a rank for each data set. As 14 methods are compared, the method with the highest classification accuracy receives rank 1, the second best receives rank 2 and so on. If the accuracies tie, the ranks are shared so that the total sum is constant $(1 + 2 + 3 + \cdots + 14 = 105)$. The total rank of a method is calculated as the mean across all 8 data sets. The total ranks and the mean accuracies of the 14 classification methods are shown in the two columns to the right of the colour matrix in Figure 4.

The colour matrix represents the classification accuracies of the methods sorted by total rank. Warm colours (brown, red and yellow) correspond to higher accuracy while cold colours (green and blue) correspond to lower accuracy. The figure reveals several interesting patterns in addition to the already discussed superior accuracy of ensembles over individual classifiers. First, the cooler colours in the last column (20s data set) indicate relatively low accuracy compared to the middle columns. This seems counter-intuitive because the frequency spectrum and the EDA and pulse averages are calculated from larger bouts of the signal, and should be less noisy. The reason for this anomaly is most likely the smaller number of data points to train the classification methods. Note that the cooler colours in the first couple of columns is not unexpected. Three- and four-second sections may be insufficient to noisy estimates of the features, hence the lower accuracy. Second, the mixture of colours in the row corresponding to

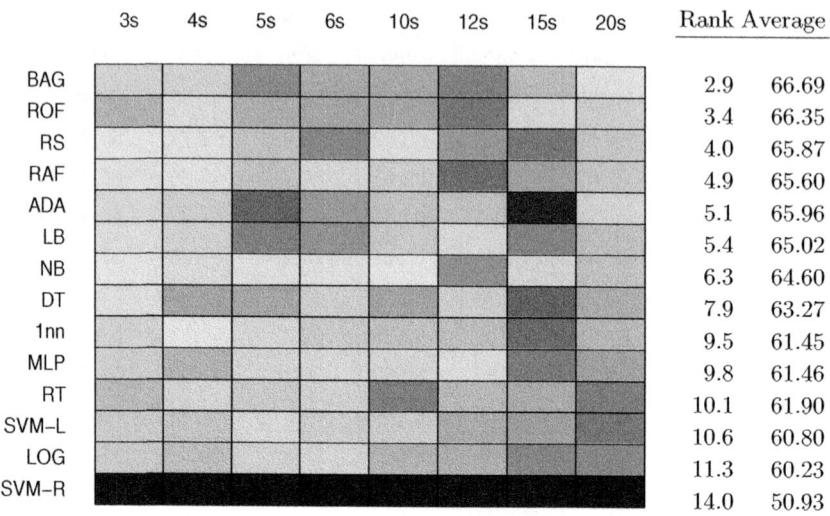

Fig. 4. Colour matrix for the classification methods sorted by their average ranks. Brown/red correspond to high accuracy; green/blue correspond to low accuracy.

AdaBoost (ADA) supports the finding elsewhere that AdaBoost's performance can vary considerably for noisy data. This row also contains the single dark brown cell corresponding to the highest accuracy of 73.8% achieved in the whole experiment.

6 Conclusion

This paper presents a case study of affective data classification coming from three biometric modalities: EEG electrode, electrodermal sensor (EDA) and pulse reader, embedded in a system called AMBER. The results indicate that positive and negative emotional states evoked by audio stimuli can be detected with good accuracy from a time segment spanning a few seconds. This work serves as a first step in a developing an inexpensive and accurate real-time emotion recognition system. Improvements on the hardware and the preprocessing of the signals are considered. We are currently working towards preparing an experimental protocol and the supporting software for gathering data from AMBER on a large scale. The new protocol will be based on a combination of visual, audio and computer-game type of stimuli.

References

1. Boss, D.O.: EEG-based emotion recognition (2006),
   http://emi.uwi.utwente.nl/verslagen/capita-selecta/CS-oude=Bos-Danny
2. Chanel, G., Kronegg, J., Grandjean, D., Pun, T.: Emotion assessment: Arousal evaluation using eEG's and peripheral physiological signals. In: Gunsel, B., Jain, A.K., Tekalp, A.M., Sankur, B. (eds.) MRCS 2006. LNCS, vol. 4105, pp. 530–537. Springer, Heidelberg (2006)
3. De Martino, F., Valente, G., Staeren, N., Ashburner, J., Goebela, R., Formisano, E.: Combining multivariate voxel selection and support vector machines for mapping and classification of fMRI spatial patterns. NeuroImage 43(1), 44–58 (2008)
4. Hardoon, D.R., Mourao-Miranda, J., Brammer, M., Shawe-Taylor, J.: Unsupervised analysis of fMRI data using kernel canonical correlation. NeuroImage 37(4), 1250–1259 (2007)
5. Ko, K., Yang, H., Sim, K.: Emotion recognition using EEG signals with relative power values and Bayesian network. International Journal of Control, Automation, and Systems 7, 865–870 (2009)
6. Kuncheva, L.I.: Combining Pattern Classifiers. Methods and Algorithms. John Wiley and Sons, N.Y (2004)
7. Kuncheva, L.I., Rodríguez, J.J.: Classifier ensembles for fMRI data analysis: An experiment. Magnetic Resonance Imaging 28(4), 583–593 (2010)
8. Kuncheva, L.I., Rodriguez, J.J., Plumpton, C.O., Linden, D.E.J., Johnston, S.J.: Random subspace ensembles for fMRI classification. IEEE Transactions on Medical Imaging 29(2), 531–542 (2010)
9. Liao, W., Zhang, W., Zhu, Z., Ji, Q.: A decision theoretic model for stress recognition and user assistance. In: Proceedings of the National Conference on Artificial Intellegence, vol. 20, PART 2, pp. 529–534 (2005)
10. Liao, W., Zhang, W., Zhu, Z., Ji, Q., Gray, W.D.: Towards a decision-theoretic framework for affect recognition and user assistance. International Journal of Man-Machine Studies 64(9), 847–873 (2006)
11. Mourao-Miranda, J., Bokde, A.L.W., Born, C., Hampel, H., Stetter, M.: Classifying brain states and determining the discriminating activation patterns: Support Vector Machine on functional mri data. NeuroImage 28(4), 980–995 (2005)
12. Nakasone, A., Prendinger, H., Ishizuka, M.: Emotion recognition from electromyography and skin conductance. In: Proc. of the 5th International Workshop on Biosignal Interpretation, pp. 219–222 (2005)
13. Petrantonakis, P., Hadjileontiadis, L.: Emotion recognition from EEG using higher-order crossings. IEEE Transactions on Information Technology in Biomedicine 14(2), 186–197 (2010)
14. Picard, R.W.: Affective computing. Technical Report 321, M.I.T Media Laboratory Perceptual Computing Section (1995)
15. Picard, R.W.: Emotion research by the people, for the people. Emotion Review (2010) (to appear)

16. Rebolledo-Mendez, G., Dunwell, I., Martínez-Mirón, E.A., Vargas-Cerdán, M.D., de Freitas, S., Liarokapis, F., García-Gaona, A.R.: Assessing neuroSky's usability to detect attention levels in an assessment exercise. In: Jacko, J.A. (ed.) HCI International 2009. LNCS, vol. 5610, pp. 149–158. Springer, Heidelberg (2009)
17. Rodríguez, J.J., Kuncheva, L.I., Alonso, C.J.: Rotation forest: A new classifier ensemble method. IEEE Transactions on Pattern Analysis and Machine Intelligence 28(10), 1619–1630 (2006)
18. Sato, J.R., Fujita, A., Thomaz, C.E., Martin, M.G.M., Mourao-Miranda, J., Brammer, M.J., Amaro Junior, E.: Evaluating SVM and MLDA in the extraction of discriminant regions for mental state prediction. NeuroImage 46, 105–114 (2009)
19. Sherwood, J., Derakhshani, R.: On classifiability of wavelet features for eeg-based brain-computer interfaces. In: Proceedings of the 2009 International Joint Conference on Neural Networks, pp. 2508–2515 (2009)
20. Takahashi, K.: Remarks on emotion recognition from biopotential signals. In: 2nd International Conference on Autonomous Robots and Agents, pp. 186–191 (2004)
21. van Gerven, M., Farquhar, J., Schaefer, R., Vlek, R., Geuze, J., Nijholt, A., Ramsey, N., Haselager, P., Vuurpijl, L., Gielen, S., Desain, P.: The brain-computer interface cycle. Journal of Neural Engineering 6(4), 41001 (2009)
22. Wang, Z., Childress, A.R., Wang, J., Detre, J.A.: Support vector machine learning-based fMRI data group analysis. NeuroImage 36(4), 1139–1151 (2007)
23. Witten, I.H., Frank, E.: Data Mining: Practical Machine Learning Tools and Techniques, 2nd edn. Morgan Kaufmann, San Francisco (2005)
24. Yasui, Y.: A brainwave signal measurement and data processing technique for daily life applications. Journal Of Physiological Anthropology 38(3), 145–150 (2009)
25. Yazicioglu, R.F., Torfs, T., Merken, P., Penders, J., Leonov, V., Puers, R., Gyselinckx, B., Hoof, C.V.: Ultra-low-power biopotential interfaces and their applications in wearable and implantable systems. Microelectronics Journal 40(9), 1313–1321 (2009)

Towards a Fully Computational Model of Web-Navigation

Saraschandra Karanam[1,*], Herre van Oostendorp[2], and Bipin Indurkhya[1]

[1] International Institute of Information Technology-Hyderabad,
Gachibowli, Hyderabad, Andhra Pradesh, India
saraschandra@research.iiit.ac.in, bipin@iiit.ac.in
[2] Utrecht University, Utrecht, The Netherlands
herre@cs.uu.nl

Abstract. In this paper, we make the first steps towards developing a fully automatic tool for supporting users for navigation on the web. We developed a prototype that takes a user-goal and a website URL as input and predicts the correct hyperlink to click on each web page starting from the home page, and uses that as support for users. We evaluated our system's usefulness with actual data from real users. It was found that users took significantly less time and less clicks; were significantly less disoriented and more accurate with system-generated support; and perceived the support positively. Projected extensions to this system are discussed.

Keywords: cognitive model, automation, support, navigation, web.

1 Introduction

This paper presents an approach towards the support of web navigation by means of a computational model. Though several studies have established the usefulness of providing support based on cognitive models to end-users, no such fully automatic tools have been developed so far for web navigation. Several existing tools are either used for evaluating hyperlink structure (Auto-CWW based on CoLiDeS) or for predicting user navigation behavior on the web (Bloodhound based on WUFIS).

For example, Cognitive Walkthrough for the Web (Auto-CWW) [1] is an analytical method (based on a cognitive model of web-navigation called CoLiDeS [2]) to inspect usability of websites. It tries to account for the four steps of parsing, elaborating, focusing and selecting of CoLiDeS. It also provides a publicly available online interface called AutoCWW (http://autocww.colorado.edu/), which allows you to run CWW online. One bottleneck is that the steps of identifying the headings and the hyperlinks under each heading in a page, designating the correct hyperlink corresponding for each goal and various parameters concerning LSA (a computational mechanism to compute similarity between two texts, described later in detail) like selecting a semantic space, word frequencies and minimum cosine value to come up

---
[*] Corresponding author.

with the link and heading elaborations need to be entered manually. Auto-CWW then generates a report identifying any potential usability problems such as unfamiliar hyperlink text, competing/confusing hyperlinks. A designer can make use of this report to make corrections in the website's hyperlinks.

Bloodhound developed by [3] predicts how typical users would navigate through a website hierarchy given their goals. It combines both information retrieval and spreading activation techniques to arrive at the probabilities associated with each hyperlink that specify the proportion of users who would navigate through it. Bloodhound takes a starting page, few keywords that describe the user-goal, and a destination page as input. It outputs average task success based on the percentage of simulated users who reach the destination page for each goal.

ScentTrails [4] brings together the strengths of both browsing and searching behavior. It operates as a proxy between the user and the web server. A ScentTrails user can input a list of search terms (keywords) into an input box at any point while browsing. ScentTrails highlights hyperlinks on the current page that lead the user towards his goal. It has been found that with ScentTrails running, users could finish their tasks quicker than in a normal scenario without ScentTrails.

Both Auto-CWW and Bloodhound are tools for web-designers and evaluators and not for supporting end-users. ScentTrails, though it is designed for supporting end-users, it makes an underlying assumption that knowledge about the website structure is assumed to be known beforehand. User can enter queries at any point of time during a browsing session and the ScentTrails system directs the user along paths that lead to his or her desired target pages from the current page. Our proposed system does not make this assumption. It takes as input from the user only the goal and a website URL and nothing else. A fully automatic model of web-navigation has many potential benefits for people working under cognitively challenging conditions. Screen readers for visually impaired persons can be made more efficient with an automated model that can read out only relevant information. Elderly people have one or more of the following problems: they can forget their original goal, or forget the outcome of the previous steps which can have an impact on their next action; they may have low mental capacity to filter unnecessary information that is not relevant to their goal; and their planning capabilities may be weak in complex scenarios [5]. For these situations, an automated model can plan an optimized path for a goal, can provide relevant information only and keep track of their progress towards the completion of the task. Naive users who are very new to the internet generally do not employ efficient strategies to navigate: they follow more of an exploratory navigation style; they get lost and disoriented quite often; and they are slow and also inefficient in finding their information. An automated model that can provide visual cues to such users can help them learn the art of navigation faster. An experienced internet user generally opens multiple applications on her or his machine and also multiple tabs in a browser. She or he could be listening to songs, writing a report, replying by email to a friend, chatting with friends on a chat-application and searching on internet for the meaning of a complex word she or he is using in a report. Under these scenarios, she or he would definitely appreciate an automated model that can reduce the time spent on one of these tasks.

In previous research [1], it was shown that the CoLiDeS model could be used to predict user navigation behavior and also to come up with the correct navigation path for each goal. CoLiDeS can find its way towards the target page by picking the most relevant hyperlink (based on semantic similarity between the user-goal and all the hyperlink texts on a web-page) on each page. The basic idea of the CoLiDeS model [2] [7] is that a web page is made up of many objects competing for user's attention. Users are assumed to manage this complexity by an attention cycle and action-selection cycle. In the attention cycle they first parse the web page into regions and then focus on a region that is relevant to their goal. In the action-selection cycle each of the parsed regions is comprehended and elaborated based on user's memory, and here various links are compared in relevancy to the goal and finally the link that has the highest *information scent* – that is, the highest semantic similarity between the link and the user's goal – is selected. For this, Latent Semantic Analysis technique is used [6]. This process is then repeated for every page visited by users until they reach the target page. This model can be used to come up with a tool in which we give the successful path back to the user and this could help the user in reducing the efforts spent in filtering unnecessary information. Thus, the tool we are developing is designed to help users in cognitively challenging scenarios. In the next section, we provide the details of our system.

2 System Details

We provide a brief description of Latent Semantic Analysis (LSA) developed by [6], which forms the backbone of our system. LSA is a machine-learning technique that builds a semantic space representing a given user population's understanding of words, short texts and whole texts by applying statistical computations, and represents them as a vector in a multidimensional space of about 300 dimensions. It uses singular value decomposition: a general form of factor analysis to condense a very large matrix of terms-documents co-occurrence into a much smaller representation. The cosine value between two vectors in this representation gives the measure of the semantic relatedness. Each cosine value lies between +1 (identical) and −1 (opposite). Near-zero values represent two unrelated texts. LSA provides many different semantic spaces ('psychology', 'biology', 'heart', for example) to represent the differences in vocabulary levels of various user-groups and terminology in different domains (http://lsa.colorado.edu). To model age differences, there are semantic spaces available based on American Grade levels: 03, 06, 09, 12 and 1st year college. LSA provides a functionality to compute similarity between a piece of text and a group of texts (one-to-many analysis) and a functionality to retrieve all those terms that are close (minimum frequency and minimum similarity measure can be specified) to a particular term from the semantic space.

Building a semantic space locally and running LSA algorithms is not our expertise. So, we plan to use the LSA server provided by the University of Colorado (http://lsa.colorado.edu). This means that we require our program to automatically fill in values in the forms in different pages of the LSA server. While we leave out the coding details, we provide a high-level description of the steps involved in running our system:

1 Take user-goal and website URL as input.
2 Extract hyperlinks from the website.
3 Elaborate the hyperlink text using LSA's 'Near-neighbor analysis'. These elaboration steps simulate the spreading-activation processes happening in the user's working memory, which are known to help in text comprehension. [1].
4 Compute semantic similarity between the user-goal and the elaborated representation of hyperlink text using LSA's 'one-to-many analysis'.
5 Output the hyperlink with the maximum similarity measure.
6 Extract the URL associated with the hyperlink in the previous step.
7 If the current page is the target page, stop the system, else go to Step 8.
8 Give the URL extracted in Step 6 as input back to Step 2.

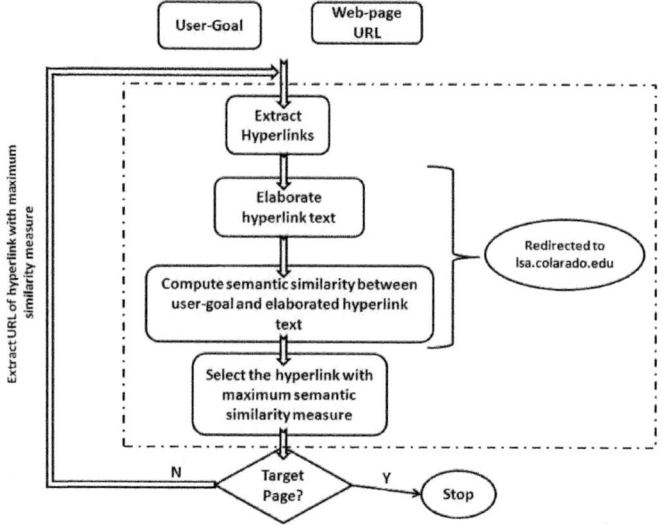

Fig. 1. Schematic representation of the steps involved in the automation system

Refer to Figure 1 for a schematic representation of these steps.

3 A Behavioral Study to Evaluate the Support Generated by the Automated System

Our aim is to study the usefulness of automated support generated by our system. Would the user find it useful? In which form should the support be provided? There is already some literature in this direction [7], where support in the form of auditory cues was found to be annoying by the participants. Here we replicate the study of [7] by using visually highlighted links. In [7], the process of inserting suggestions was not done automatically. We do this by creating two versions of a mock-up website: a control condition without any visual highlighting and a support condition with the model-predicted hyperlinks highlighted in green colour. We hypothesize that the support will be found useful by the participants; and their navigation performance in terms of time, number of clicks taken to finish the task, accuracy and overall disorientation will significantly improve.

3.1 Method

Participants. Nineteen participants from International Institute of Information Technology-Hyderabad and five participants from Jawaharlal Nehru Technological University, Hyderabad participated in the experiment. All were computer science students. A questionnaire with six multiple-choice questions on Human Body, the topic of our mock-up pages, was given to the participants. Correct answers were scored as 1 and wrong answer as 0. Individual scores for the participants were computed by adding their scores for all the questions. All our participants scored low, so we can safely assume that they had low prior domain knowledge.

Material. A mock-up website on the Human Body with 34 web pages spread across four levels of depth was used. Eight user-goals (or tasks), two for each level, which required the users to navigate, search and find the answer, were designed.

Design. We had two conditions: a control condition, where no support was provided; and a support condition, where support was provided in the form of highlighted links. These conditions are shown in Figures 2 and 3, respectively.

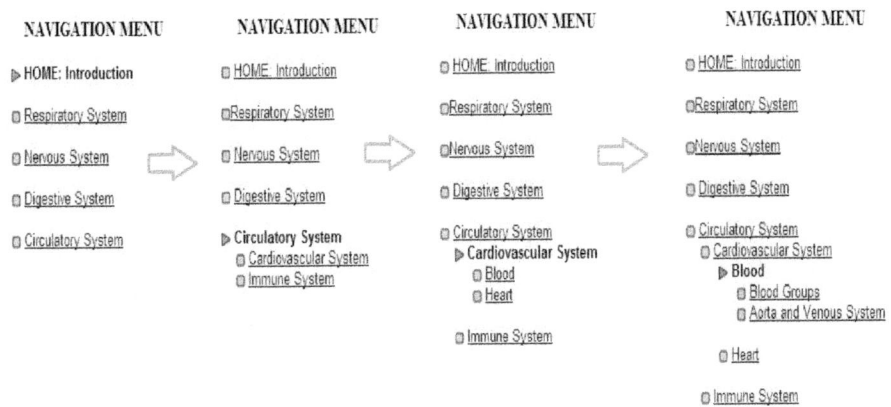

Fig. 2. Navigation format on different pages in the control condition

The automated system based on CoLiDeS described in the previous section was run on the eight different information retrieval tasks. The results of the simulations were paths predicted by the system. Based on these results, the model-predicted paths for each goal were highlighted in green color (See Figure 3). It is important to emphasize that this support is based on the automated model, which is based on CoLiDeS: computation of semantic similarity between the goal description and hyperlink text. We used a between-subjects design: half the participants received the

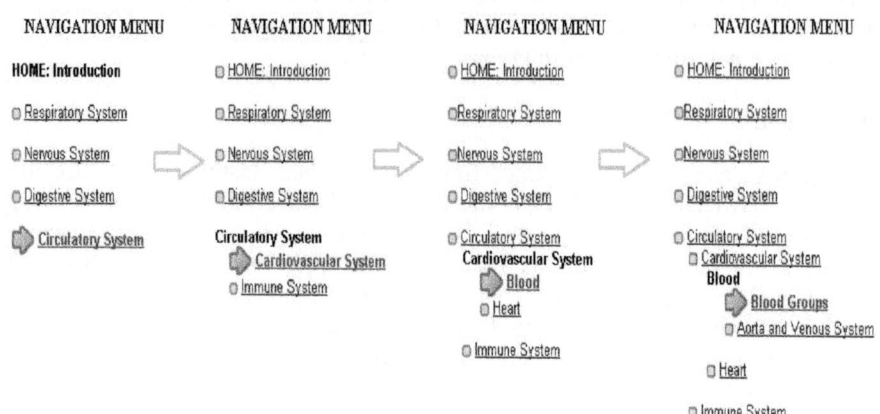

Fig. 3. Navigation format on different pages in the support condition

control condition and the other half the support condition. The dependent variables were mean task-completion time, mean number of clicks, mean disorientation and mean task accuracy. Disorientation was measured by using Smith's measure [8]

$$L = \sqrt{((N/S - 1)^2 + (R/N - 1)^2)}$$

Where, R = number of nodes required to finish the task successfully (thus, the number of nodes on the optimal path); S = total number of nodes visited while searching; N = number of different nodes visited while searching. Task accuracy was measured by scoring the answers given by the users. The correct answer from the correct page was scored 1. A wrong answer from the correct page was scored 0.5. Wrong answers from wrong pages and answers beyond the time limit were scored 0.

Procedure. The participants were given eight information retrieval tasks in random order. They were first presented with the task description on the screen and then the website was presented in a browser. The task description was always present in the top-left corner, in case the participant wished to read it again. In the control condition, participants had to read the task, browse the website, search for the answer, type the answer in the space provided and move to the next task. In the support condition, the task was the same except that they got a message reminding them that they are moving away from the model-predicted path if they did not choose the model-predicted hyperlink, but they were free to choose their path.

3.2 Results

Mean task-completion time. An independent samples t-test between control and support conditions was performed. There was a significant difference in mean task-completion times between the control and support conditions $t(22)=2.67$, $p<.05$. Figure 4 shows the means plot. Participants took significantly less time to complete their tasks in the support condition when compared to the control condition.

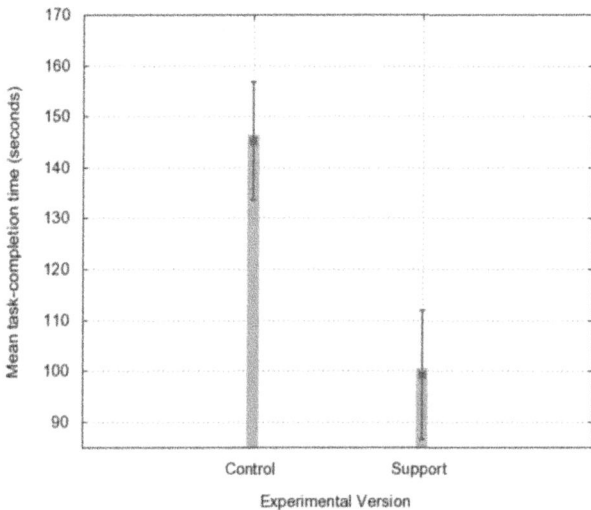

Fig. 4. Mean task-completion time in control and support conditions

Mean number of clicks. An independent samples t-test between control and support conditions with mean number of clicks as dependent variable was performed. There was a significant difference in number of clicks between control and support conditions t(22)=5.47, p<.001. Figure 5 shows the means plot. Participants took significantly less number of clicks to reach their target page in support condition when compared to control condition.

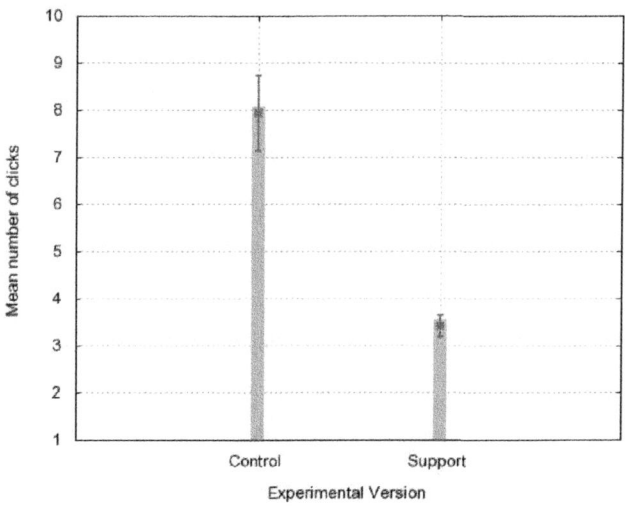

Fig. 5. Mean number of clicks in control and support conditions

Mean disorientation. A similar independent samples t-test between control and support conditions with mean disorientation as dependent variable reveals a significant difference t(22)=8.202, p<.001. When support was provided there was a very significant decrease in disorientation than when there was no support. See Figure 6 for the mean disorientation in the control and support conditions.

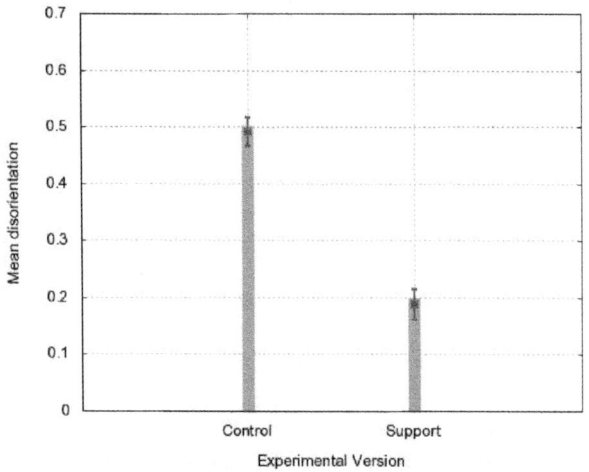

Fig. 6. Mean disorientation in control and support conditions

Mean task accuracy. Next, independent samples t-test was performed between the control and support conditions with mean task accuracy as dependent variable. The difference was found to be highly significant t(22)=-3.112, p<.01. Participants were more accurate in the support condition compared to the control condition. Support helped them not only to reach their target pages faster but also to answer the question more accurately. See Figure 7 for mean task accuracies in both conditions.

Follow-up analysis with task complexity. We did a follow-up analysis including task complexity as extra independent variable. Task complexity was defined in terms of the depth of the target page corresponding to each goal. Thus, we had 4 levels of complexity for 4 levels of depth. Briefly, the main results are that the mean task-completion time was significantly less for the first three levels of complexity in the support condition when compared with the control condition. Mean number of clicks was significantly less in the support condition when compared to the control condition for all four levels of complexity. Mean disorientation was significantly less in the support condition for second, third and fourth levels of complexity. Task accuracy was significantly higher in the support condition in first and third levels. Overall, the main pattern is that the support had positive effects, sometimes particularly in more complex situations (e.g. the reduction of disorientation with more complex tasks).

Finally, by means of a brief questionnaire consisting of 7-point Likert scales we examined the perception of participants regarding the support. The statements included were: "If there were a switch off button for support, it would be useful"

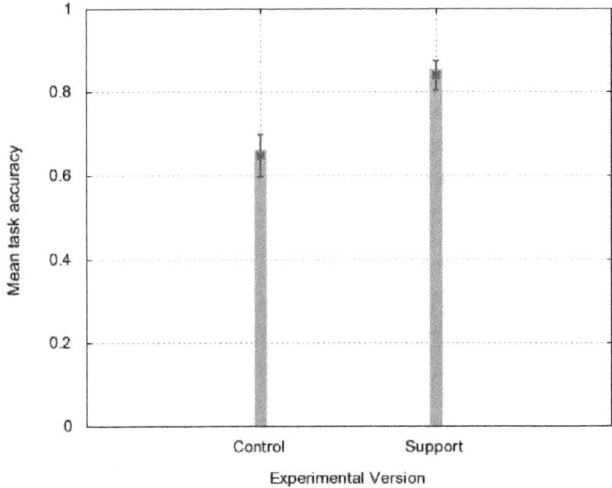

Fig. 7. Mean task accuracy in control and support conditions

($M=3.3$, $SD=2.3$), "Learning to use the site was easy" ($M=5.6$, $SD=1.9$), "Support helped me reach my target page quicker" ($M=6.3$, $SD=1.4$) and "Support made it very easy to navigate" ($M=5.6$, $SD=2.3$). These results indicate that participants had in general a (very) positive opinion about the support mechanism.

4 Conclusions and Discussion

We took the first steps towards filling a gap in the domain of cognitive models of web-navigation: lack of a fully automated support system that aids the user in navigating on a website. This is important as it illustrates the practical relevance of developing these cognitive models of web-navigation. We provided the details of our prototype. We evaluated our system with an experiment and found a very positive influence of model-generated support on navigation performance of the user. The participants were significantly faster, took significantly less number of clicks and were significantly less disoriented and more accurate when support was provided than in the no-support control condition. All in all, the main conclusion is that the support seems to have had a significant positive influence on performance.

However, our system is preliminary and we developed this to illustrate the positive effects of such a system. Several improvements can be made to the existing system, some of which will be discussed next. The current system uses only semantic similarity between the user-goal description and the hyperlink text to predict the correct hyperlink. Several other parameters can be incorporated: like background knowledge of the user in relation to the hyperlink text, frequency of usage of the hyperlink text by the user, and whether there is a literal matching between the user goal and the hyperlink text (such as, if the user is looking for 'heart diseases' and the hyperlink also says 'Heart Diseases') [9]. The path adequacy concept and backtracking rules of CoLiDeS+ can be taken into account [7]. Semantic information from pictures, if included, could make the predictions even more accurate as shown in [10]. For simplicity, and also to establish a proof of concept, we ran our model only on our mock-up website. Real websites provide more programming challenges like handling image links, handling the terms that are not present in the semantic space etc. Though we introduced the paper with cognitively challenging scenarios, in our behavioral study, we used a conventional scenario. It would be interesting to replicate the study under such scenarios of heavy cognitive load and look at the results again. We expect a much higher significant improvement in performance under such conditions. Also the perception of usefulness of the support system under such conditions would be very high. Which form of support is the best? Visually highlighting correct hyperlinks as we did in this study? Or re-ordering the hyperlinks in the order of their similarity measure with respect to the user-goal or using a bigger font? The current version of our system does not give feedback if the user is already on the target page. The user is left pondering over questions such as: Is this the target page? Should I navigate further? Other questions are: How intrusive should the support be? Should it take over the user-control over the navigation completely? When does it get annoying? These are some research questions open for investigation. We are currently working already on incorporating additional parameters when computing information scent and plan to answer these questions in our future studies.

References

1. Blackmon, M.H., Mandalia, D.R., Polson, P.G., Kitajima, M.: Automating Usability Evaluation: Cognitive Walkthrough for the Web Puts LSA to Work on Real-World HCI Design Problems. In: Landauer, T.K., McNamara, D.S., Dennis, S., Kintsch, W. (eds.) Handbook of Latent Semantic Analysis, pp. 345–375. Lawrence Erlbaum Associates, Mahwah (2007)
2. Kitajima, M., Blackmon, M.H., Polson, P.G.: A comprehension-based model of Web navigation and its application to Web usability analysis. In: Proceedings of CHI 2000, pp. 357–373 (2000)
3. Chi, E.H., Rosein, A., Supattanasiri, G., Williams, A., Royer, C., Chow, C., Robles, E., Dalal, B., Chen, J., Steve, C.: The Bloodhound Project: Automating discovery of web usability issues using the InfoScent Simulator. In: Proceedings of CHI 2003, pp. 505–512. ACM Press, New York (2003)
4. Olston, C., Chi, E.H.: ScentTrails: Integrating browsing and searching on the web. ACM Transactions on Computer Human Interaction 10(3), 1–21 (2003)
5. Kitajima, M., Kumada, T., Akamatsu, M., Ogi, H., Yamazaki, H.: Effect of Cognitive Ability Deficits on Elderly Passengers' Mobility at Railway Stations - Focusing on Attention, Working Memory, and Planning. In: The 5th International Conference of The International Society for Gerontechnology (2005)
6. Landauer, T.K., Foltz, P.W., Laham, D.: An introduction to latent semantic analysis. Discourse Processes 25, 259–284 (1998)
7. Juvina, I., van Oostendorp, H.: Modeling semantic and structural knowledge in Web-navigation. Discourse Processes 45(4-5), 346–364 (2008)

8. Smith, P.: Towards a practical measure of hypertext usability. Interacting with Computers 8, 365–381 (1996)
9. Kitajima, M., Polson, P.G., Blackmon, M.H.: CoLiDeS and SNIF-ACT: Complementary models for searching and sensemaking on the Web. In: Human Computer Interaction Consortium (HCIC), Winter Workshop (2007)
10. Karanam, S., van Oostendorp, H., Puerta Melguizo, M.C., Indurkhya, B.: Integrating semantic information from pictures into cognitive modeling of web-navigation. Special Issue of European Review of Applied Psychology (in press)

Stairway Detection Based on Single Camera by Motion Stereo

Danilo Cáceres Hernández, Taeho Kim, and Kang-Hyun Jo

Intelligent System Laboratory, School of Electrical Engineering,
University of Ulsan, Ulsan 680-749, Korea
{danilo,thkim,acejo}@islab.ulsan.ac.kr

Abstract. In this paper we are proposing a method for detecting the localization of indoor stairways. This is a fundamental step for the implementation of autonomous stair climbing navigation and passive alarm systems intended for the blind and visually impaired. Both of these kinds of systems must be able to recognize parameters that can describe stairways in unknown environments. This method analyzes the edges of a stairway based on planar motion tracking and directional filters. We extracted the horizontal edge of the stairs by using the Gabor Filter. From the specified set of horizontal line segments, we extracted a hypothetical set of targets by using the correlation method. Finally, we used the discrimination method to find the ground plane, using the behavioral distance measurement. Consequently, the remaining information is considered as an indoor stairway candidate region. As a result, testing was able to prove its effectiveness.

Keywords: Stairway segmentation, Gabor filter, Maximum distance of ground plane, Line segments extraction, Stair candidate region.

1 Introduction

Autonomous stair climbing navigation has been studied in computer vision. Nevertheless, most of the publications define in a priori the actual position of the stairway in an image in order to apply the stair recognition algorithm [2]-[6]. In this paper the authors present a stairway detection algorithm for determining stair localization from image sequence processing, without any prior information about the position of the stairs. As mentioned above, this stage is necessary in order to perform the climbing process in unknown environments for autonomous systems.

There are many types of stairs. For example, there are simple stairs, which consist of one straight piece with or without an intermediate landing. Other types of stairs are formed by several straight sections with intermediate landings for a change in direction if so desired. Fundamentally, stairs can be described as structures that follow a series of steps or flights of steps for passing from one level to another. Based on this property, we tried to extract from given images only the set of the longest line segments according to the stairway edges. In an image, this is denoted as a "set of lines which are parallel to each other". The detection of this set of lines is important in describing the localization of the candidate region of the stairs in an imaging sequence. Combining this set of line segments with the information of the dynamics of the scene, the proposed algorithm was able to determine the candidate area of the stairs within the image plane.

The authors of several studies have addressed the problem of stair climbing [2]-[6]. Nevertheless, in most of the cases, robots were placed in front of the stairways, and, in some cases, human operators moved the robot towards the stairways region. Consequently, in most of these approaches the localization problem still remains. To this end, the proposed method presents a strategy for solving the stairway localization problem in autonomous robotics. On the other hand, some related works combine brightness information with 3-D data from a stereo camera system. In [3], the authors worked with the correlation-based stereo by knowing the disparity. In addition they also estimated the ground plane by using the Least Median of Square. As a result they created an algorithm for the detection and 3D localization of stairways. Nevertheless, the process mentioned above required a computational complexity.

2 Algorithm Description

Our proposed algorithm for indoor stairs segmentation has six steps which are: (1) extracting frames, (2) finding the maximum distance of the ground plane (MDGP), (3) extracting the line segments, (4) extracting the region of interest (ROI), (5) finding correspondences, and (6) ground discrimination. From the last step we will define the stair candidate.

2.1 Extracting Frames

Using our algorithm, we extracted frames in a short time interval that started just before the earliest detection from the video capture sequence. Due to the fact that frames are collected in a short time interval, there is a temporally close relation between frames. Consequently, from the video image sequences, both the spatial and temporal information is extracted in order to know the dynamics of the scene. Our method was based on the assumptions that stairways are located below the true horizon. Consequently, in the coordinate system of the single camera, axis Z is aligned with the optical axis of the camera. Axes X and Y are aligned with axes x and y of the image plane. In Fig.1 P is a point in the world at coordinate (X,Y,Z), the projection of this point into the image plane is denoted R(xc,yc).

2.2 Finding the Maximum Distances of the Ground Plane

The proposed method consists mainly in extracting the information surrounding the ground plane projected onto the image plane. The MDGP is the maximum distance that the proposed method is able to compute from the camera position to the horizon line onto the image plane by using the Pythagorean trigonometric identities. When given an image, our goal is to compute the MDGP according to the following set of equations (1, 2, 3):

$$d = \frac{h}{\tan(\delta + \alpha)} \qquad (1)$$

$$\alpha = \tan^{-1}\left(\frac{y - y_c}{f}\right) \qquad (2)$$

$$f_{(pixel)} = \frac{f_{(mm)} \times Img_{(pixel)}}{CCD_{(mm)}} \qquad (3)$$

where d is the distance estimation between the camera and the target object in the image, h is the height of the camera above the ground, δ is the angle between the optical axis of the camera and the horizontal projection, α is the angle formed by scanning the image from top to bottom with respect to the center point of the image ($R(x_c, y_c)$ in Fig. 1), y is the pixel position on the y axis, yc is the pixel located in the center of the image, f is the focal length in pixels and *mm*, Img is the image width, and CCD is the sensor width.

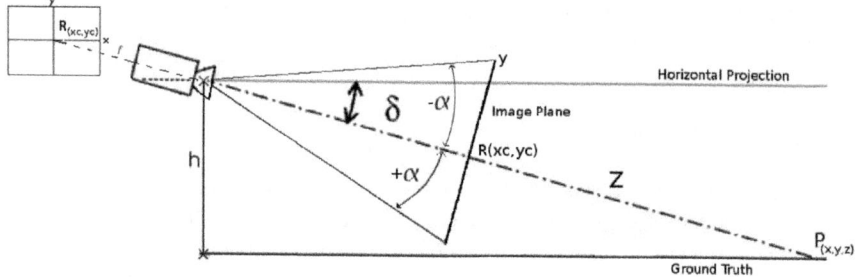

Fig. 1. Distance measuring by motion stereo based on single camera

After scanning the image from top to bottom, the algorithm was able to find a point of discontinuity (see Fig. 2). This discontinuity causes an ambiguity in the distance value when |-α| is more than or equal to δ. The point is highly related with MDGP. By finding the pixel in which the discontinuity happened the proposed algorithm distinguished where the ground plane candidate area was.

2.3 Extracting Line Segments

This step consisted of extracting the line segments of the stairs, by applying the Gabor Filter [7, 8, 9]. Gabor Filters are the most widely used texture feature extractors. The Gabor Filter is made of directional wavelet-type filters, or masks, and consists of a sinusoidal plane wave of particular frequency and orientation, modulated by a Gaussian Envelope. Using different values for wavelength, variance of the Gaussian

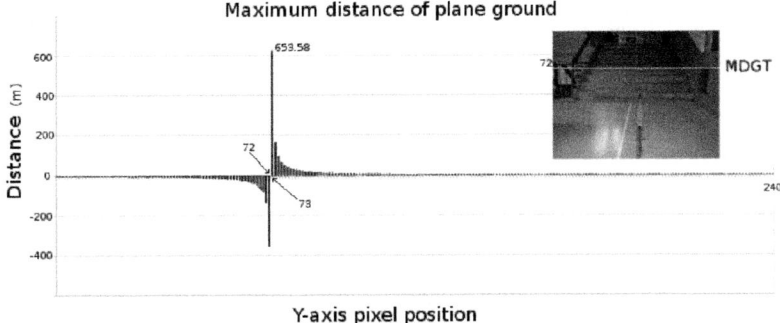

Fig. 2. Finding the maximum distances of the ground plane. The graph above shows the pixel value where the discontinuity occurs. For $\delta = 9.3$, the discontinuity occurs between the pixel value 72 and 73 on the y-axis. Consequently the image shows the ambiguity in the distance value from the 72 pixel value.

and orientation, we experimentally verified a set of Gabor Filters in order to get the best response. Eq. (4) shows the real part of the Gabor Filter:

$$G_{\lambda,\theta,\varphi,\sigma} = \exp^{-\left(\frac{x'^2 + \gamma y'^2}{2\sigma}\right)} \cos\left(2\pi \frac{x'}{\lambda} + \varphi\right) \tag{4}$$

$$x' = x\cos\theta + y\sin\theta \tag{5}$$

$$y' = -x\sin\theta + y\cos\theta \tag{6}$$

where x and y specify the position intensity value along the image, λ represents the wavelength of the sinusoidal factor, θ represents the orientation, φ is the phase shift, σ is the standard deviation of the Gaussian envelope along the x and y axes, and γ is the spatial aspect ratio, and specifies the ellipticity of the support of the Gabor function. The main idea of this section is to get an estimate of information from the stairs, such as the orientation angle of the line candidates and the number of line segments of stair candidates. This set of parameters is used in order to define the stair candidate area. First, the longest line segments have to be extracted from the binary image obtained after applying the Gabor Filter with a length $l \geq 20$, where l is the euclidean distance between the endpoints of the line segments. Second, after extracting the longest segmented lines the proposed algorithm estimated the orientation angle in the image from the previous step. This process is done in order to estimate the localization of the candidate stair area with respect to the coordinate system of the camera. Angles which are smaller than -3 degrees refer to the left, angles between -3 and 3 degrees refer to the center, and angles wider than 3 degrees refer to the right, according to the horizontal angle of view. From the remaining angles, the mean value was calculated in order to estimate the angle orientation of the stairs. Third, the initial point for each line segment were extracted, in order to project a line using the estimated angle of orientation and to generate the data set of pixel targets position at time t (PTP). As a result, the proposed algorithm was able to extract the set of line segments from a sequence of images.

2.4 Extracting the Region of Interest

In this section we propose an automatic target system detection. This process consists of two steps. First, using the dataset of point on PTP, from the current image frame (from now time t) a target area that includes the pixel target position were extracted, called region of interest (ROI). Due to the fact that frames are collected in a short time interval, there is a temporally close relation between frames. For the next image frame (from now time $t+1$), the possible target area where the pixel target position must appear were extracted. As mentioned above, we consider the data set of pixel target position as our target at time t (ROI1), as well as at time $t+1$ (ROI2). The set of ROI1 is confined in a small block with a 20x10 pixels size, and the set of ROI2 is confined in a block with a 40x20 pixels size.

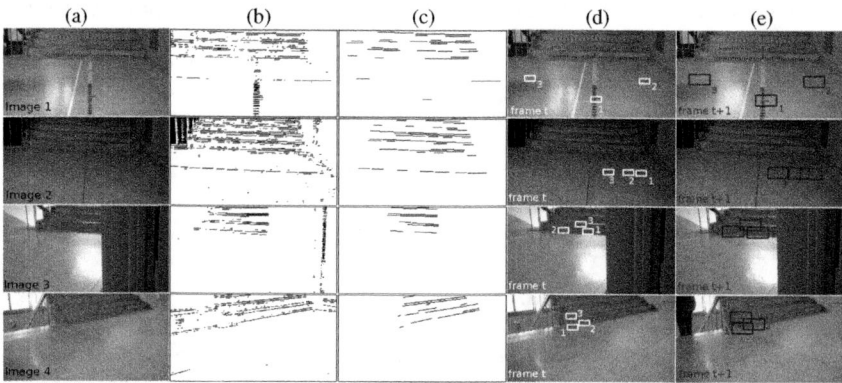

Fig. 3. Region of interest at frame t and frame t+1 stages. (a) Grayscale input image. (b) Result of binarization process from Gabor filter. (c) Extracting the longest line segments. (d),(e) Extracting the area of interest for ROI1. The white block represents the ROI1, using a 20x10 pixel size for each block extracted at time t; the numbers show the order of extraction. The blue block represents the ROI2, using a 40x20 pixel size for each block extracted at time t+1; the numbers show the order of extraction.

2.5 Finding the Correspondences

In this step, we propose to find the image correspondence by cross-correlation on the basis that the set of frames is acquired in a short time period. By measuring and finding the best similarity between ROI1 and ROI2, we proposed to extract from the image frame at time $t+1$ the new position of the pixel target position at time $t+1$. The relationship between the set of sROI_1 and sROI_2 can be expressed by the following equation (7):

$$rs = \frac{\sum_i (x_i - mx)(y_i - my)}{\sqrt{\sum_i (x_i - mx)^2} \sqrt{\sum_i (y_i - my)^2}} \qquad (7)$$

where *rs* is the cross correlation coefficient, x_i is the intensity of the i-th pixel in image ROI1, y_i is the intensity of the i-th pixel in image ROI2, *mx* and *my* are the means of intensity of the corresponding images. For each ROI1 and its respective ROI2 we generated a matrix, which included all the correlation coefficients computed for every displacement of the ROI1 into the ROI2 area. In this matrix, the maximum value of the coefficient correlation represents the position of the best matches between the ROI1 and the ROI2. By finding the correspondences between them, extraction of the new pixel target position for the frame at time *t+1* was attainable.

2.6 Estimating the Ground Plane

From the set of line segments, the algorithm removes those candidate segments that can represent ground plane information based on the concept that stairways can be described as objects with a set of lines segments with a certain height level. After finding the correspondences, the algorithm extracted a set of pixel target points for two different frames. Using this information, those line segments that showed a distance discrepancy between the frame at time *t+1* and the frame at time *t* according to the camera displacement are removed. In other words, the displacement of the camera is highly related with the displacement of the target in an image sequence. The difference between the frame at *t+1* and the frame at time *t* has to be almost the same as the distance of the camera displacement. Then, the distance estimation for each pixel target point in exposure time is calculated. This step is done according to eq. (1, 2, 3). Table 1 shows the estimation result.

By computing the mean for the data set of absolute errors, we were able to remove those sets of lines in the frame at time *t+1* if the mean value was more than 10% of

Table 1. Probability first loop results in 4 different data set images with different properties

Image	No. Areas	P_A	P_L	P_P	P_C
1	2	0.11	0.15	0.09	0.12
		0.77	0.85	0.91	**0.84**
2	3	0.11	0.13	0.09	0.11
		0.05	0.02	0.01	0.03
		0.70	0.85	0.90	**0.82**
3	1	0.31	1.00	1.00	**0.77**
4	4	0.12	0.34	0.11	0.19
		0.03	0.01	0.01	0.02
		0.04	0.09	0.03	0.06
		0.49	0.56	0.85	**0.63**

Note No. Areas is the number of stairway candidates into the image plane, P_A is the probability per each area, P_L is the probability value per each area, P_P is the probability of the pixel numbers per each area and P_C in bold is the best stairway candidate in the image.

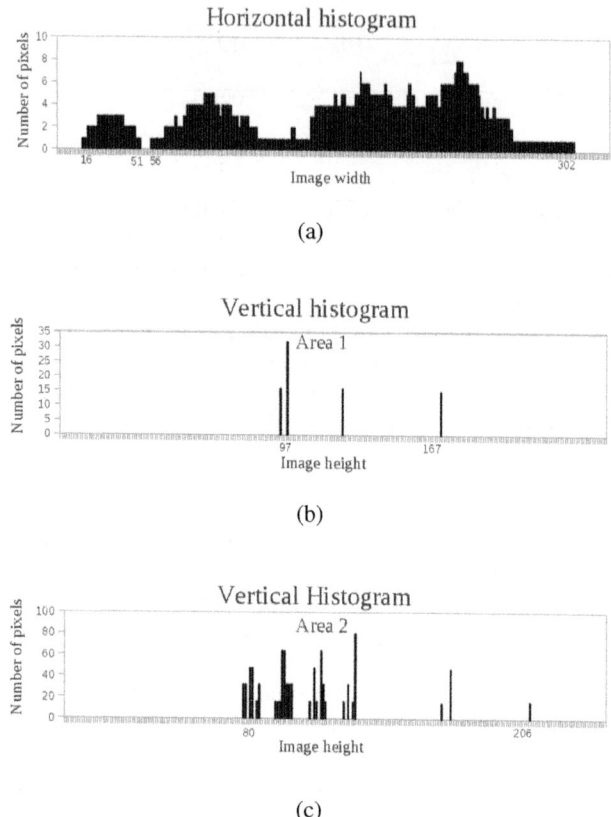

Fig. 4. Horizontal and vertical position histogram. (a) By scanning the image using the horizontal histogram position we identify the candidate stairways areas. In this image there are two stairways candidates which were detected by the process. (b) Vertical histogram position result for area 1. (c) Vertical histogram position result for area 2. The number of lines and pixels in each candidate area are extracted by applying the vertical histogram position.

the displacement of the camera between frames. This process was done with the assumption that a value of 10% of the displacement of the camera between frames can be a target point with a certain height level above the ground plane. After verifying the value and determining the removal of those line segments in the frame at time $t+1$, we scanned the image by using a horizontal and vertical histogram position in order to determine the distribution of the set of line segments (see Fig 4). This step is done through the analysis of the data that was extracted specifically for this process, such as the number of density areas, as well as the number of lines and pixels per each density area. Fig. 5 shows these steps according to eq. (8)

$$P(A) = \frac{N_A}{N} \tag{8}$$

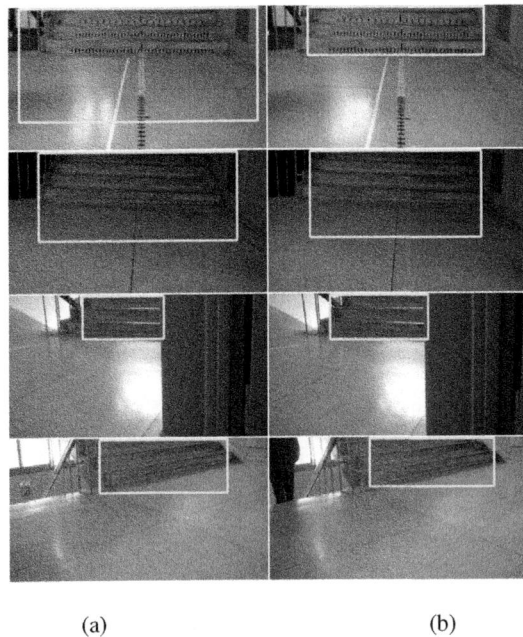

(a) (b)

Fig. 5. Stairway detection process result of the first two image sequences used in the first loop. (a) Stairway candidate area at time t after extracting the ROI1 and ROI2. (b) Stairway candidate area at time t+1. Segmented result after determining the removal of those line segments for which the distance difference is less than or equal to 10% of the camera displacement in the frame at time t+1.

where N_A is the number of areas, line segments, and pixels in A, and N is the total number of areas, line segments, and pixels in the image. The process mentioned so far was recursively repeated until we extracted, from the set of targets, relevant information for describing stairways (see table 1). The process stops at the exact moment when the mean value of the distance discrepancy between targets is not the same as the camera displacement.

2.7 Stairway Detection

The purpose of this step is to extract from the image the candidate stairway region. Once we get this point the proposed algorithm is able to describe the area where the stairway candidate area is confined using the maximum and minimum values of the last frame which detect a height information. Fig. 5 shows the candidate stairway.

From the detected area, we extract the distance and the angle between the coordinate system of the camera and the candidate region. Table 2 shows us the distance and the angle estimation result with respect to the camera coordinate system. This process of stairway detection is relevant for the implementation of autonomous stair climbing navigation and passive alarm systems intended for blind and visually impaired people. In consequence, the proposed algorithm can determine the localization of stairways through a given sequence of images given by motion stereo.

Table 2. First loop results in 4 different data set images with different properties

Targets	Blocks	\multicolumn{3}{Image 1}			Image 2			Image 3			Image 4	
		rs	de(m)	dd(m)	rs	de(m)	dd(m)	rs	de(m)	dd(m)	rs	de(m)	dd(m)
1	ROI1	0.97	1.553	0.048	0.88	1.996	0.040	0.88	4.269	0.165	0.86	3.327	0.102
	ROI2		1.505			1.956			4.104			3.225	
2	ROI1	0.73	2.145	0.066	0.88	2.016	0.040	0.74	4.447	0.178	0.84	3.809	0.067
	ROI2		2.079			1.976			4.269			3.742	
3	ROI1	0.76	2.241	0.050	0.80	2.036	0.040	0.92	5.916	0.310	0.94	4.640	0.100
	ROI2		2.192			1.996			5.609			4.542	
Average				0.055			0.040			0.218			0.090

Note that rs is the cross correlation coefficient, de is the distance estimation, and dd is the distance difference.

3 Experiment Result

In this section, we will show the ending results of the experiment. All the experiments were done on Pentium(R) Dual-Core CPU E5200@ 2.5 GHz with 2 GB RAM under Emacs environment. We used 320x240 images. Table 1 and Table 2 shows the results for the first loops, Table 3 shows the computational time required per loop iteration. Figure 5 shows the result of the different groups of data sets. The depicted results are due to the implementation of the distance measurement to estimate the ground plane, in combination with a Gabor Filter in order to find the high density of line segments in the candidate area. The experiment described a system that can detect the localization of stairways by the combination of the edges on a planar motion tracking.

Table 3. Stairway candidate localization for the first loop results in 4 different data set images with different properties.

Image	Real dist. (m)	Est. dist (m)	Relat. Error	Est. Angle	Time (sec)
1	3.75	3.67	2.13%	-0.61	0.07
2	2.20	2.06	6.36%	-0.83	0.07
3	4.25	4.18	1.65%	-5.02	0.08
4	5.25	5.08	3.24%	7.96	0.09

Note this information is done using as reference the coordinate system of the camera. Real dist. is the real distance in meter, Est. dist is the distance estimation, Relat. error is the relative error, Est. Angle is the angle estimation. Time is the computational time.

4 Conclusion

In this paper we presented relevant information for stairs detection and localization without any a priori information about the position of the stairways in the image. This approach was done through the extraction of the stairway features by analyzing the edges of the stairways belonging to the horizon projected in 2D space. Before the edge analysis was applied, the presented approach had allowed the assessment of the maximum distance of ground plane. By using this approach the authors presented a method to identify the projected horizon in 2D space. This process is important at the time of reducing the computational time for developing a real time application. By using the Gabor filter, the algorithm can extract the line segments of the stairways showing a robustness to the noise, such as the illumination condition (see Fig. 3(a) and 3(b)), where most of the line segments are located above the ground plane. The contribution of this paper is the fact that the proposed algorithm gave us an estimate of information about the stairways, such as the localization with respect to the coordinate system of the camera (see table 3). This information is important and necessary in order to localize stairways in unknown environments. It is also a fundamental step for the implementation of autonomous stair climbing navigation and passive alarm systems intended for blind and visually impaired people. As a future work, we will improve the performance by adding information about the camera rotation in 3D world (such information is important because nonplanar surfaces affect the result) and we will also improve the accuracy of the distance estimation result.

References

1. Hernandez, D.C., Jo, K.-H.: Outdoor Stairway Segmentation Using Vertical Vanishing Point and Directional Filter. In: The 5th International Forum on Strategic Technology (2010)
2. Aparício Fernandez, J.C., Campos Neves, J.A.B.: Angle Invariace for Distance Measurements Using a Single Camera. In: 2006 IEEE International Symposium on Industrial Electroncis (2006)
3. Cong, Y., Li, X., Liu, J., Tang, Y.: A Stairway Detection Algorithm based on Vision for UGV Stair Climbing. In: 2008 IEEE International Conference on Networking, Sensing and Control (2008)
4. Lu, X., Manduchi, R.: Detection and Localization of Curbs and Stairways Using Stereo Vision. In: International Conference on Robots and Automation (2005)
5. Gutmann, J.-S., Fucuchi, M., Fujita, M.: Stair Climbing for Humanoid Robots Using Stereo Vision. In: International Conference on Intelligent Robots and System (2004)
6. Se, S., Brady, M.: Vision-based detection of stair-cases. In: Fourth Asian Conference on Computer Vision ACCV 2000, vol. 1, pp. 535–540 (2000)
7. Ferraz, J., Ventura, R.: Robust Autonomous Stair Climbing by a Tracked Robot Using Accelerometer Sensors. In: Proceedings of the Twelfth International Conference on Climbing and Walking Robots and the Support Technologies for Mobile Machines (2009)
8. Barnard, S.T.: Interpreting perspective images. Artificial Intelligence 21(4), 435–462 (1983)
9. Weldon, T.P., Higgins, W.E., Dunn, D.F.: Efficient Gabor Filter Design for Texture Segmentation. Pattern Recognition 29, 2005–2015 (1996)
10. Lee, T.S.: Image Representation Using 2D Gabor Wavelets. IEEE Transactions on Pattern Analysis and Machine Intelligence 18(10) (1996)
11. Basca, C.A., Brad, R., Blaga, L.: Texture Segmentation.Gabor Filter Bank Optimization Using Genetic Algoritnms. In: The International Conference on Computer Tool (2007)
12. Deb, K., Vavilin, A., Kim, J.-W., Jo, K.-H.: Vehicle License Plate Tilt Correction Based on the Straight Line Fitting Method and Minimizing Variance of Coordinates of Projection Points. International Jounal of Control Automation, and System (2010)

Robot with Two Ears Listens to More than Two Simultaneous Utterances by Exploiting Harmonic Structures

Yasuharu Hirasawa, Toru Takahashi, Tetsuya Ogata, and Hiroshi G. Okuno

Graduate School of Informatics, Kyoto University, Kyoto, Japan
{hirasawa,tall,ogata,okuno}@kuis.kyoto-u.ac.jp

Abstract. In real-world situations, people often hear more than two simultaneous sounds. For robots, when the number of sound sources exceeds that of sensors, the situation is called *under-determined*, and robots with two ears need to deal with this situation. Some studies on under-determined sound source separation use L1-norm minimization methods, but the performance of automatic speech recognition with separated speech signals is poor due to its spectral distortion. In this paper, a two-stage separation method to improve separation quality with low computational cost is presented. The first stage uses a L1-norm minimization method in order to extract the harmonic structures. The second stage exploits reliable harmonic structures to maintain acoustic features. Experiments that simulate three utterances recorded by two microphones in an anechoic chamber show that our method improves speech recognition correctness by about three points and is fast enough for real-time separation.

1 Introduction

Since people have increasing opportunities to see and interact with humanoid robots, *e.g.*, the Honda ASIMO [1], Kawada HRP [2], and Kokoro Actroid [3], enhanced verbal communication is critical in attaining symbiosis between humans and humanoid robots in daily life. Verbal communication is the easiest and most effective way of human-humanoid interaction such as when we ask a robot to do housework or when a robot gives us information.

Robots' capabilities are quite unbalanced in verbal communication. Robots can speak very fluently and sometimes in an emotional way, but they cannot hear well. This is partially due to many interfering sounds, *e.g.*, other people interrupting or speaking at the same time, air-conditioning, the robot's own cooling fans, and the robot's movements. Even if robots have a higher-level of intelligence, they cannot respond absolutely because they cannot correctly hear and recognize what a human is saying.

Robots working with humans often encounter an *under-determined* situation, *i.e.* there are more sound sources than microphones. One good way to deal with many sounds is to use sound separation methods for preprocessing. However, conventional methods, such as Independent Component Analysis (ICA) [4] and beamformer [5], cannot handle under-determined situations well; ICA requires microphones of the same number of sound sources and beamformer requires more. Although using many microphones improves performance, there are some undesired aspects; space is needed for microphones, cost is incurred, performance may saturate, and the computational costs will become too expensive to attain real-time processing. Thus, it is critical to cope with under-determined situations by using a small number of microphones with small computational cost.

This paper is focused on under-determined speech separation that can be used for a simultaneous speech-recognition system. We focus on Automatic Speech Recognition (ASR) for robots, although most conventional methods developed so far have focused on sound source separation for humans. Yılmaz et al. [6] use a time-frequency mask estimated by histogram clustering assuming that at most one speech signal is dominant in each time-frequency region. Lee et al. [7] estimate a mixing matrix and speech signals concurrently using a maximum a posteriori estimation. Bofill et al. [8] introduce a two-stage approach, that is, an estimation of a mixing matrix followed by that of the speech signals. They assume that each time-frequency region contains a number of speech signals that does not exceed the number of microphones [9].

These methods exploit the sparseness in the power distribution of voices. For each time-frequency region, only a few sound sources are *dominant*; that is, they have relatively high power in each frequency bin at each frame. Under this assumption, we can handle under-determined speech separation as if there are less speech signals than microphones by ignoring non-dominant utterances. Needless to say, dominant source estimation is critical for these methods because a single misestimation causes two types of severe degradation. The first is a lack of spectra in the misestimated source because the power of non-dominant sources is regarded to be zero. The second is noise leakage from other sources that is derived from the residual power of the misestimated source.

In this paper, we propose new constraints that exploit harmonic structure in order to improve the accuracy of dominant source estimation. Our method is based on the conventional L1-norm method, which enables the efficient handling of many sources. By adding the new constraints, we can maintain acoustic features and improve ASR results.

2 Under-Determined Simultaneous Speech Separation

2.1 Problem Settings for Under-Determined Speech Separation

The problem settings for speech separation used in this paper are as follows.

Input M mixtures of N simultaneous speech signals
Output   Each speech signal of N talkers
Assumption Mixing matrix $\boldsymbol{H}$ is known, and $N > M$

We add a supplementary explanation to the assumption. Since our method is based on the conventional L1-norm method [8], that requires mixing matrix H, our method also requires mixing matrix H. To satisfy this requirement, we can measure the transfer function in advance. If this is impossible, we can use the method to estimate the mixing matrix from the observed signals [10].

2.2 Requirements for Under-Determined Speech Separation

There are two main requirements for under-determined speech separation.

1. Ability to handle a large number of talkers
   When the number of talkers increases, each time-frequency region contains more and more dominant sources. This means that a separation method must not have too strict assumption about the number of sources in each region.
2. Reduction of distortion in acoustic features
   As shown in Fig. 1, our separation results are used for ASR. Since the speech-recognition module uses acoustic feature values for easy recognition, our separation method must maintain these values of the original speech signals. Since we use the Mel-Frequency Cepstrum Coefficient (MFCC) [11] as the feature value, separating the following two areas is especially important. The first is time-frequency regions that have high power, and the second is time-frequency regions that are low in frequency.

2.3 Conventional L1-Norm Speech Separation Method

Let us first describe the speech mixing process. We modelize the sound-transfer function as a linear time-invariant function. Using a Short-time Fourier Transform (STFT), the speech-mixing process can be written as follows:

$$x = \sum_{j=1}^{N} h_j s_j = Hs \tag{1}$$

in which s_j is a speech signal of talker j, s is a vector equal to $[s_1, s_2, ..., s_N]^T$, h_j is a vector of transfer functions from talker j to all microphones, x is the

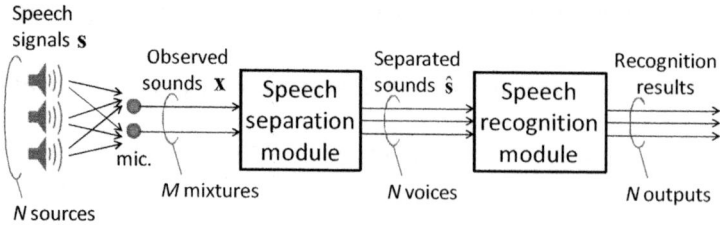

Fig. 1. Simultaneous speech-recognition system

observed signals, and $\boldsymbol{H}$ is a matrix that consists of $\boldsymbol{h}_j$. Note that we omit frequency bin index f and time frame t in this paper because separation is done independently in each time-frequency region.

Now, we will explain the conventional L1-norm speech separation method [8]. When the norm of the complex amplitude of each talker's speech follows the same Laplace distribution independently, the logarithm of joint probability is expressed as follows:

$$\log p(\boldsymbol{s}) = -\lambda \sum_{k=1}^{N} |s_k| + C \qquad (2)$$

in which λ is a positive parameter of the Laplace distributions, and C is a constant value. After an observation, since we can use Eq. (1), the logarithm of posterior probability can be written as follows using another constant C':

$$\log p(\boldsymbol{s}|\boldsymbol{x}) = \begin{cases} -\lambda \sum_{k=1}^{N} |s_k| + C' & (\boldsymbol{x} = \boldsymbol{H}\boldsymbol{s}) \\ -\infty & (otherwise) \end{cases} \qquad (3)$$

When all elements are real numbers, this is a Linear Programming (LP) problem, and we can show that $\boldsymbol{s}_{opt}$, which maximizes Eq. (3), has at most M non-zero elements. When we define $K = \{k_1, k_2, ..., k_M\}$ as a set of indices of these non-zero sources, we can also prove that the optimum separation result $\boldsymbol{s}_{opt}$ is represented as $[\hat{s}_1, \hat{s}_2, ..., \hat{s}_N]^T$ using the following equations:

$$\hat{\boldsymbol{s}}_K = \boldsymbol{H}_K^{-1} \boldsymbol{x}, \qquad (4)$$
$$\hat{s}_i = 0 \qquad \forall i \notin K, \qquad (5)$$

in which $\hat{\boldsymbol{s}}_K = [\hat{s}_{k_1}, \hat{s}_{k_2}, ..., \hat{s}_{k_M}]^T$ and $\boldsymbol{H}_K = [\boldsymbol{h}_{k_1}, \boldsymbol{h}_{k_2}, ..., \boldsymbol{h}_{k_M}]$.

Using this knowledge, a maximum a posteriori problem can be written as a combinatorial optimization problem as follows. This means that estimating K,

$$K_{opt} = \underset{K}{\operatorname{argmin}} \sum_{i=1}^{M} |\hat{s}_{k_i}| \qquad (6)$$

inwhich
$$\hat{\boldsymbol{s}}_K = \boldsymbol{H}_K^{-1} \boldsymbol{x} = [\boldsymbol{h}_{k_1}, \boldsymbol{h}_{k_2}, ..., \boldsymbol{h}_{k_M}]^{-1} \boldsymbol{x} \qquad (7)$$
$$K = \{k_1, k_2, ..., k_M\} \qquad (1 \leq k_1 < ... < k_M \leq N) \qquad (8)$$

which is the set of dominant sources, is directly connected to the separation accuracy of this method.

Since our method uses a time-frequency expression, elements are usually complex numbers. In this case, this is not a LP problem but a Second-Order Cone Programming (SOCP) problem. Thus, the above combinatorial optimization does not theoretically maximize Eq. (3). However, Winter et al. [12] demonstrate that this combinatorial optimization can be solved much more quickly than a strict solution and that the solution is similar to a strict one even when elements are complex numbers. Thus, in this paper, we use the above combinatorial optimization to obtain the solution.

We will now discuss the problem that occurs when we use this method in a simultaneous speech recognition system. As we stated in subsection 2.2, we need a separation method that has the ability to handle a large number of talkers and reduces distortion in acoustic features. Since this method can handle at most M dominant sources in each time-frequency region, this method satisfies the first requirement. However, it does not satisfy the second requirement because the accuracy of the dominant source estimation is about 50 to 60%, and this poor estimation accuracy causes a lack of spectra and noise leakage, which greatly distort the acoustic features. To improve the system's ASR results, it is necessary to improve the accuracy of the dominant source estimation.

3 Under-Determined Speech Separation Using Constraints on Harmonic Structure

3.1 Outline of Our Method

We focus on the harmonic structure of speech sounds and propose a separation method that has new constraints on harmonic structure. There are three main reasons we focus on the harmonic structure.

1. Harmonic structure has an overtone structure, thus we can consider the relationship between frequencies, though conventional L1-norm method cannot.
2. We can extract a harmonic structure relatively easily because it does not overlap frequently since speech signals have sparse power distribution.
3. Harmonic structure has high power and is contained in low-frequency areas, thus it should be separated correctly in order to maintain acoustic features.

Figure 2 outlines our proposed method. First, we use a conventional L1-norm method explained in subsection 2.3 and obtain tentatively separated sounds. Second, we extract the harmonic structure from these sounds. Finally, we separate the speech mixtures again, with constraints that exploit harmonic structure. We will refer to these two separations as the first and second separations. The first separation is done only to obtain the tentative sound used to extract the harmonic structure, and the second separation is to obtain the final output sounds.

Our method divides the extraction of the harmonic structure into two phases. The first phase is to estimate the fundamental frequency, and the second is to estimate the harmonic structure. The reason we use this two-phase approach is to improve robustness for extracting the harmonic structure from the tentative separated sound containing a lack of spectra and noise leakage since it is the output sound of conventional L1-norm method.

Another arrangement of our method is to reduce the additional computational cost of introducing the second separation. Speech separation needs to be done quickly because our system will be implemented for real-time use in robots. Since the second separation is similar to the first separation, we can reuse the calculated results to reduce the additional computational cost.

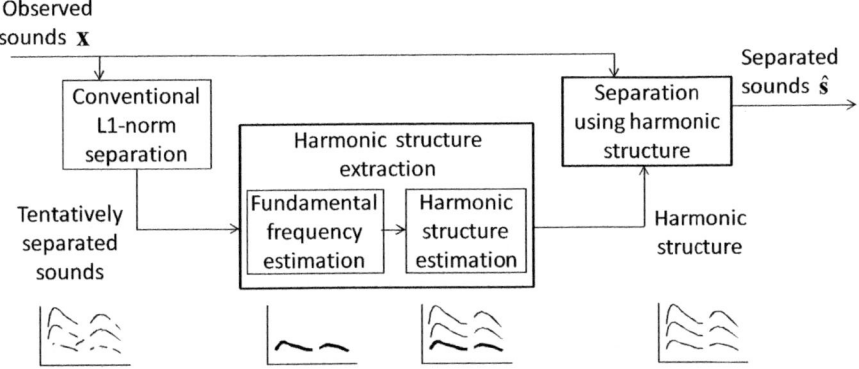

Fig. 2. Outline and flow of our method

3.2 Separation with Constraints

This subsection introduces the most essential part of our method, shown by the upper right block in Fig. 2: the separation method when the harmonic structure is already known. First, we define P as a set of sources having a harmonic structure in one time-frequency region. Since the harmonic structure has high power, we can add the constraint that the set of dominant sources K must include P. In other words, $P \subseteq K$ must be true. Please note that if there are more than M harmonic structures in one time-frequency region, $|P| > M$, the above constraint cannot be satisfied. In that case, we use another constraint that states that sources without a harmonic structure must not be included in K, i.e. K must be included in P ($P \supset K$).

Using these constraints, the separation problem can be written as follows.

$$K_{opt} = \underset{K}{\operatorname{argmin}} \sum_{i=1}^{M} |\hat{s}_{k_i}| \tag{9}$$

in which

$$\hat{s}_K = \boldsymbol{H}_K^{-1}\boldsymbol{x} = [\boldsymbol{h}_{k_1}, \boldsymbol{h}_{k_2}, ..., \boldsymbol{h}_{k_M}]^{-1}\boldsymbol{x} \tag{10}$$

$$K = \{k_1, k_2, ..., k_M\} \quad (1 \leq k_1 < ... < k_M \leq N) \tag{11}$$

$$P \subseteq K \quad (|P| \leq M) \tag{12}$$

$$P \supset K \quad (|P| > M) \tag{13}$$

The difference between this combinatorial optimization problem and the one discussed in subsection 2.3 is the existence of Eqs. (12) and (13). When we take into consideration the time-frequency region which does not have harmonic structure ($P = \phi$), these constraints do not take effect; thus, we can reuse the separation results from the first separation. In addition, even when we consider time-frequency that contains harmonic structure, we can reuse the calculation results of the matrix operations in Eq. (10).

4 Experiments

To determine the improvements obtained with our method, we carried out two experiments using synthesized sounds. Table 1 lists the experimental conditions, and Fig. 3 shows the arrangement of microphones and loud speakers. To determine the proper STFT frame length, we use the results of a previous experiment by Yılmaz et al. [6].

Our evaluations use two kinds of measurements. The first is a Signal-to-Noise Ratio (SNR) to check whether our method can accurately separate mixed speech signals, and the second is ASR correctness to check whether the output sounds maintain acoustic feature values and are suitable for ASR.

Additionally, we also evaluate calculation time, which is important for real-time human-robot interaction. The following is the result. Table 2 shows the calculation time taken to separate 198 mixtures, which equalled 1218 seconds. To solve the LP problem, we implement a program in Matlab that follows subsections 2.3 and 3.2. To solve the SOCP problem, we use the SeDuMi toolbox with the CVX with default settings.

Table 1. Experimental conditions

N, M	3 talkers (same loudness), 2 microphones
Sampling frequency	16 kHz
Impulse response	Recorded in anechoic chamber
Sound sources	JNAS 200 sentences (males and females)
STFT window size / shift size	1024 points (64 ms) / 256 points (16 ms)
Speech recognizer	Julius 3.5 fast
Acoustic model	PTM triphone, 3 state HMM
Language model	Statistical model, 20k words
Acoustic feature	MFCC 12 + ΔMFCC 12 + ΔPow
Analysis window size / shift size	400 points (25 ms) / 160 points (10 ms)

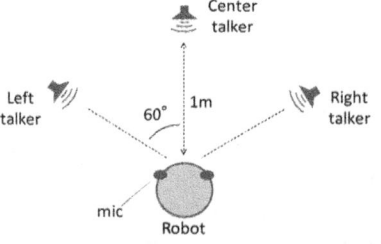

Fig. 3. Arrangement of mics and talkers

Table 2. Separation time and RTF

Method	Time(s)	RTF
Conventional LP	266	0.23
Proposed method	376	0.31
SOCP solver	31686	26.0

As Table 2 shows, although the proposed method needs to separate a sound mixture twice, it is not much slower than conventional LP thanks to reusing the calculated results as mentioned in subsection 3.1. Here, SOCP, which is a theoretically justified method, takes a very long time to separate the sound mixture, and its Real Time Factor (RTF) is much higher than one, meaning that we cannot use this method in real-time.

4.1 Results of Experiments

Figure 4 shows the original signal, the separation results with conventional method, the estimated harmonic structure, and the separation results with the proposed method. It shows only the low-frequency region because most of the estimated harmonic structure is in the low-frequency region. The black in the lower left figure represents the estimated harmonic structure, and the black and white represents high and low power in the other three figures, respectively.

The results of conventional method 4(b) indicate that there are some spectra lacking in the black circles and some noise leakage in the black rectangles. However, in the results obtained with the proposed method seen in 4(d), the spectra in the black circles are recovered, and noise leakage in the black rectangles is

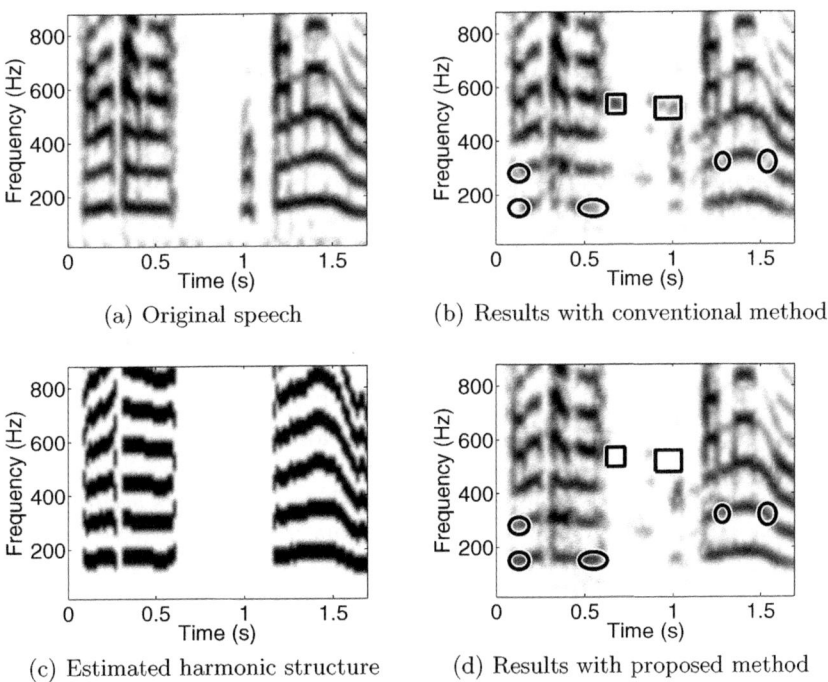

Fig. 4. Spectrograms and estimated harmonic structure

Table 3. Average SNR (dB) and ASR correctness (%)

Method	SNR			ASR correctness		
	Left	Center	Right	Left	Center	Right
(a) Conventional method	6.1	5.5	5.7	68.6	64.3	66.1
(b) Proposed method	**6.4**	**5.9**	**6.0**	**71.1**	**66.9**	**68.2**
(c) Optimum harmonic	6.7	6.2	6.4	79.3	74.0	77.7
(d) Optimum in all TF	7.0	6.3	6.7	86.8	85.9	86.3
(x) SOCP solver	6.2	5.7	5.8	68.3	65.4	65.8

reduced. This means that our method improves the accuracy of dominant source estimation and reduces interference.

Table 3 lists the average SNR and average ASR correctness of each talker. In this table, "(c) Optimum harmonic" represents the condition in which harmonic structures are given in all time frames, and "(d) Optimum in all TF" represents the condition in which dominant sources are given in all time-frequency regions; thus, this is the upper bound under our experimental condition. Note that the ASR correctness of clean speech is about 93%.

As this table shows, "(b)Proposed method" outperforms "(a) Conventional method" in both measurements. In addition, the table shows the results of "(x) SOCP," a theoretically justified solution, are as good as "(a) Conventional method."

4.2 Discussion

This experiment demonstrated that our method improves separation for both measurements of SNR and ASR correctness. This means that our method, which uses harmonic structure constraints, can estimate dominant sources correctly and maintain acoustic feature values.

In Table 3, the difference between "(b) Proposed method" and "(c) Optimum harmonic" shows that we can improve our method more by at most 8 points if we estimate harmonic structure more accurately. Also, the difference between "(c) Optimum harmonic" and "(d) Optimum in all TF" shows that there is room to improve our method; e.g., add new constraints for high frequency regions.

Finally, we also find that a solution using SOCP solver has almost no advantage under our experimental condition in the sense that its accuracy and is almost same to and its computational complexity is more than the conventional method. Of course, SOCP solver can solve more generalized problem, so we need to weigh its computational complexity against the benefits.

5 Conclusion and Future Work

We proposed a speech separation method that can be used to achieve a simultaneous speech-recognition system. Since the conventional L1-norm methods have the problem of inaccurate dominant source estimations, the separation results from the L1-norm methods lack spectra and contain noise leakage. Thus, the ASR results for these sounds are not good.

In this paper, we focused on the fact that the harmonic structure has a high-power overtone structure, and we proposed constraints on harmonic structure. More concretely, first, we used a conventional L1-norm method and extracted the harmonic structure. Second, we separated the speech mixtures again with the constraint that the harmonic structure is always powerful. The experiment revealed that our method improved the correctness of ASR by about three points and that it is fast enough to use as a real-time separation system.

As future work, we need to check the robustness of this method, *i.e.* separation accuracy for when a given mixing matrix H is not exactly correct, which often happens when we pre-measure the mixing matrix. In addition, we also want to tackle the reverberation environment. Even though our experiments were carried out using impulse responses in an anechoic chamber, developing a method that works properly in a standard reverberation room is essential for robots that work in real environments.

Acknowledgments. Part of this study was supported by a Grant-in-Aid for Scientific Research (S) and the Global COE Program.

References

1. Hirose, M., Ogawa, K.: Honda humanoid robots development. Philosophical Trans. A 365(2007), 11–19 (1850)
2. Akachi, K., Kaneko, K., et al.: Development of humanoid robot HRP-3P. In: Proc. Humanoids 2005, pp. 50–55 (2005)
3. MacDorman, K.F., Ishiguro, H.: The uncanny advantage of using androids in cognitive and social science research. Interaction Studies 7(3), 297–337 (2006)
4. Hyvärinen, A., Oja, E.: Independent Component Analysis: Algorithms and Applications. Neural Networks 13(4-5), 411–430 (2000)
5. Griffiths, L., Jim, C.: An alternative approach to linearly constrained adaptive beamforming. IEEE Trans. on Antennas and Propagation 30(1), 27–34 (1982)
6. Yılmaz, O., Rickard, S.: Blind separation of speech mixtures via time-frequency masking. IEEE Trans. on Signal Processing 52(7), 1830–1847 (2004)
7. Lee, T.W., Lewicki, M.S., et al.: Blind source separation of more sources than mixtures using overcomplete representations. IEEE Signal Processing Letters 6(4), 87–90 (1999)
8. Bofill, P., Zibulevsky, M.: Underdetermined blind source separation using sparse representations. Signal processing 81(11), 2353–2362 (2001)
9. Li, Y., Cichocki, A., Amari, S.: Analysis of sparse representation and blind source separation. Neural Computation 16(6), 1193–1234 (2004)

10. Li, Y., Amari, S., Cichocki, A., Ho, D.W.C., Xie, S.: Underdetermined blind source separation based on sparse representation. IEEE Trans. on Signal Processing 54(2), 423–437 (2006)
11. Davis, S., Mermelstein, P.: Comparison of parametric representations for monosyllabic word recognition in continuously spoken sentences. IEEE Trans. on Acoustics, Speech and Signal Processing 28(4), 357–366 (1980)
12. Winter, S., Sawada, H., Makino, S.: On real and complex valued L1-norm minimization for overcomplete blind source separation. In: Proc. WASPAA 2005, pp. 86–89 (2005)

Author Index

Abreu, Rui II-416
Adibuzzaman, Mohammad I-135
Aguilar Ruiz, Jesús I-1
Aguilar-Ruiz, Jesus Salvador II-367
Ahamed, Sheikh Iqbal I-135
Ahmad, Mumtaz I-19
Ahriz, Hatem II-436
Álvarez-García, Juan A. II-48
Andrushevich, Aliaksei II-459
Ángel Gómez-Nieto, Miguel II-396
Arredondo, Tomás V. II-183

Barber, Rick I-285
Barhm, Mukhtaj Singh II-511
Basile, Teresa M.A. I-275
Batteux, Michel I-186
Benferhat, Salem I-49
Ben Hamza, Abdessamad II-68
Bentahar, Jamal II-37, II-68
Bernardi, Giulio II-79
Bernardino, Anabela Moreira II-469
Bernardino, Eugénia Moreira II-469
Bosse, Tibor II-566
Briano, Enrico II-58
Brooks, Philip W. II-312
Buchanan, Bruce G. I-176
Bullard, Kalesh S. II-328
Bustillo, Andres I-199

Candel, Diego C. II-183
Card, Stuart W. II-296
Carson-Berndsen, Julie II-426
Castro, Carlos I-79
Cerruela García, Gonzalo II-396
Cesta, Amedeo II-79, II-216
Chandrasekaran, Muthukumaran II-347
Che, Chan Hou II-276
Chen, Chun-Hao I-95
Chen, Pei-Yu I-125
Chen, Qiao II-237
Cho, Chin-Wei I-242
Christy, Thomas I-317
Chumakov, Roman II-153

Cortellessa, Gabriella II-79
Craw, Susan II-436
Crawford, Broderick I-79
Creixell, Werner II-183

Dague, Philippe I-186
Damarla, Thyagaraju I-69
D'Amico, Rita II-79
Deb, Kaushik II-163
De Benedictis, Riccardo II-79
de Haro-García, Aida II-357
del Castillo-Gomariz, Rafael II-376
Díaz, Daniel II-216
Di Mauro, Nicola I-275
Dong, Liang I-176
Dssouli, Rachida II-37

el-Khatib, Khalil II-511
El-Menshawy, Mohamed II-37
Embley, David W. I-253

Faghihi, Usef II-27
Fan, Jinfei I-59
Felfernig, Alexander I-105
Ferilli, Stefano I-275
Ferreiro, Susana I-199
Fezer, Karl F. II-328
Fiani, Philippe I-186
Forouraghi, Babak II-302
Fournier-Viger, Philippe II-27
Fujita, Hamido II-21
Fujita, Katsuhide II-501
Fumarola, Fabio I-285
Funatsu, Kimito I-115

Gao, Qi-Gang I-146
Gao, Xiang II-193, II-256
Gao, Yang I-207
García-Pedrajas, Nicolás II-357, II-376, II-386
Garza-Castañón, Luis I-10
Garza Castañón, Luis Eduardo I-29
Glimsdal, Sondre II-532
Gómez-Pulido, Juan Antonio II-469
Gonzalez, Jesus A. I-39

Author Index

González-Abril, Luis II-48
Gonzalez-Sanchez, Alberto II-416
Gorritxategi, Eneko I-199
Granmo, Ole-Christoffer II-522, II-532
Grzenda, Maciej I-232
Guillén, Deneb Robles I-29

Han, Jiawei I-285
Haque, Munirul I-135
Hasan, Chowdhury I-135
Haselager, Willem I-295
He, Guang-Nan I-207, I-220
Hernández, Danilo Cáceres I-338
Hernandez-Leal, Pablo I-39
Hindriks, Koen I-295
Hindriks, Koen V. II-556
Hirasawa, Yasuharu I-348
Hong, Tzung-Pei I-156
Hoogendoorn, Mark II-566
Huang, Weili II-276

Ibarguengoytia, Pablo H. I-39
Indurkhya, Bipin I-327
Isak, Klaus I-105
Ito, Takayuki II-501

Jo, Kang-Hyun I-338, II-163
Jonker, Catholijn I-295
Jonker, Catholijn M. II-120, II-556

Kancherla, Kesav II-446
Kane, Mark II-426
Kaneko, Hiromasa I-115
Karanam, Saraschandra I-327
Kawaguchi, Shogo II-501
Kawsar, Ferdaus I-135
Kesim Cicekli, Nihan II-406
Kim, Taeho I-338
Kitahara, Tetsuro II-1
Klapproth, Alexander II-459
Klein, Michel C.A. II-98, II-130, II-566
Koh, Jia-Ling I-242
Krishnamoorthy, Mukkai I-253
Kuncheva, Ludmila I. I-317

Lamirel, Jean-Charles I-19
Lamiroy, Bart I-264
Le, My Ha II-163
Létourneau, Sylvain I-165
Levin, Mark Sh. II-459

Li, Ning I-220
Lim, Andrew II-193, II-237, II-246, II-256, II-276, II-286
Lin, Chun-Wei I-156
Lin, Frank Yeong-Sung I-125
Lin, Wen-Yang I-95
Lopresti, Daniel I-264
Lozovyy, Paul II-319
Luque Ruiz, Irene II-396

Ma, Li II-302
Malerba, Donato I-285
Mall, Raghvendra I-19
Mandl, Monika I-105
Mansoor, Sa'ad P. I-317
Massie, Stewart II-436
Mehrotra, Kishan. G. I-69
Mena Torres, Dayrelis I-1
M'Hallah, Rym II-226
Mills, Ian II-436
Minamikawa, Atsunori II-89
Mirenkov, Nikolay II-11
Mogles, Nataliya II-130
Mohan, Chilukuri K. I-69
Monfroy, Eric I-79
Montecinos, Mauricio I-79
Morais, Hugo II-490
Morales, Eduardo F. I-39
Morales-Menendez, Ruben I-10, I-29
Mosca, Roberto II-58
Muecke, Karl II-266
Mukkamala, Srinivas II-446

Nadig, Karthik II-347
Nagy, George I-253
Nkambou, Roger II-27

Oddi, Angelo II-216
Ogata, Tetsuya I-348, II-1
Okuno, Hiroshi G. I-348, II-1
Oommen, B. John II-522
Oon, Wee-Chong II-246, II-256
Oppacher, Franz I-59
Ormazábal, Wladimir O. II-183
Ortega-Ramírez, Juan A. II-48
Ozono, Tadachika II-173
Öztürk, Gizem II-406

Pagani, Marco II-79
Pan, Dan I-146

Pandhiti, Swetha II-328
Paquet, Eric I-85
Pérez-Rodríguez, Javier II-357, II-386
Pierce, Iestyn I-317
Pinto, Tiago II-490
Polyakovskiy, Sergey II-226
Potter, Walter D. II-312, II-328
Powell, Brian II-266
Praça, Isabel II-490

Qin, Hu II-193, II-256
Qu, Yan II-328
Qureshi, Faisal Z. II-511
Qwasmi, Nidal II-511

Ramaswamy, Srini II-336
Ramirez-Mendoza, Ricardo I-10
Rapin, Nicolas I-186
Rasconi, Riccardo II-216
Rasheed, Khaled II-347
Revetria, Roberto II-58
R-Moreno, M. Dolores II-216
Rodrigues, Fátima II-490
Rodríguez, Juan J. I-199
Rodríguez Sarabia, Yanet I-1

Sánchez-Pérez, Juan Manuel II-469
Santiesteban-Toca, Cosme Ernesto
    II-367
Scheidegger, Carre II-203
Schumann, Anika II-480
Seth, Sharad I-253
Shah, Arpit II-203
Shang, Lin I-207
Shintani, Toramatsu II-173
Shiramatsu, Shun II-173
Simon, Dan II-203, II-319
Singh, Gurmeet I-69
Smith, Reid G. I-176
Soria-Morillo, Luis M. II-48
Soto, Ricardo I-79
Sucar, L. Enrique I-39
Sugawara, Kohei II-21
Swezey, Robin M.E. II-173

Takahashi, Toru I-348, II-1
Testa, Alessandro II-58
Thomas, George II-319
Thomson, Richard II-436
Threm, David II-336
Tiberio, Lorenza II-79

Tiihonen, Juha I-105
Titouna, Faiza I-49
Treur, Jan I-306, II-109, II-130, II-542,
    II-566
Tsai, Chun-Ming II-143
Tudón-Martínez, Juan Carlos I-10

Umair, Muhammad II-109

Vale, Zita II-490
van de Kieft, Iris II-120
van der Laan, Yara I-306
van der Wal, C. Natalie II-566
van Gemund, Arjan J.C. II-416
van Lambalgen, Rianne II-98
van Oostendorp, Herre I-327
van Riemsdijk, M. Birna II-120
van Wissen, Arlette II-130
Vega-Rodríguez, Miguel Angel II-469
Viktor, Herna Lydia I-85
Villar, Alberto I-199
Visser, Wietske II-556

Wan, Wei II-68
Wang, Shyue-Liang I-156
Watanobe, Yutaka II-11
Wei, Lijun II-286
Wei, You-En I-95
Wen, Ya-Fang I-125
Weng, Yujian II-246
Weninger, Tim I-285
White, Tony I-59
Wiggers, Pascal I-295
Wilson, Nic II-480
Woo, Byung-Seok II-163

Yamakawa, Nobuhide II-1
Yang, Chunsheng I-165
Yang, Kuo-Tung I-156
Yang, Yu-Bin I-207, I-220
Yen, Hong-Hsu I-125
Yokoyama, Hiroyuki II-89
Yoshioka, Rentaro II-11
Yu, Liguo II-336

Zeng, An I-146
Zhang, Xuan II-522
Zhang, Yao I-207, I-220
Zhu, Wenbin II-193, II-237, II-246,
    II-276, II-286

7

GPSR Compliance

The European Union's (EU) General Product Safety Regulation (GPSR) is a set of rules that requires consumer products to be safe and our obligations to ensure this.

If you have any concerns about our products, you can contact us on ProductSafety@springernature.com

In case Publisher is established outside the EU, the EU authorized representative is:

Springer Nature Customer Service Center GmbH
Europaplatz 3
69115 Heidelberg, Germany

Batch number: 09478804

Printed by Printforce, the Netherlands